THE CONVENTION ON BIOLOGICAL DIVERSITY AND DEVELOPING COUNTRIES

ENVIRONMENT & POLICY

VOLUME 25

The titles published in this series are listed at the end of this volume.

The Convention on Biological Diversity and Developing Countries

by

G. Kristin Rosendal

The Fridtjof Nansen Institute,
Lysaker, Norway

KLUWER ACADEMIC PUBLISHERS
DORDRECHT / BOSTON / LONDON

A C.I.P. Catalogue record for this book is available from the Library of Congress.

ISBN 0-7923-6375-2

Published by Kluwer Academic Publishers,
P.O. Box 17, 3300 AA Dordrecht, The Netherlands.

Sold and distributed in North, Central and South America
by Kluwer Academic Publishers,
101 Philip Drive, Norwell, MA 02061, U.S.A.

In all other countries, sold and distributed
by Kluwer Academic Publishers,
P.O. Box 322, 3300 AH Dordrecht, The Netherlands.

Printed on acid-free paper

Printed in the Netherlands.

Contents

PREFACE AND ACKNOWLEDGEMENTS

I would like to express my appreciation to the Norwegian Research Council and the Fridtjof Nansen Institute (FNI) for financial support, and to Director Willy Østreng (FNI) for moral support. Thanks also to Director Desmond McNeill at the Centre for Environment and Development (SUM, UiO) for letting me spend an inspiring semester at SUM. I am also very grateful to Susan Høivik at PRIO for invaluable language consulting.

For the empirical part of this study I am indebted to a large number of people who have shared their views and knowledge with me. In particular, I would like to thank biologist Bente Herstad, then at the Norwegian Ministry of Environment, who gave me the first introduction to this issue-area in 1988, and who has kept up my interest in the political and scientific aspects of biodiversity ever since (by way of digression). Special thanks also to Dr. Tewolde B. Egziabher, the Director of the Ethiopian National Conservation Secretariat, for granting me several invaluable interviews during my trips to Addis Ababa. Many thanks also to Jan P. Borring at the Norwegian Ministry of Environment for reading through and giving me very useful advice on various parts of the manuscript.

For the theoretical part of the study I am particularly grateful to professor Arild Underdal at the University of Oslo for inspiring and guiding me throughout the research process. I would also like to give special thanks to Steinar Andresen at FNI, who has provided constant support and advice.

At FNI I am also grateful to a number of people who have given me valuable comments and helped me in my work: Regine Andersen, Gaudenz Assenza, Stian Reklev Grytvik, Erling Hagen, Torunn Laugen, Kari Lorentzen, Maryanne Rygg, Ann Skarstad, Gro Skaaren-Frystro, Øystein B. Thommessen and Jørgen Wettestad. Special thanks to Øystein and Torunn for their patience and invaluable support throughout the time of writing this book, and thanks to Anne Skjæret for moral support.

The responsibility for this text remains of course with me.

Lysaker, February, 2000

G. Kristin Rosendal

ABSTRACT

First part of analysis: International level – negotiations and output

Background: There is a widespread consensus that the loss of biodiversity represents a threat to the most basic human needs, essentially in food and medicine. There is similarly little doubt that the greatest variety of species and genetic diversity, as well as the greatest losses of this diversity takes place in developing countries in the South. This situation has put the spotlight on the need for biodiversity conservation in tropical countries. At the same time, the rapid developments within in the new biotechnologies have brought the economic potentials of biodiversity to the attention of politicians and corporate interests worldwide. Along with this attention came the resolve in many OECD countries to strengthen national patent legislation in order to secure private rights to genetic resources – the raw materials of the food and pharmaceutical industries. This environmental threat coupled with the huge economic potential of the biological resources gave rise to a vehement North - South controversy during the international biodiversity negotiations.

The Convention on Biological Diversity (CBD) seeks to respond to both of these conditions, as it combines the principles of conservation, sustainable use and equitable sharing of benefits derived from use of biological resources. The principles of equitable sharing of benefits and equal responsibility for conservation costs render the CBD largely as a victory of the South – at least a symbolic victory.

Theoretical approach: The negotiation process leading up to the CBD is analysed from three perspectives from international relations theory – a power-based, an interest-based and an ideational (normative/cognitive) perspective. The aim is to examine whether these may shed light on the negotiation result, and subsequently the implementation process (Chapters 3 to 6).

Findings: The finding that 'the South' actually did win through with their claims seemed rather unprecedented in the history of international negotiations and agreements. It suggested that the time, place and procedures for negotiating environmental treaties make a difference. At the same time, it could mean that 'the North' has little intention to follow up their part of the deal. The manner in which related aspects of the issue has been, and is being dealt with, in overlapping and partly competing international fora may still carry structural effects, negating efforts to achieve the objectives of the CBD. Dominant actors may still have their way, as they may have greater impact on choosing the "right" forum for advocating their interests. The Trade-Related Aspects of Intellectual Property Rights (TRIPs) under the World Trade Organisation will be up for review in the next negotiation round, and is likely to reaffirm the strongly conflicting positions of the parties in the biodiversity issue. In this arena, concern for competitiveness in the biotechnology sector may take precedence over concerns for improved conservation and equitable sharing of benefits in the biodiversity issue-area. In practical terms, this would mean that 'Northern' governments are less likely to establish relevant legislation, such as providing the requirement in patent legislation that all genetic material must be obtained on mutually agreed terms and on prior informed consent.

Increasing the potential for successful implementation, however, some important users are changing their behaviour in the direction prescribed by the CBD: There are some indications that central actors in international biodiversity transactions – major seeds corporations, botanical gardens and scientific institutions – are choosing to adhere to the new rules and principles provided by the CBD on equity and fair sharing in genetic resources transactions. This trend is, however, far more pronounced among private actors than within the public sector. Private seeds collection agencies may be more open to seeing the potential competitive advantage of building a fair and above-board image in germplasm transactions.

Second part of analysis: National & International links in domestic implementation

Background: An underlying premise in this analysis is that a symbolic victory alone is far from sufficient for starting to resolve the problems within the biodiversity issue-area. This has been my rationale for examining whether and how the principles of the CBD seem to be gaining strength by being translated into practical politics. I have also argued that this translation into national policies is equally important in 'the North' and in 'the South'. Concurrent action is a necessary condition for progress in the issue-area, as countries would not have the will or ability to solve the problems single-handedly.

Hence, I asked what could be expected in terms of such domestic legislation. The developing countries were expected to have an incentive for action due to their victory in the negotiations, but basically their adherence would depend on other parties complying with the principle of equitable sharing. The developed countries can hardly be pressured to adhere to their obligations. This is partly due to the inherent nature of international relations – with no central international authority to force states to fulfil their international duties. Moreover, the developed countries have few incentives to comply, as they have access to breeding material (seeds) through international gene banks and appear as the stronger part in bioprospecting deals. Hence, the South has limited sanctions to induce such adherence.

Theoretical approach: In this second part of the analysis, a domestic implementation model is developed. This is done partly for the examination of how national biodiversity policies correspond to the international agreement, and also with a view to control for domestic factors. Finally, the links between international and domestic level factors are examined by proposing and focusing on four mechanisms through which an international institution may affect domestic policies: through legitimacy and perceived obligations (moral mechanism), through the establishment of incentive and enabling mechanisms (material), through stabilising mutual expectations among the implementing states (mutual reassurance) and through strengthening domestic groups and individuals in the civil society (empowerment).

Findings: The analysis revealed a rather high level of consistency between the CBD and Ethiopian biodiversity policies. I then went on to ask how this consistency could be explained by institutional mechanisms. I was particularly interested in whether the CBD had changed the established way of i) viewing international exchange of and access to seeds among users and owners, and ii) formulating conservation projects. I found that the CBD was regarded as a useful and fair framework for linking domestic biodiversity policies, especially with regard to controlling access to plant genetic resources (seeds). On the other hand, the expectations for external financial transfers for biodiversity projects in the wake of the CBD were very high and probably exaggerated. The money provided by external donors was increasing, but often tied to conditions that seemed to negate the Ethiopian government's efforts to follow up some of the objectives of the CBD. Hence, I found evidence that the international institution have had an impact on domestic policies through both moral and material mechanisms – effects that are enhanced by domestic factors, such as a high level of domestic commitment to fulfil these international obligations. This opportunity is hardly utilised, however, as the mutual reassurance mechanism is still lacking (Chapters 7 to 9).

With regard to domestic implementation in owner countries, the lessons from Ethiopia are broadly optimistic: The biodiversity issue links up with long-standing domestic traditions – national sovereignty over biological resources, equitable sharing of benefits from use of biological resources – and is hence enhancing a strong domestic commitment and awareness. Hence, the biodiversity issue is quite easily translated as a priority in domestic policy-making. On the negative side, traditional approaches to biodiversity conservation among donors may obstruct recipients' efforts to adhere to the CBD: Wildlife conservation still takes precedence over on-farm, in-situ plant genetic resources projects. Wildlife conservation projects and policies are more likely to be one step removed from local and indigenous community participation, and are more likely to be dominated by standard approaches within development aid. This consistent northern conservation bias remains a central constraint on the international commitment to share responsibility for CBD implementation. It is important to co-operate with owner countries (principally recipients) in defining projects that fit into their own policy agenda, in order to secure commitment for domestic implementation.

List of Tables

List of Boxes

ABBREVIATIONS AND ACRONYMS

CBD	Convention on Biological Diversity
COP	Conference of the Parties
CGIAR	Consultative Group on International Agricultural Research
CGRFA	Commission on Genetic Resources for Food and Agriculture (FAO)
DC	developing country
EFAP	Ethiopian Forestry Action Plan
EPA	Environmental Protection Agency
EPRDF	Ethiopian People's Revolutionary Democratic Front
FAO	United Nations Food and Agricultural Organization
G-77	Group of 77
GEF	Global Environment Facility
GATT	General Agreements on Tariffs and Trade
GRAIN	Genetic Resources Action International (previously ICDA)
IARCs	International Agricultural Research Centres
IEA	International Environmental Agreements
ICDA	International Coalition for Development Action (now GRAIN)
IBPGR	International Board for Plant Genetic Resources (now IPGRI)
IPGRI	International Plant Genetic Resources Institute (previously IBPGR)
IPR	Intellectual Property Rights
IUCN	World Conservation Union
LIC	Low Income Countries (World Bank definition)
MNRDEP	Ministry of Natural Resources Development and Environmental Protection
NCS	National Conservation Strategy
NEPA	National Environmental Protection Authority
NGO	Non-governmental organisation
NIC	Newly Industrialised Countries
OECD	Organisation for Economic Co-operation and Development
PA	Peasant Association
PGR	Plant Genetic Resources
RAFI	Rural Advancement Foundation International
REST	Relief Society of Tigray
TRIPs	Trade Related Aspects of Intellectual Property Rights
UNCED	United Nations Conference on Environment and Development (Rio Conference)
UNCHE	United Nations Conference on Human Environment (Stockholm Conference)
UNDP	United Nations Development Programme
UNEP	United Nations Environmental Programme
UPOV	Union for the Protection of New Varieties of Plants
WIPO	World Intellectual Property Organisation
WTO	World Trade Organization
WWF	World Wide Fund for Nature

PART ONE: INTRODUCTION AND ANALYTICAL APPROACH

There is a grandeur in this view of life, with its several powers, having been originally breathed into a few forms or into one; and that, whilst this planet has gone cycling on according to the fixed law of gravity, from so simple a beginning endless forms most beautiful and most wonderful have been, and are being evolved.

Charles Darwin, 1859, *The origin of species*

CHAPTER 1

INTRODUCTIONS

1.1 The Issue and the Main Research Question

Recent decades have witnessed an upsurge in transborder environmental problems, as well as a whole series of international arrangements set up to meet these challenges. Hence, it becomes pertinent to ask how well these international environmental agreements (IEA) have succeeded in fulfilling their intended tasks.

Clearly, the overall structural power relations and the interests of negotiation parties account for much of what may be achieved in international environmental co-operation. On the other hand, however, the very way in which negotiations are organised may have an independent effect on the formation and implementation of international agreements. Contrary to the international structural division of power and interests, institutional factors may be amenable to short and medium term manipulation. If that is so, greater attention to conscious design of the negotiation forum may make it possible to enhance the chances of reaching international agreement as well as increasing the potential for successful implementation. Today a growing academic field has begun focusing on institutional factors and how they affect the negotiation of IEAs and their implementation. The broad theme for investigation in this study is the formation and implementation of a global environmental agreement in a North-South context:

> *This study will focus on how institutional factors at the international level may influence the formation process of an international environmental agreement (IEA), and subsequently affect relevant policies in a developing country (IEA implementation).*

Institutions are broadly defined as "sets of rules of the game or codes of conduct that serve to define social practices, assign roles to the participants in these practices, and guide the interactions among occupants of these roles" (Young, 1994:3). A sub-group of institutions is labelled international *regimes,* defined as "implicit or explicit principles, norms, rules and decision-making procedures around which actors' expectations converge in a given issue-area" (Krasner, 1982:185). The terms "regime" and "institution" are frequently applied interchangeably, as I will also do (see section 1.2.1). *Institutional factors* include obvious functions provided by a regime, such as information, monitoring and the use of incentives, but may also involve more subtle aspects. The focus on institutional factors rests on the assumption that rules and procedures may have an independent role in moulding behaviour and establishing expectations among the negotiating parties. For instance, procedures for participation and agenda setting may prove to have an independent impact on the achievements of international collaboration. The term *implementation* will be used here to indicate a deliberate effort by national authorities to follow up their international commitments in domestic policies within the specific issue-area. "Negotiation output" or "IEA objectives" will be used to denote the item which is to be implemented (see section 1.2.3).

The rapid loss of *biological diversity*[1] represents a considerable global environmental problem. International efforts here have been most comprehensively expressed in the international Convention on Biological Diversity, Rio 1992 (hereafter: the CBD or the BioConvention.) While the main bulk of biological diversity is situated in tropical countries in the "South", it is primarily the developed countries that have economic and (bio)technological capacity to extract the full commercial benefits from these resources: The benefits include large revenues from sectors such as agricultural biotechnology and plant-based pharmaceuticals. The negotiations on biodiversity have thus concerned distribution as well as conservation issues. This is the basis for my choice of studying the creation and implementation of the CBD in a North-South perspective, and with a *developing country* as the main case.[2] *Ethiopia* is an interesting case for studying biodiversity conservation and use for two main reasons. First, there is the unique biological diversity of the country.[3] Secondly, there is the fact that Ethiopia has long conducted a deliberate policy aimed at conservation of, and control over, its own genetic resources.[4]

Against this backdrop, I propose to investigate the following research question which I believe may shed light on the broad thesis question:

---

1 Biological diversity is defined as the diversity between and within species (genetic diversity) and ecosystems world-wide (Convention on Biological Diversity, May 22, 1992, Nairobi).

2 There are several ambiguities concerning the concepts of developing and developed/industrialised countries, or the "South" and the "North". The concepts tend to obscure obvious and significant political, economic and cultural differences within the two groups. As large parts of the international debate on biodiversity have been formulated as part of a North-South conflict, and also for reasons of clarification, I shall nevertheless keep to this admittedly simplified dictonomy. In various parts of the study, the differences between countries within these crude categories will be addressed, however. This was also an item for discussion during the CBD negotiations, as the treaty makes specific use of these concepts in designating duties and rights to the parties. It was resolved by applying the UN definition for developed and developing countries as well as defining a third category for the "East-bloc": "economies in transition".

3 Ethiopia has long been regarded as a centre of diversity for a number of important food crops, like wheat, barley, cabbage and onions, in addition to a very large variety of types of coffee.

4 Genetic resources are the hereditary material (genes) in all animals, plants and micro-organisms; the concept refers to the genetic diversity within and between species.

How have international institutional factors had an impact on the formation of the CBD (regime creation) and subsequently affected domestic implementation of the CBD objectives in Ethiopia (regime impact)?

While my hypothesis is that regimes do matter, the broad null-hypothesis is that domestic policies in Ethiopia would not have been significantly different in the absence of the CBD. In the analysis I shall seek to move beyond the question of whether regimes matter, and examine how – through what causal mechanisms – institutional factors may have had an impact.

The focus of analysis will thus be on the creation and impact of this specific international environmental agreement. The analysis will be conducted on two levels and in two main parts: *Regime formation* at the international level, and *Regime impact* at the domestic level. Throughout the analysis I shall seek to identify causal mechanisms linking activities at the international and national level.

Regime formation

The analysis of *regime formation* at the international level involves several theoretical and methodological steps. I start with the formation of the CBD, rather than using the CBD itself as the point of departure, because I assume that the factors influencing the negotiation process and output at the international level will have an impact also in the domestic implementation phase. We need to start with some ideas about what kinds of output at the international level would enhance the likelihood of domestic implementation in a developing country. States might be presumed to be more *willing* to follow up their international obligations if they see it as being in their interest to do so, or if they regard the agreement as legitimate. Alternatively, the output may be something dictated by more powerful states. Finally, national compliance with international regulations may come about as a result of external factors, such as economic or technological developments.

This case in point is Ethiopia – a developing country (DC). Existing international structural power relations would hardly predict a breakthrough for DCs interests in IEAs.[5] If, however, regimes matter this could mean that institutional factors have facilitated such a breakthrough: This could be implied for instance by the regime having facilitated learning and by the regime containing incentive mechanisms. In order to be able to offer predictions as to the kind of influence the international agreement may have brought to bear on domestic implementation, I will need to determine whether and how it has come to coincide with DC interests. There are three possible results in this regard:

---

5 For the main part of this study I shall apply the terms "developing countries" (DCs) and "the South" interchangeably. In some instances I will use the term "Low Income Countries" (LIC), adopted from the World Bank. LIC will be used to distinguish the low income countries from the newly industrialised countries (NIC.) This distinction is made clear in Chapter 5. The more inclusive term, "DC", seems at present to be the most simple, broadly used, and intuitive suitable term available; I shall address the need to employ other terms as I go along. For the "North", I shall in addition use the terms "developed" or "industrialised countries", or OECD countries, interchangeably.

- First, there may be *no correspondence* between the CBD and DC interests. Even in that case, there may still be a scope for institutional factors having influenced the output,[6] but the regime will be less likely to affect domestic implementation in DCs.
- Second, the CBD objectives may *happen to coincide* with the interests of DCs to a greater extent than was the case under pre-regime conditions, through a coincidence of compatible interests. If so, the potential effects of institutional factors on output are probably limited, but there may still be regime effects in the implementation phase.
- Third, the CBD output may represent a *real breakthrough* for DC norms and interests, "against the odds". If so, there is ample reason to question the role of institutional factors.

In speaking of a *breakthrough* for DC norms and interests, we must bear in mind that this is more a question of degrees, rather than an either/- or situation. The rationale for introducing the concept of breakthrough is to have some kind of standard by which to measure the CBD output. The idea of a common set of DC norms and interests must, of course, be handled with caution.

Furthermore, in order to examine the possible role of institutional factors in the regime formation process, I will need to control for exogenous factors. Drawing on a range of sources within international relations theories, I shall study regime formation from various analytical perspectives, each giving rise to different analytical inferences about what kind of result (output) to expect. The theoretical background and analytical framework for the analysis of regime formation will be presented and developed in Chapter 3. Regime formation is analysed in Chapters 4 and 5.

A central aim in this study is to use established theory to promote and further develop an analytical framework for studying implementation of international environmental agreements (IEA) in developing countries (DCs). Most studies of the formation and effects of international environmental institutions (and these too are still in their early phases) have been directed towards OECD countries.[7] Bringing in

---

6 While the bulk of regime theory is occupied with studying what and how institutional factors may promote agreement among actor interests, institutional factors may be conductive in strengthening actor controversies, as well.

7 IEA effectivness is being studied at Dartmouth College: "*The effectiveness of International Regimes*"; at the Institute of Marine Studies, University of Washington; Institute of Political Science, University of Oslo and the Fridtjof Nansen Institute (FNI): "*The Effectiveness of International Resource Cooperation*"; and at the Centre for International Affairs, Harvard University: "*The Effectiveness of International Environment Institutions*". The FNI is also involved in research connected with the environmental research programme of the European Union. See e.g. David Vogel & Veronica Kuhn, 1987, "The comparative study of environmental policy: a review of the literature", in Dierkes M. et al. (eds.) *Comparative Politics Research: Learning from Experience*, (Aldershot: Gower).

the North-South perspective may call for a somewhat different approach, and possibly lead to some refinements in established theoretical propositions.

Regime impact

The other main focus of this study concerns (i) *whether* Ethiopian biodiversity policies are in compliance with the commitments under the CBD, and if so, (ii) *what factors* may account for this consistency and whether it is a result of deliberate efforts to *implement* the CBD. In order to talk about regime impact we must be able to trace the consistency back to a causal effect by institutional factors.

i) The first question regarding *measurement* (*whether consistency*) must deal with the following issues:

- A crucial task is to trace and establish correspondence between the CBD objectives and national biodiversity policies. Ethiopia has long pursued deliberate policies relating to conservation and use of plant genetic resources.[8] Consistent policies may, or may not, have been enacted *prior* to the establishment of an international agreement in order to *anticipate* and ensure a position in the forefront of international developments. Perhaps governments and national industries seek competitive gains, for instance by adjusting to stricter regulations in advance. Technological adjustment may give a competitive edge to industries, as would be the case in the phasing out of, and finding substitutes for, CFC in the ozone issue-area. Similarly, by enacting "good" conservation strategies in anticipation of a new Bioregime, a government may seek to mark itself as a trustworthy recipient of additional funding from the international donor community for domestic environmental programmes. Hence, the analysis will have to distinguish existing policies in the issue-area from newer policies which may somehow be related to the CBD.
- On a similar note, we are faced with a contrafactual issue: Would domestic legislation have been different in the absence of the IEA? This points to the need for diachronic comparison, to study the development of conservation policies in Ethiopia over time, and through different phases. Can we pinpoint related changes or improvements in domestic policy legislation in conjunction with, or coinciding with, the introduction of CBD? In that case, then it must be possible to identify specific formulations indicating such covariance in international and national policy outputs.
- Conversely, we must also allow for a time-lag between the conclusion of an IEA and the examination of whether policies are in compliance.

---

8 If no such effects can be identified, there is obviously little sense in discussing the influence of international regimes on domestic implementation. This is not a specific problem related to implementation in developing countries. Whether our case is a DC or an OECD country, there will probably be some policies already existing in the issue-area, prior to the advent of the IEA.

ii) If we find that Ethiopia has been following commitments under the CBD, the next questions concern whether and how this may be ascribed to impact from the regime. What are the main factors *accounting* for consistency between Ethiopian biodiversity policies and the CBD? Is domestic policy the result of deliberate implementation efforts?

- First, it would seem that some degree of coincidence of interests may be a necessary, if not sufficient, condition for states to be willing to implement their environmental obligations under an IEA. A perfect coincident of interests would, on the other hand, present great methodological difficulties to the analysis of whether or not the regime might have had an impact.
- Second, even though interests are perfectly matched, a country may well still choose to free-ride with regard to IEA obligations. On the other hand, it seems reasonable that a country that sees its own preferences reflected in an IEA would be more likely to fulfil its obligations, than if the opposite was the case.
- Third, domestic factors must be controlled for. This involves checking whether any covariance between the CBD objectives and Ethiopian biodiversity policies may be put down to relevant and significant changes in domestic factors. If domestic policies are in compliance with the objectives in an IEA, this may be caused by factors unrelated to the international institution. There may, for instance, be domestic factors, such as economic or technological developments, which account for goal attainment (compliance). Domestic factors will also give an indication of the *ability* to conduct implementation. Clearly, states vary in terms of their capacity to formulate and enforce environmental policies, hence, an understanding of relevant domestic factors is essential in order to evaluate domestic implementation activities. Low implementation activity in a country with high implementation capacity must rank lower than low implementation activity in a country with low implementation capacity.
- Fourth, if international regimes have an impact on domestic policies, it must furthermore be possible to link such impact to domestic factors. Hence, I will examine whether and how domestic implementation activities can be identified and linked through causal mechanisms to international institutional factors (i.e. related to the establishment of the CBD).

If we are to identify the causal mechanisms linking factors at the international and national level, we need a firm grasp of the most important domestic factors affecting environmental policies. Furthermore, in order to control for external factors that might account for national compliance with international standards, I will also have to check for possible significant and relevant changes in domestic factors.

What sources can we draw on in developing such an analytical framework? In Chapter 7 I shall explore the lessons from implementation literature for contributions to the construction of an analytical framework. This will involve various methodological problems as to establishing linkages between theoretical concepts and empirical data, and; more specifically, it raises the question of the transferability of concepts between developed and developing countries. Another important methodological concern is the quality and amount of data needed to substantiate findings of causal mechanisms and links.

In order to focus on implementation in a North-South perspective, I have chosen a partly explanatory, partly exploratory *case-study* as the main methodological approach. Considerable uncertainty still attends variables and causal mechanisms in this area. The case-study approach would appear appropriate, since it allows for an in-depth analysis of causal relationships through the method of *process-tracing*. The first part of the study involves the challenge of compressing and handling large quantities of theoretical and empirical data. Conversely, the second part must deal with a situation where empirical data are relatively scarce, as it is difficult to gain access to documentation of the policy process in a country like Ethiopia. These aspects, along with other methodological problems, are addressed in Chapter 2. We will return to them more specifically as the analysis proceeds.

In one of my daughter's nursery books, there is a story of the crow who has just finished doing his spring-cleaning, by dint of heaping all his belongings in a big pile on the floor. When asked how he will manage to locate the stuff at the bottom of the pile, he replies with great equanimity: "Don't worry, I put everything on top". This is very much what I would like to do. In studies such as these, a major problem is often felt to be: "How to present a coherent view of all the items within the pile at the same time?" In reality, however, the problem is related more to how to present the various components in as clear and logical a manner as possible. Before proceeding to the presentation and development of analytical frameworks, I consider it necessary to prepare the ground by elaborating on (1) the dependent variables, and (2) the issue-area itself:

1. As noted, the two dependent variables are, – *institution (regime)* and *implementation*. It is important to notice that the regime constitutes the dependent variable in the first analysis (regime formation). In the next step of the analysis the regime constitutes an independent variable, as this second analysis is concerned with possible regime effects (domestic implementation). In order to measure implementation, we must have a firm idea as to what is to be implemented. It is essential to distinguish institutional factors influencing regime formation, from the rules and regulations constituting the regime to be implemented (see section 1.2.1). The next step is to clarify what constitutes implementation activities in this study, which actors it refers to and the timeframe included in the study (sections 1.2.2-4).

2. Secondly, a brief introduction of biodiversity as an issue-area is long overdue here. We also need to view it in comparative perspective with other issue-areas in global environmental negotiations. An important point relating to the evaluation of regime formation, as well as the measurement of domestic implementation activities, is the question of what may be expected. Among the

large number of issues negotiated on the international arena, some problem areas will obviously prove more easy to handle than others (section 1.3.)

1.2 The Dependent Variables: Institutions and Implementation

In this section I will seek to clarify the following:

1. What are "institutions"? How may we distinguish between the international institution as an independent factor affecting implementation, and the policy programme of the institution that is to be implemented?
2. Why should we study implementation at all?
3. How should we delineate the study of implementation, conceptually and time-wise?
4. How may we define "implementation" in the case of biodiversity conservation and use?

1.2.1 International institutions

This study is basically concerned with a subgroup of institutions, the international regime, although the two terms are often used interchangeable. In Stephen Krasner's well-known definition, international regimes are "implicit or explicit principles, norms, rules and decision-making procedures around which actors expectations converge in a given issue-area" (Krasner, 1982:185). Keohane has offered a somewhat simpler definition which sees regimes as "institutions with explicit rules, agreed upon by governments, that pertain to particular sets of issues in international relations" (Keohane, 1989:4). The way we define regimes obviously has a bearing not only on how we distinguish them from other entities, but also on how we distinguish the item to be implemented. Krasner's definition highlights that, in order to talk about a regime, there must be some degree of normalisation of rules as well as some degree to which actors' expectations converge in a specific issue-area.

Krasner's definition provides several concepts on which to hinge the classification of regimes:

- *Principles* denote general goals and causal beliefs, while *norms* describe general rights and obligations (Krasner, 1982:186). We may distinguish between principles and norms on the basis of whether they set up homogenous or differentiated roles among the participants, and whether they assign property rights aimed at internationalisation, nationalisation or free access (Levy et al., 1995:275).
- *Rules* provide concrete and explicit regulations for action, in terms of prohibitions, requirements and permissions concerning, e.g., standards, timetables and targets.

- *Procedures* include decision-making in terms of voting systems, the organisational set-up[9] and various compliance mechanisms (Levy et al., 1995:277). This is the classification I will employ in describing the negotiation output of the CBD in Chapter 5; it will also be useful for making comparison with other international conventions.

We also need to distinguish between the institutional aspects of the regime and the character of any explicit organisation that may accompany it. In Oran Young's definition (1982b:18), "an *institution* is a recognised pattern of behaviour; an *organization* is a material entity possessing attributes such as an office, personnel, a budget, equipment, and so forth". Essentially, *regimes* are issue-specific and do not have the capacity to act, whereas an *organisation* is equipped with staff capable of acting and need not be restricted to one issue-area (Hansenclever et al., 1996:179). We may illustrate the relationship between the two by saying that a widely recognised institution will be more likely to be surrounded by a strong organisational set-up, or that a strong organisational set-up may increase the impact of the institution. On the other hand, an institution may eventually become bogged down by organisational overload. This could happen if bureaucratic red-tape takes precedence over active policy-making and implementation.

Hence, with reference to this study, the United Nations Food and Agricultural Organization (FAO) is clearly an organisation, but it contains several institutions, such as the Undertaking on plant genetic resources. The General Agreement on Tariffs and Trade (GATT) is an institution, and one which has recently strengthened its organisational aspects by its transformation into the World Trade Organization (WTO). Likewise, the CBD is treated as an institution, but it has organisational attributes, as exemplified by the Secretariat and various subsidiary bodies. Ciba-Geigy and Monsanto (transnational corporations involved in *inter alia* seeds), on the other hand, are organisations.

Studying the effects of international institutions on implementation generally implies a focus on the institutional mechanisms that may affect actor behaviour. Such mechanisms may include incentives, sanctions, monitoring and information; they may point to the authority and legitimacy bestowed by the institution, and they may involve the effect on behaviour from regular negotiations. While I am interested in the effects of such mechanisms, I also want to focus on how such mechanisms come to be set up in the first place. Obviously, an international regime cannot be any stronger than the negotiating parties wanted it to be.

In studying implementation we are thus faced with a situation where the item to be implemented (the objectives and mechanisms of an IEA, a policy-output or a programme), may also affect the process of implementation itself. Implementation

---

9 Krasner delimits his definition of procedures to: "decision-making procedures are prevailing practices for making and implementing collective choice" (1982:186). I take the definition to include also the institutional set-up, like the secretariat and scientific bodies attached to the regime, as these will be part of the decision-making procedures of the institution.

studies often draw a distinction between the content of the policy to be implemented and the context in which the policy is to be implemented. Both of these are assumed to affect implementation (see Chapter 7). Hence, we must distinguish between the international institution (or regime) as a context, and the content of the international agreement (or output or programme).

On the basis of Krasner's definition of regimes, Levy et al. (1995) reason that when normalisation is low, we may talk about "tacit regimes", when convergence of expectations is low we may talk about "dead letter regimes", and when both have a high score we may conclude that we are dealing with a "classic regime". Furthermore, in Krasner's definition it is possible to distinguish between *implicit* norms and rules generated by the regime, and the *explicit* regulations (more or less stringent, clear rules) to which states have committed themselves. The *implicit* norms may affect state behaviour, and the *explicit* regulations prescribe goals for state behaviour in a specified issue-area. In line with Jørgen Wettestad's advice (Wettestad, 1995), I shall distinguish between the two as *structural* regime rules and *regulatory* regime rules. The *structural* part of a regime include the procedural rules (participation, decision-making, compliance mechanisms) whereas the *regulatory* part include the principles, norms and regulations emanating from the regime.[10] Both these sets of rules may have bearings on implementation, even though it is of course the latter that is to be implemented. It seems to me that Krasner's definition makes it easier to distinguish between structural and regulatory rules. This also illustrates the difference between the study of regime formation on the one hand – which entails its structural and regulatory rules, and on the other hand the process of implementing the regulatory rules into practical politics.

Accordingly, the policy objectives set out in the CBD text constitute the regulatory rules, the subject for implementation. In the absence of clear targets, however, it may be difficult to distinguish between the different types of *regulatory* rules. On the one hand there are the regulations or specific policy goals to be implemented, and on the other the principles and norms constituting part of the expectations associated with and generated from the regime. The objective of *fair and equitable sharing* set out in the CBD is one example of a rule of such dual nature. It constitutes a regime expectation that may affect implementation; at the same time, we want to examine whether this objective is being implemented by the Convention parties. In this respect, the vague regulatory rules of the "Bioregime" may constitute a difficult case, compared to other international institutions, when it comes to measuring implementation. On the other hand, even though the CBD lacks quantitative standards, we should be able to identify activities that are "moving in the right direction", also with a view to increasing equitable sharing.

A further example may shed light on the often blurred relationship between structural and regulatory rules, or between mechanisms and objectives. This concerns the common use of *national reporting* as a monitoring mechanism in IEAs.

---

[10] Oran Young makes a similar distinction between the substantive core (rights and rules), and the procedural component of a regime (Young, 1982b).

This mechanism has been set up in order to enhance implementation, in other words it is a means and not an end in itself. On the other hand, when the parties agree to submit their national reports, this is clearly part of the implementation process itself: Hence, national reporting can be seen both as a mechanism and as an objective.

1.2.2 Why study implementation?

Two main reasons stands out here: practical necessity, coupled with a growing academic interest. Both at the national and international level, implementation studies are confronted with methodological hurdles.

Along with the proliferation of international environmental agreements, there has been a growing academic concern with factors influencing the creation and operation of international environmental regimes.[11] While the main bulk of research to date has been devoted to the creative phases,[12] there is now *a marked shift towards the question of operation.* In several environmental issues, the time now seems ripe for studying implementation of the objectives established by the institution. A consequence of turning to the operational effectiveness of international agreements is the need to examine implementation at the national level, as well as the links between the national and international arenas.

Paralleling the development in domestic policy studies, it has become clear to students of international environmental agreements that the decision, or *output, is not the end of the story*: The classical perception, as expressed in Weber's "ideal bureaucracy" and Taylor's "Scientific Management", held that implementation will happen "automatically", given the appropriate policies. The next phase involved an increasing understanding of the complexities associated with implementation (Simon, 1947; Etzioni, 1961), accompanied by a widespread disillusionment regarding the capacity of states to plan and transform their societies (Wildavsky, 1982).

At both levels, attention has traditionally focused on the causal factors leading to the policy output. This may partly be due to the substantial methodological problems linked to the study of implementation. The questions of *when* to start (and when to stop) studying implementation, whether to measure implementation success in terms of *behavioural change* or genuine *problem solving*, and how to identify *relevant*

---

11 Haas, Keohane & Levy (1993) take institution to mean, "persistent and connected sets of rules and practices that prescribe behavioral roles, constrain activity, and shape expectations. They may take the form of bureaucratic organizations, regimes (rule-structures that do not necessarily have organizations attached), or conventions (informal practices)". In *Institutions for the Earth* (Cambridge, MA: MIT Press).

12 So also with regard to the biodiversity issue. The author has previously analyzed the negotiation phase leading up to the CBD, i.a. in the report: *International conservation of biological diversity: The quest for effective solutions*, FNI, R:012-1991 (Lysaker: Fridtjof Nansen Institute).

actors[13] in the implementation process, should serve to indicate the considerable challenges involved in this endeavour.

The concepts of *compliance* and *implementation* have often been used in an intermittent way in the literature. Recently there seems to be converging agreement on their use, namely to apply the following distinction:

- *Implementation* is used to indicate the actual activities pursued by governments in order to fulfil their international commitments. In this case, it makes sense to question whether institutional factors have had an impact on behaviour change in the implementing country. At the same time, implementation activities do not necessarily lead to actual compliance.
- *Compliance* refers exclusively to whether policies are in accord with international standards or not. It does not refer to the causality behind that output. For instance, a prescribed and actual 30% reduction in emissions of certain pollutants could have come about even if the government made no active efforts whatsoever: The reduction could be due to economic recession or to technological developments. Hence, there may be perfect compliance without any intentional implementation efforts on the part of governments.

Implementation has been defined by van Meter and van Horn (1975) as "those actions by public and private individuals or groups that are directed at the achievement of objectives set forth in prior policy decisions". Implementation is also often understood in a broader sense, including "activities that move in the right direction", even though it may fall short of specific, rule-oriented *compliance*.[14] The *intentional* element pointed out by van Meter and van Horn ("directed at") is important for distinguishing between implementation and compliance. This is also important for the way in which implementation will be understood in this study: "deliberate efforts by national authorities to follow up their international commitments in domestic policies within the specific issue-area". In the identification and evaluation of these efforts, I shall also be looking for "activities moving in the right direction".

Neither compliance nor implementation will necessarily lead to actual *problem solving*, as the international regulations prescribed may have been insufficient. This brings us to the next question: How to approach and delimit the study of implementation?

---

13 The debate about who constitutes relevant actors is primarily linked to the top-down / bottom-up controversy. The study of IEA implementation is primarily a top-down project, but this should not lead to the neglect of non-central decision-makers (see Chapter 7).

14 This is what Marc A. Levy, Oran R. Young & Michael Zürn calls "the political approach" to the study of regime consequences. "The political approach" focuses on behavioral change, while other approaches may look for regime effects in terms of problem-solving, economic efficiency or justice. Marc A. Levy, Oran R. Young & Michael Zürn, (1995).

1.2.3 How to study implementation?

The whole implementation process may be conceived of as a series of *heuristic stages* at different levels. Here we must note that these are ideal type phases which are not intended to imply causality nor rigid chronology in the policy-making and implementation process:

1. Problem perception: *some degree of consensus that some kind of problem exists, which may be improved by international co-operation (getting to the agenda)*
2. Negotiations: *international negotiation process (regime formation)*
3. Output: *international environmental agreement (IEA)*
4. International implementation: *further international negotiations on i.a. protocols, enforcement and financial mechanisms*
5. Domestic implementation in three stages:
   - 5a. Domestic implementation output: *ratification, legislation establishing domestic policies and programmes aimed at following up international commitments*
   - 5b. Domestic implementation outcome: *enforcement of these domestic policies leading to corresponding behaviour change in target groups at sub-national level*
   - 5c. Impact: *genuine environmental problem-solving.*[15]

An essential problem in the study of implementation of international environmental agreements is associated with the choice of dependent variable: whether to stop at *domestic legislation in agreement with international goals* (5a), or at *behaviour change in agreement with* policy goals by target groups (5b), or to go on to study *impact* in terms of *effective problem solving* (5c). The third one (5c) is clearly premature as well as beyond the scope of this study: First, the CBD is too immature, both in terms of the short time since its conclusion, and because it is largely a framework convention, which does not allow for a study of its effectiveness in terms of problem solving. Second, measuring "real" ecological problem solving is generally believed to fall outside the purview of social scientists. Such measurement necessitates a broad involvement of natural science expertise, and this will still, in most cases, involve a great deal of uncertainty.

The study of the two other alternatives (5a and 5b) is also complicated, as IEA objectives often lack tangible standards by which activity may be measured. Furthermore, looking for institutional effects only in terms of legislation and behaviour change, rather than real problem solving, raises the question of whether this is enough to conclude that regimes "matter". To this Haas, Keohane and Levy (1993) reply: "For international institutions to make a difference environmentally,

[15] The steps in this "implementation process" are not necessarily linked by any causal connections. Problem solving may come about without regime-activated behaviour change, as e.g. developments in the technological field may render regime-prescribed activities redundant. In that case the merits can obviously not be ascribed to the regime.

they must spawn political change, and therefore it is appropriate to judge them according to how well they do so" (1993:397).

In light of the short time-span and the vague objectives of the CBD, implementation will be studied primarily by looking at the national policies and legislation which have been initiated and enforced (5a), and which may be said to be "activities that move in the right direction".

The main indicator of *output implementation* will be harmonisation or integration of the CBD objectives in national legislation. This again is indicated by:

1. follow-up in the form of conservation plans and programmes,
2. symbolic language (i.e. using the same terms),
3. policy instruments aimed at conservation activities, and
4. related institutional changes in government agencies.

This does not completely solve the measurement problems, as there will still be uncertainty regarding what activities are really headed in the right direction, and also about how far these activities must go in the right direction in order to count as implementation activities (more about this in section 1.3). As pointed out by Chayes and Chayes (1993), the acceptable level of compliance will vary across regimes, times and occasions, and partly depend on the significance that parties place on others' performance. Evaluating successful implementation of international environmental agreements is not a straightforward matter, for several reasons:

- First of all, there is no one set of standard criteria for evaluating whether implementation has been successful or not. Any evaluation is thus basically a matter of individual perceptions, for instance whether an output is evaluated in terms of ecological objectives, in terms of its economic efficiency, in terms of its social justness, or with a view to the political feasibility of implementing it.
- Moreover, the individual evaluation of implementation success will depend on whether aspirations were high or low to begin with.

As mentioned in the introduction, the evaluation of compliance and implementation is very often complicated by existing policies in an issue-area – policies enacted prior to the actual adoption and entry into force of an IEA. Such policies may or may not have been enacted in order to anticipate and be in the forefront of international developments. This easily leads to a contrafactual analysis, asking whether domestic biodiversity legislation would have been different (stronger/weaker) in the absence of the CBD. In order to establish some link between the CBD and domestic output, it must be possible to identify changes in domestic policy legislation in conjuncture with, or as a result of the CBD. This was a central point during my interviews in Ethiopia, reported in the second main part of this study.

By looking at the content of policy and the political and institutional context in which these policies are being developed, we may also begin to address the potential for *implementation outcome* (level 5b). In the next section I shall briefly present the activities associated with outcome in terms of expected target group behaviour. Here it is also important to keep track of what is happening at the international level (4). As yet, the CBD is largely a framework convention, and the parties are still engaged in filling it with practical activities. This ensuing international bargaining is likely to have implications for the national policy process, but the development of domestic legislation may also feed back into the international arena: For instance, it is conceivable that a country may stand out as a role model for other countries in terms of bold legislative proposals. Thus, activities at the various levels will to some extent follow in chronological order, but there will also be considerable scope for feedback mechanisms between them: Implementation at one level may (and is clearly intended to) affect implementation at the next one, and this must be taken into account in the analysis. On the other hand, it seems clear that, at this point in time, implementation at level 5b has not come far.

My delineation (choice of focus) of the dependent variable also has implications in terms of *when to start and when to stop the study of implementation*. As far as the time-span (*when*) is concerned, this is assumed to pull in two directions. Whether our focus is on output, outcome or problem solving, certainly some time must pass after a policy decision is made for any effects to be noticeable. If the time horizon is cut too short, we may neglect the potential for learning and adapting to new standards. On the other hand, execution of policy decisions may dwindle over time, as new ideas capture attention (Sabatier, 1986).[16] A relevant finding in the Ethiopian case-study is that despite the short time-span, the CBD is already being used to legitimise and spur national legislation and policy formation. The time-span available for this study is about five years (1992-1997). If the main goal is to study implementation in terms of outcome, the general advice is that five to ten years may be needed (Sabatier, 1986). Since my focus is on output, I believe that five years will suffice for the purpose of this study.

In the next section, I present the main objectives of the CBD and the implementation activities which may be associated with following up these objectives.

1.2.4 How to conceive of implementation in biodiversity conservation and use?

In order to evaluate implementation, it must thus be possible to identify new legislation, programmes or agencies as a result of the IEA, and furthermore ask whether these activities seem to be heading in the right direction.

---

[16] In order to capture possible peaks and declines in implementation performance, the ideal would obviously be to conduct repeated measurements. This phenomenon is no less present in environmental issues, as public opinion has had a tendency to ebb and flow quite substantially in this sector. It must also be assumed that the less resources that are available for implementation in the particular policy area, the more vulnerable it will be to such fluctuations.

What kind of domestic policy (output) and what kind of behaviour change (outcome) may be associated with implementing the objectives envisaged in the Biodiversity Convention?

The CBD text has no obvious focal points. There are three main objectives:

- conservation of biological diversity,
- sustainable use of its components, and the
- fair and equitable sharing of the benefits arising out of utilisation of genetic resources.

The first two objectives pertain to a large array of interrelated sectors, as the main threats to biodiversity – habitat destruction, overexploitation and displacement by introduced species – are associated with extensive human activities such as industry, agriculture and forestry.[17] As regards action on these specific sectors, the CBD leaves it to the member states to enact more detailed regulations. Such regulations may involve specific costs to specific domestic actors, for instance in restrictions on land use for local people living close to wildlife reserves. In developed countries there may be restrictions also in terms of control of biotechnological enterprises. Policies directed at enhanced wildlife management must address local administrative bodies, and then again local communities of people, throughout the country.

The third objective of the CBD may be easier for DCs to follow, as general benefits may more easily be seen to outweigh the costs. The CBD states that each country has the sovereign authority to determine access to its genetic resources, that access to genetic resources requires *prior informed consent* and must be on *mutually agreed terms*, and that a country providing genetic resources is entitled to benefit from the commercial use of its resources. The Convention envisages three basic mechanisms by which a country may benefit from external use of its genetic resources:

- participation in the research using the resources,
- receiving technology which embodies or utilises the resources, and
- sharing in a fair and equitable way the benefits arising from the commercial utilisation of the genetic material or resource.

National policy formulation in this area may thus be directed at enacting legislation which secures for the country sovereign rights over its own genetic resources. This may take the form of establishing agencies to control external access to genetic resources, like domesticated plant seeds and medicinal plants, and formulating the "terms of trade".

This section has provided a first glance of the BioConvention in relation to domestic politics. In the following, I shall explain how the subject can be viewed in

---

[17] See e.g. the Red Data Books of IUCN. See also E.O. Wilson, (1992) *The Diversity of Life* (Cambridge, MA.: Harvard University Press).

an international setting by contrasting biodiversity with other global environmental issues.

1.3 Biodiversity as a Global Environmental Problem

Among the great number of IEAs, clearly some issues are more easily solved than others.[18] In this section I shall briefly contrast the issue-area of biodiversity with other global and regional agreements, primarily those dealing with pollution control. Here, as in other sections, I make these comparisons for illustrative purposes and not in a systematic manner.

A common example of an easy-to-solve issue-area involves the international efforts to combat *ozone depletion*: This *global pollution issue* was in the later phases characterised by a low degree of political contention and fairly simple technical solutions with relatively low costs to the parties concerned. As large industrial corporations did manage to develop substitutes for CFCs, and thus achieved a comparative edge, they emerged as important environmental allies in the push for stricter regulations. All countries are vulnerable to ozone depletion, while the costs of phasing out CFC gasses have been relatively low. On this basis ozone has generally developed into a "benign" issue.

Climate change, on the other hand, has been characterised as a "malign" issue-area: It is far more complex, involving great and unevenly distributed costs in restricting and changing behaviour in central sectors of society, such as transport and energy. Biodiversity resembles climate change in this respect. It involves the important sectors of agriculture and forestry, as well as the biotechnology, pharmaceutical and chemical industries. The capabilities for utilising biological resources economically are unevenly distributed, and so is the international distribution of biodiversity itself.

Efforts to find common solutions to combat the depletion of the *ozone*-layer were also greatly boosted by various critical scientific findings about ozone holes ("exogenous shocks") (Litfin, 1994). By contrast, loss of biodiversity and climate change, which also takes place in a gradual manner, hardly achieve the same shocking headlines. With respect to tropical forests, there is also an element of *scientific disagreement* present. Scientists from Brazil, Malaysia and Thailand are known to operate with smaller figures for deforestation than do the World Bank, international research institutes, and various environmental NGOs. Likewise, estimates of species extinction are hampered by uncertainty with regard to the number of species as a whole. This is not to say that the scope of extinction is insignificant. It is estimated that extinction due to destruction of habitats is taking place at a rate several thousand times as fast as what would be "normal" under "natural" conditions (Wilson, 1988).

---

[18] The number is uncertain, ranging between one and two hundred depending on whether i.a. protocols and bilateral agreements are included.

The causal relationship between, for instance, “toxic” pollutants and degradation of the sea is hardly disputed. In contrast to most *regional pollution regimes*, however, in the biodiversity area the links between causes and effects are much more complex. This leads back to the first aspect, pointing to the great number of sources leading to biodiversity loss, and indicating that almost every sector of human activity is involved and may be affected by efforts to address the problem.

About 95 % of the world’s terrestrial biodiversity is found within the *jurisdiction of sovereign states*. This aspect distinguishes the CBD from other IEAs like *ozone*, the *global climate* and the *Law of the Sea* (UNCLOS). While the CBD concerns what are essentially domestic resources, many other IEAs address resources outside the national jurisdiction of any one state. On the other hand, it is true that the climate issue also involves high costs for important domestic sectors.

Moreover, *most of the world’s terrestrial species biodiversity is located in developing countries*. The greatest losses of biodiversity are expected to take place in areas with the highest species diversity – the tropical forests of Latin America, Asia and Africa. Wetlands, coral reefs and coastal areas in many parts of the world are also considered to be under severe threat. It is also in many tropical regions that we find the so-called Vavilov centres: the “centres of origin” and diversity for the world’s food crops (see Box 5.6). Hence, policies to mitigate biodiversity loss will frequently place the main burden on developing countries. Moreover, biodiversity conservation will often have a short-term adverse effect on sectors such as agriculture and forestry, on which these countries are heavily dependent. Further exasperating this picture is the fact that while the main owners of biological resources are developing countries, the main users are the developed countries. The central conflict arises from the following situation:

- The huge economic potentials of genetic resources, due to progress in the new biotechnologies[19], have largely coincided with the realisation that biological resources are endangered. About 75 % of the world’s terrestrial species is found in what may generally be termed the South. At the same time, it is the industrialised countries who possess the economic and (bio)technological capacity to utilise the full economic value of genetic resources on a commercial basis.[20] The international debate on biodiversity is thus concerned not only with

---

[19] “Old biotechnology” refers to the use of micro-organisms (like yeast) to transform a product, e.g. brewing beer and baking bread, while “the new biotechnologies” refers to activities like tissue culture and recombinant-DNA (r-DNA) techniques. Such genetic engineering is a technology used to alter the genetic material of living cells through direct interference with the genome in order to make them capable of producing substances or performing functions alien to the unmanipulated cells.

[20] Countless useful genetic variations of global significance have emanated at the local farm and rural community level. One example is the yellow dwarf virus resistance gene found in Ethiopia, on which California’s US$ 160 million annual barley crop depends, as well as the high-lysine gene in sorghum, also of Ethiopian origin. (UNDP project document, ETH/93/G31 - A Dynamic Farmer-Based Approach to the Conservation of Ethiopia Plant Genetic Resources.)

conservation, but also with how to distribute the profits from utilisation of biological resources.

In the Montreal Protocol on the Ozone Layer, the developing countries achieved unprecedented acceptance for *technological and financial transfers* from the North. This success was largely possible because the developed world had more to lose, the developing countries were able to put up a credible threat about not joining the Convention, and the actual price tag was after all acceptable for the North. On a similar note, the Montreal Protocol on the Ozone Layer has been cited as a role model for a biodiversity convention. This international agreement is unique in that it sets out to deal with problems of pollution that have yet to inflict their damage.

The aspects which have been briefly introduced in this section will be subject to more detailed analysis in Chapter 4. Before turning to the development of the analytical framework, I shall now present my methodological approach, as well as indicate some of the more general methodological problems raised in this study.

CHAPTER 2

METHODOLOGICAL DELIBERATIONS

2.1 Choice of Methodological Approach

The study takes as its point of departure different theoretical perspectives from which analytical inferences will be developed deductively. These inferences will be compared to the empirical findings along the way, allowing analytical feedback to the theoretical perspectives. Applying the method of *process tracing* (detailed case-study) makes it possible to examine multiple observations, and to assess which causal mechanisms have been activated (King et al., 1994). Against this backdrop, I have chosen a methodology roughly described as a descriptive and explanatory case-study within the qualitative domain. This choice involves of course the trade-off between depth and scope – the single case allowing for greater focus on causal mechanisms, while the comparative approach may provide more room for testing hypothesis.

With regard to the international negotiations leading to the CBD, data material is at an acceptable (manageable) level. I have been following that process for the last eight years, staying in contact with the (primarily Norwegian) negotiators, and receiving the related documents on a continual basis. There is sufficient data to provide thick description of the various phases of the process, necessary for the process-tracing method. As regards the Ethiopian case-study, however, things have been a bit more problematic, as data material here have been less easily available and less plentiful. Beside a theoretical interest, an additional practical reason for choosing Ethiopia was that I could link up with a team from the University of Oslo engaged in research in Ethiopian natural resource management. The team has a fairly well established network of contacts in Ethiopia, thus facilitating access to informants and documents. I was also familiar with some of the central Ethiopian negotiators myself, from my established contact with the biodiversity issue in international negotiations.

The data from Ethiopia have been gathered through non-structured and open elite interviews as well as written documentation. As national legislation in the issue-area is a focal point in the analysis, the documentation is primarily in the form of formal legal texts and ministerial background documents providing input in the legislation process. Another important focus is the development of conservation programmes, and for this purpose relevant documentation from non-governmental and intergovernmental organisations is included.[21]

2.2 Implications of Methodological Approach

While the research question and theoretical approach goes a long way in defining the possible scope for methodological choices, the methodology will in turn have implications for what one finds. This section will explore some of these possible implications, relating them to my own study of regime formation and implementation of the CBD in Ethiopia.

Construct validity

Are we actually measuring or investigating what we intend to measure and investigate, in line with what our theoretical perspective prescribe? The importance of having a clear idea about how to deal empirically with the analytical concepts applied in a study (construct validity) was the background for going into detail on *implementation* and *institution / regime* in the preceding chapter.[22] In this particular case, degree of *implementation* could simply have been defined as the degree of correspondence between the policy objectives in the CBD and domestic biodiversity policy in Ethiopia. When we define and limit implementation to *intentional activities* moving in the right direction, this enables us to *demarcate it from compliance* with objectives, which may happen coincidentally due to e.g. external economic or technological developments.

Construct validity is often cited as a serious obstacle to applying concepts developed for studies in OECD countries to studies in developing countries. By choosing, in Chapter 7, to draw on the framework worked out by Martin Jänicke (1995), I can limit these obstacles: Jänicke's model was built as an analytical framework for the study of environmental policies in developing countries; he later went on to see whether the same concepts could be applied in analysing environmental policies in developed countries as well.

There may be specific problems related to *construct validity in the interview situation,* as different values on individual factors such as sex, age and cultural background may confuse or limit intersubjective understanding of concepts. In this case, many of my informants, like myself, have long been familiar with the international biodiversity issue; judging from our dialogues we do seem to "speak the same language". This situation of shared knowledge may, however, lead into the opposite pitfall – unconditional agreement. A typical bias is the tendency to perceive biodiversity mostly in terms of a North-South issue, and in effect toning down

21 Some of my respondents wished to remain anonymous. I prepared a confidential list of interviewees in Appendix 1 in the dissertation version of this book (Rosendal, 1999).

22 Like *power* and *state*, *regime* is a conceptual creation, and neither can be observed as they really are (Kratochwil & Ruggie, 1986:764).

potential domestic conflicts. Hence, it has been important to include NGO members and regional level administrative staff, many of whom have not been involved in the domestic biodiversity legislation process, nor were they acquainted with the CBD.

Internal validity: Control for third variables

Internal validity relates to our efforts to avoid spurious effects when examining causal relationships between the factors. Each time a causal relationship is suspected, or an inference about what caused an event is made, the researcher also risks overlooking alternative, rival explanations. To identify causal mechanisms requires causal inference, and hence involves the need to define causality.[23] The causal effect itself can be defined *counterfactually* as the difference between actual outcome (the value on the dependent variable with the presence of the independent variable) and expected result (the value on the dependent variable without the independent variable) (King et al., 1994: 81): "What would the output have been like in the absence of the international institution?"

In my study, I operate with several explanatory perspectives, trying out one after another. The empirically based pattern (between an independent and the dependent variable) is then compared to several theoretically predicted patterns. The essence of the process-tracing method is that these predictions or inferences are made repeatedly through the various stages of the process under examination, thus increasing the number of observations within the single case study. When patterns coincide, internal validity is deemed to be strengthened. Nevertheless, there is always the possibility that some other set of factors constitutes the "real" explanation.[24]

Making inquiries about how international environmental agreements may be successfully implemented, basically implies a desire that this should come about. Had we re-phrased the question, and instead asked how states may succeed in defecting and becoming free riders vis-à-vis international treaties, or how local communities might free themselves from centrally directed policies, we might have ended up with a different set of findings with regard to implementation.[25] Another

---

23 A causal mechanism describes the link between an independent and dependent variable. A methodological implication pertains to my efforts in Tables 3.1 and 3.2 to sort out the main independent variables and causal pathways in the analytical framework of this study: One scholar's independent variable may be another's causal mechanism. Hence, the ordering in the tables may be questioned, but I still consider it necessary to include them.

24 For instance, my interviewees agree that the CBD is already being used to legitimise and spur national legislation and policy formation. An alternative explanation is that there are economic or technological developments, or an upsurge in domestic pressure groups, or even external pressure, which may be the real factors accounting for the apparent boost in policy activity. The theoretical propositions with which I operate are primarily *probabilistic* in nature. Hence, when the "right" conditions are found to be present without being followed by the anticipated event, this can be taken to weaken the theoretical proposition – but it does not necessarily refute it.

25 This latter point relates to the top-down/bottom-up debate in implementation theory which is discussed in Chapter 7.

belief may be the inherent fairness or goodness of the norms furnished by international institutions, as the origins of such norms are sometimes traced back to epistemic communities or grassroots movements: This potential bias will be addressed along the way in my study.

External validity: Generalisation

Lijphard (1975) has argued that what is lost in terms of statistical generalisation and control for third variables in single or multi case-studies compared to statistical method, may be gained by lesser problems of validity and reliability. In order to generalise, we must be able to make causal inferences regarding the relationship between independent and dependent variables. It may be argued that virtually all studies of international regime formation are to some extent marred by selection bias with regard to the dependent variable – as they all focus on regimes which actually have come into existence, rarely if ever controlling for issue-areas for which no international regime did emerge. In effect, the selection of cases regarding *regime formation* has basically very little variation on the dependent variable, although it may be argued that the output may be more or less suited to address the problem for which the regime was intended to solve. The outlook, in terms of drawing causal inferences, is somewhat better for the study of the implementation phase (*regime effects*), as there is obviously great variation among regimes and within regimes as to how they score on national implementation.

Bernauer (1994) argues about the importance of defining the universe from which our case or samples are drawn, regardless of the scope for generalisation. Generalisations in my case concerns both other *countries* and other *issue-areas*. There may be sizeable differences between countries as to the implications for domestic implementation of the CBD objectives. The Ethiopian case-study might be of scant relevance for implementation in most OECD countries, but it may not be unique as far as implementation in developing countries is concerned. Moreover, there are intrinsic differences between *issue-areas* for which international regimes are created. The type of issue-area may have implications for the kind of output to be expected. The scope for generalisation between issue-areas and countries is dealt with in section 9.6.

PART TWO:

THE CONVENTION ON BIOLOGICAL DIVERSITY – REGIME FORMATION AND IMPLEMENTATION

"When I use a word", Humpty Dumpty said, in a rather scornful tone, "it means just what I choose it to mean – neither more nor less."

"The question is", said Alice, "whether you can make words mean so many different things."

"The question is", said Humpty Dumpty, "which is to be master – that's all. When I make a word do a lot of work like that," said Humpty Dumpty, "I always pay it extra."

Lewis Carroll, *Alice's Adventures in Wonderland; Through the Looking Glass*

CHAPTER 3

THEORETICAL BACKGROUND AND ANALYTICAL APPROACH

Once a deductive approach has been chosen, the next step involves a careful examination of relevant theoretical contributions. Studies of international relations, international negotiations and, more specifically, international environmental agreements, differ quite substantially in terms of theoretical point of departure and analytical framework. Differences between scholars are not so much due to rigorously opposing views regarding the main factors at play, as to differences in focus, as authors cultivate and analyse particular aspects of international relations. Hence, to some extent it will make sense to classify the contributions as different theoretical schools, but it will also be a matter of distinguishing between different foci. A related and important aspect of these frameworks involves the process of identifying and defining fruitful analytical concepts.

Here we need to start by presenting a general picture of the main theoretical perspectives, in order to identify their main orientations. This will prove useful for narrowing down and identifying relevant contributions with a view to constructing an analytical framework for this study. Taking Stephen Krasner's three main models within regime theory as a point of departure, we can use the following classification:[26]

1. Realist regime model (power-based perspectives);
2. Modified regime model (interest-based perspectives);
3. Ideational regime model (norms and idea-based perspectives).[27]

---

[26] Krasner (1982) uses the classifications: realist regime model (power-based), modified regime model (interest-based) and Grotian regime model (norm-based). In my view, the Grotian term could just as well have been applied to the second category. Compared to the ideal types employed by the English school on thinking about international society; Hobbes (realism), Grotius (rationalism) and Kant (idealism), the classification applied by Krasner seems confusing. Hence, I chose to apply the term ideational in order to illustrate that within this third category the main focus is on norms and ideas.

[27] This also roughly corresponds to Arild Underdal's (Underdal, 1993) typology, in which he stresses the importance of understanding the intended scope of validity of different approaches / schools:

i) neo-realism (power-based) explanations are mainly designed to account for the weight the different actors will carry in the aggregation phase;

ii) institutionalism (interest-based) explanations, which are developed to specify what constitutes rational choices in situations characterised by interdependence; and

iii) cognitivists (knowledge based) explanations which focus on formation and impact of values, beliefs and ideas. See also Arild Underdal (1998a).

Two important points concerning these models/perspectives:

- While none of them disputes the importance of actor preferences, power relations and the ground rules laid down by international institutions in accounting for outputs of international decision-making processes, the perspectives vary as to which factors are seen as having the most significant impact in explaining variation in output.
- There are many ways of dividing up these and related contributions in "meta-analytical" frameworks. Related distinctions could for instance have been based on public international law, international relations theory and policy analysis.[28] The choice and ordering of the broader theoretical field will, like the choice of the more specific analytical framework, depend largely on the specific question in focus. At the same time, it is clear that the choice of theoretical approach will to some extent affect the answers revealed by the study.

In section 3.2, these three perspectives, their main factors and main propositions, will be further developed, to enable us to make inferences about the kind of output to expect. The analysis in Chapters 4 and 5 will seek to reveal their applicability in explaining the various stages of the negotiation process and finally the international output – the CBD. The next step will be to make inferences about the implications of this first evaluation for the subsequent implementation process.

3.1 Theory Sources: The Growth of International Relation Theory

In earlier times, the subject-matter of international relations was dealt with by law, history and philosophy. As a reaction to World War 1, idealism (or liberalism) dominated the approach to international relations. The subject developed in the two powerful victors, the UK and the USA, both of whom wanted as little change in international order as possible. The mainstream view was that the ideals of democracy and international diplomacy could help in combating misunderstandings like those that had brought about WW1 (Hollis & Smith, 1990). E.H. Carr (1962) attacked this *liberal idealism* for being utopian and normative; and Morgenthau advocated *realism* as a theory which tries to understand international politics as it actually is, rather than as people would like to see it (1960:17).

A second world war ensued; after that, much of the post-WWII period was dominated by the *realist* school, which held that international law would rarely be able to play more than an epiphenomenal role in the ordering of international

[28] These distinctions are employed by Joyeeta Gupta (1997). See also Anne-Marie Slaughter Burley (1993): Slaughter Burley relies on the distinction between Realism, Institutionalism and Liberalism in her examination of the field. As most of the propositions inferred from public international law (*see* Slaughter Burley, 1993:220) may be found within the various branches of regime theory (Modified and Grotian), I shall limit my analytical framework to contributions within this broad field.

relations.[29] In this view, the international system is dominated by national interests and the struggle for power. States are perceived as essentially unitary, rational actors (URA model), acting in a context of anarchy, and with state autonomy as their overriding concern. The lack of a supranational authority in an anarchic international system severely restricts co-operation among states, even when there may be common interests. States engaged in international collaboration will be worried about cheating, and also about *relative gains*: the fear that other parties might gain more from the arrangement than they do (Grieco, 1988). Visualised as ideal types at two opposing ends of a continuum, realists conceived of moral and ethics as a product of power and politics, reflecting the interests of dominant powers – rather than as the product of universal ethics, as the idealists had claimed.[30]

The proponents within each school would, however, tend towards a more nuanced view of international relations: Central to the realist rejection of ideals as a guiding principle for understanding international politics, was the historical lesson that ideals and morals were just as likely to be used to legitimise evil as to pursue good. In this light, the realist project may be seen as an effort to "look the truth of power straight in the face" (Morgenthau, 1960:16). The continuum of high- and low-salience issue-areas is also important for grasping the applicability intended for the theory of political realism. The realists, in the old Hobbesian style, nevertheless offered a pessimistic view of the prospects for international co-operation, as cheating would always be an option. This meant that they would tend to view H0 as the most likely proposition concerning regime effects.[31]

A major challenge to realism came in the 1970s, as core realist assumptions came under attack, and attention was directed to interdependence and the positive role of international institutions in providing order in international relations. These scholars focused on *regimes*: the principles, norms, rules and decision-making procedures that pattern state expectations and behaviour in specific issue-areas. The regime concept was (re)launched in modern international relations theory by Ruggie (1975), who differentiates between three levels of institutionalisation: *epistemic communities,* which are purely cognitive; *international regimes,* which consist of a set of mutual expectations and generally agreed-to rules; and *international organisations* (1975:569). The regime tradition argued that the realists could not explain why international regimes persisted after the perceived decline in US hegemony. The notion of "hegemony" was the resort of the realists to explain how

---

[29] See Carr (1962); Morgenthau (1960); and Waltz (1959).

[30] The realists set out to establish a science, and thus a positivist approach. There was great belief in natural science in US and among immigrants from war in Europe. Still, the realists were criticised by the behaviouralists (like Morton Kaplan) for being normative in their definition of power, balance of power and national interest – because none of these terms could be defined "objectively" – as well as for ignoring domestic factors.

[31] H0 is the null-hypothesis saying that regimes do not matter.

regimes could after all emerge and be sustained. This turning point is largely associated with Robert Keohane's *After Hegemony* (1984).[32]

A large body of academic work found testimony that regimes enhance the likelihood of co-operation between states by reducing the costs of making transactions consistent with the regime, by reducing the incentives to cheat, by enhancing the value of reputation, by providing information, and by facilitating monitoring. In *some* international issue-areas, mutual interests could thus override the fear of relative gains and loss of autonomy and provide a basis for development of international regimes (Zacher & Sutton, 1991). These scholars then set out to study how improved institutional design might maximise effective international collaboration and compliance with international obligations. By approaching international law in terms of efficiency and transparency, rather than the moral obligations of the idealists, regime theorists thus provided a theory instrument to deduce that rules and norms do matter.

While the realists assume that power and interests go a long way in explaining the output of international relations, the main thrust of the regime school was to add the significance of institutional factors in modifying interests, and hence to focus on the interplay of interests and institutions in international relations.

The above is, of course, a simplification: both the regime and the later institutionalist tradition include a diversity of approaches to international relations, as they seek to cultivate and focus on different factors believed to affect these relations. The majority may still be placed within a Rational Choice tradition: Actors (states) have a set of goals (international and domestic); they then choose the best actions in relation to these goals, as well as in relation to what they perceive as constraints and opportunities. The mechanism linking goals and actions is rationality. A smaller outlet focuses on norms, ideas and learning as secondary important mechanisms in addition to rationality. While most start out with the hypothesis that institutions matter, the interest-based approach studies compliance in line with the interests of the parties, looking at cost/benefit calculations, whereas the ideational approach highlights learning processes and the generation and acceptance of common international norms and ideas, such as the benefits of co-operation: *Pacta sunt servanta*. In other words, the former is basically *utilitarian* in perspective, and the latter is *primarily non-utilitarian*. With "primarily" I want to indicate that as the ideational perspective also embraces the possibility of learning about the instrumental utility of ideas, not only the normative effect, these may also fall within a utilitarian tradition. On the same note, I have chosen the term *ideational* rather than *cognitive*, so as to include both learning processes (cognitive) and normative aspects.

One "branch", which we may call "neoliberal institutionalism", retains a larger part of the realist propositions such as the URA model and a modified version of the anarchic state of international relations. The URA model is a central element in

---

[32] See also Robert O. Keohane & Joseph N. Nye (1977).

game theory, which is a tool employed by realists as well as regime scholars.[33] Within this general, interest-based approach, focusing on interest constellations and issue-specific power, we also find the more recent contributions from the Tübingen team.[34] These contributions emphasise *problem structure*, seeing the characteristics of an issue-area as important determinants in regime formation; and *situation structure*, examining in detail strategic interest constellations (Hansenclever, 1996:187–91). Institutionalism has been criticised for still regarding states as predominantly *black boxes* or *billiard balls* in the international system. In Slaughter Burley's words, this view has long traditions: "The inability to differentiate between different types of states on the basis of regime type has been implicit in international law since Grotius" (1993:226). This view is grounded in the principle of sovereign states, as equal and identical, and in the principle of non-intervention.

There have, however, been various efforts to open the black box of the state also within regime theories. The theory of *interdependence* (Keohane & Nye, 1977; Krasner; 1983) points to increasing linkages among national economies, thereby making the distinction between domestic and international societies less clear-cut. The growing interdependence between states is seen as making international relations less anarchic and reducing the autonomy of states. Still, power remains an important factor in analysing international relations, as interdependence is often asymmetric. Such asymmetric interdependence is also apparent in the differences in sovereignty among states. This power is not necessarily the sole prerogative of the state, however. As argued by Susan Strange (1989), power has gone from states to markets; this then means that environmental treaties do not matter, as governments are not fully in control of their national economies. On the one hand, it is by no means clear that all states have lost structural power to other sources of authority – whether to transnational corporations or to some amorphous international network of scientists. However, "we are keenly aware that it [the world, involved in a market system] has serious problems of inequality, of environmental damage, of financial instability, of large-scale migration, unemployment and urban decay; and that for all of these problems, the territorially-bound authority of states is totally incapable, unsuited and unmotivated to resolve" (1989:242).

Another "branch" is *transnationalism,* identified with such names as Ernst Haas (1982) and Ronnie Lipschutz (1996), and also parts of Oran Young's (1982a) work. In this literature states are still central, but no longer necessarily seen as the dominant actors in international relations. Transnationalism points to the role of non-state actors such as multinational corporations, grassroots movements and scientific communities in formulating and advocating common norms and ideas in specific sectors of international relations. Thus, a transnational or global civil

---

33 Diverging from total anarchy, the Prisoners' Dilemma may for instance be resolved by a tit-for-tat strategy in iterated games; i.e. in repeated games each actor adheres to its promises as long as the other party does. See Robert Axelrod (1984).

34 A group of researchers at the Tübingen University, led by Volker Rittberger, has over the past decade made central contributions to the previously US-dominated regime debate.

society emerges, made up of i.a. epistemic communities and grassroots movements. In this perspective, the realist notions of anarchy and states acting solely in their own self-interest are challenged by what may be termed an "ideational" perspective: The perceived need to work together towards common global welfare. Agreement is more likely when there is consensual knowledge, when there are shared policy norms, and where strong epistemic communities assist the regime.[35] This model is less well developed or specified than rational choice models: Apart from the basic role ascribed to consensual knowledge, there are few predictions about under what conditions agreement is most likely to be successfully implemented, or what the content is likely to be. As pointed out by Arild Underdal (1998a:23), the process of learning is more elusive than choice, and it is rather a supplement than an alternative to rational choice models.

Furthermore, it is the ideational perspective that helps to close the gap between basically rational-choice approaches and what is known as a discursive (or reflectivistic or constructivist) approach in the social sciences. For instance, in analysing the ozone negotiations, Karen Litfin (1994) takes as her point of departure the notion of epistemic communities, and goes on to discuss from where do norms and knowledge emanate. In her view – and basic to discourse tradition – power and knowledge cannot and should not be separated. In effect, this relationship becomes the major focal point of these analysis: "interests must be problematized as arenas of political struggle that should be formulated in light of contending knowledge claims" (Litfin, 1994:2). The point about this intrinsic power/knowledge relationship is clearly important when it comes to issues involving the environment, where science and politics are often subtly intertwined. I will comment on this as I proceed, without, however, entering into lengthy ideological debates about the "real" preferences that constitute normative or cognitive expressions.

Most branches of regime theory have long been largely dominated by US scholars; comparison with the mainstream UK approach to international relations may shed some more light on its attributes and its utility for this study. A common notion within regime theory and what is known as the *English school* (or *international society*) is the assumption of a set of commonly accepted norms and rules among the contracting parties in an issue-area or the members of international society, respectively: This is what may be called an "ideational perspective" (Evans & Wilson, 1993). Although the comparison between specific regime-areas and the often much wider idea of *international society* may appear far-fetched, these traditions certainly share the rejection of the Hobbesian (realist) view that lack of a common sovereign prevents the possibility of international rule-making and rule-enforcement (Bull et al., 1990:54). This is indeed an important element in the

---

35 An "epistemic community" is defined as a professional group that believes in the same cause-and-effect relationships, uses the same truth tests to assess them, and shares common values. The members also share "consensual knowledge" from which they convert facts to policy-relevant conclusions. The power resource of such a group is its authoritative claim to knowledge. See section 3.2.3.

assumption that regimes matter. Admittedly, as Oran Young points out, regimes may affect actor behaviour even in the absence of common social norms, by way of modifying the utilitarian (cost/benefit) calculus of actors.[36] Likewise, in what Stephen Krasner (1982) has termed the realist approach to regime theory, states may still be regarded as unitary actors engaged in power maximisation. Still, in the most widely applied definition by Stephen Krasner (1982:185), regimes generate rules and norms which may affect the activities of actors; "implicit or explicit principles, norms, rules and decision-making procedures around which actors expectations converge in a given area of international relations". In both traditions, states are still regarded as the central actor, but domestic and transnational actors must also be considered. Both approaches emphasise interdependence, but while regime theory considers this a novelty, the English school takes it for granted. Regime theory scholars reject concepts of natural law and a common morality in international society as identified with the English school, restricting their conception of common norms to specific issue-areas. The outwardly most striking difference lies in the methodology: The approach to regime studies is here far more positivistic and behaviouristic, although in the end they also often resort to interpretation rather than observed data only (Kratochwil & Ruggie, 1986:764).

Summing up, our approach to understanding and explaining international collaboration may take its point of departure in rational choice calculations based on overall structural power and interests relations, the interplay between interests and institutions; or collaboration may be interpreted in terms of the acquisition of common norms. Against this backdrop, the analytical framework will be constructed in line with Krasner's three perspectives on regime formation and effects:

1. Realist regime model (power-based perspectives);
2. Modified regime model (interest-based perspectives);
3. Ideational regime model (idea and norm-based perspectives).

These three models are not clear-cut. Indeed, they are more likely to be compatible than competitive, as they illuminate different aspects of international relations and may be "activated" at different phases of the processes of regime formation and regime operation. The first two models dominate studies of international relations and both may be placed in the rational choice tradition, while the third, ideational, model is less widely used. Providing structural explanations, models within the rational choice tradition are considered to be more parsimonious and to be applicable in extensive studies where there is a large number of actors and little uncertainty regarding interest and power relations, i.e. a low degree of complexity (Underdal, 1984). These models thus seem more suited to account for the negotiation phase and the output of international negotiations. In order to tackle

---

[36] This would be a case of "tacit regime", with low degree of formality and a high degree of converging actor expectations. Other regime types have been classified as "dead letter regimes", with high formality and low degree of convergence; and "classic regimes", featuring a high score on both dimensions. (See Levy et al., 1995:272)

the problem of parsimony, it seems logical to start out with systemic factors at the international level as a first cut: How much can be explained by looking at overall structural power and interest relations within a specific issue-area, viewing states as unitary, rational actors? The next step would be to apply the ideational perspective of norms and learning processes and search for pattern-matching with the empirical material from that perspective. Finally, one may wish to open the black box and draw on domestic factors – which are, however, more likely to be idiosyncratic (Zürn, 1993). One obvious rationale for drawing an analytical distinction between the three models, rather than lumping all explanatory factors together in one mega-model, is to gain greater clarity in the analysis and presentation of the case.

In the following two sections I relate the three models more closely to international environmental politics and biodiversity. I begin with a few general remarks about the applicability of the models, and then turn to some examples from the literature.

3.1.1 The models: scope of applicability and approach to global environmental problems

How does the realist model relate to environmental policies? It may be argued that transborder environmental problems fall within the low salience category, as co-operation is imperative yet does not put national security issues at risk.[37] Hence, a realist approach to this issue-area could be regarded as irrelevant, in the sense that states would not treat the issue as one pertaining to, or possibly threatening, their national interests. I will present three arguments for including a realist approach in this study: First, it may be argued that environmental issues are in the process of gaining increasingly high salience among governments, as important domestic sectors are becoming the targets of international environmental regulations. Second, several studies find evidence that states do accept and comply with international environmental agreements to a greater extent than would have been predicted on the basis of "national interests" alone.[38] Third, I believe that the realist perspective will give rise to inferences which may shed light on various stages of the negotiation process. As we are dealing with global relations, power can be expected to remain an important factor as there is likely to be asymmetry between the parties involved.

How do the models relate to the specific issue-area of biodiversity? The *realist* view would have a strong case in arguing that, in an issue-area characterised by such factors as scientific uncertainty, complexity, strong and conflicting interests, a global geographical scope and resources largely under national jurisdiction would make successful co-operation highly unlikely. In view of the international structural

[37] See also Kratochwil & Ruggie, who argue that the rational-choice approach of regime theorists is applicable to both high and low issue-areas, including both areas of conflict and co-operation (Kratochwil & Ruggie, 1986:762).

[38] See Haas, 1990; Wettestad & Andresen, 1991; Rosendal, 1991; Haas et al., 1993; Keohane & Levy, 1996; and Young, 1997.

power relations, this model would be suitable for highlighting possible hegemonic activity in the negotiation phase.

From an *interest*-based perspective, the issue-specific traits of biodiversity would clearly be seen as challenges, but not necessarily insurmountable ones. First, through analysing the actor interests and their cost/benefit calculations, we can identify the scope for package deals with something in it for everyone. Second, proper *institutional design* could go a long way toward finding and tapping the integrative potential in the issue-area. Regimes may thus have an impact on behaviour when uncoordinated individual action would have led to sub-optimal outcomes.

What are the implications of ideational perspectives like transnationalism for successful IEA (International Environmental Agreement) implementation? The notions of epistemic communities and global civil society seem implicitly to assume that domestic and transnational groups are inherently "better" (e.g. for solving environmental and developmental problems) than states.[39] A more cautious approach would be to avoid such assumptions, and inquire explicitly whether such transnational actors may help local communities, or leave them more open to external pressure. Another question which remains elusive within this approach concerns the relationship and distinction between consensual knowledge ("scientific facts") and common norms.

The *ideational* model would point to the scope for learning processes, again primarily with a view to the integrative aspects of the issue-area, but I would argue also due to the specific role of DCs. This perspective would indicate both these aspects as a basis for epistemic communities or grassroots movements generating common norms. When the student of international relations turns to cases of truly global environmental problems, however, the underlying ideational assumptions become strained. Here it is also questionable whether one can identify a "specific issue-area" in which to study regime formation and operation. When an increasing number of sectors in society must be engaged in solutions to global environmental problems, the development of common norms and ideas within an international regime will be hard to pin down. As the geographical scope of an international environmental arrangement increases, the "commonness" between the actors' norms and values is diluted. Is it then still pertinent to apply these kinds of assumptions when looking at implementation in developing countries? I will return to this question, both in the discussion of international and domestic factors in two subsequent sections.

---

[39] Ronnie D. Lipschutz (1996) argues that "environmentalists" within global civil society are "helping to change the ideational frameworks that support one set of constructions of social reality by replacing old intersubjective rationalities and ethics with new ones. Advocates of biodiversity do more than just bring animals, plants and bugs into the charmed circle of human ethics and law; they fundamentally change society's understanding of the relationship *between* humanity and Nature." (p. 6).

3.1.2 Studies of IEA effectiveness: What are their "projects"?

The study of implementation of international environmental agreements, and more specifically, the potential impact of international institutions in this process, is fraught with theoretical and methodological problems. Studies in this field are rapidly expanding, but they differ quite substantially in terms of theoretical point of departure and framework – making cumulative knowledge problematic and limiting the scope for comparison and generalisation. Analyses differ with regard to the measurement and specification of the *dependent variable*: the focus may be on implementation and behaviour change, social learning, or environmental problem-solving. The former approaches are often associated with the question of whether institutions matter, and are characterised by *contrafactual* analyses concerning the situation without the institution. The latter is concerned with the *optimal* situation, and the extent to which specific problems are actually solved. As argued in section 1.2.3, the time-frame of this study as well as its single-disciplinarity speaks in favour of choosing the former approach. I shall hence focus on those factors which address compliance and implementation, mainly in terms of national output, rather than being concerned with impact/problem solving effectiveness (see section 1.2). Attention will also be given to factors specific to the negotiation process, as these may have an indirect impact on the implementation process – for instance with regard to how interests have been aggregated and expressed in the international output.

Studies also differ regarding *measurement* of the output to be explained, the institutional effect on the dependent variable, and as to developing and testing theories about successful implementation: some through generalisation of "laws", and some by theory development through identification of causal mechanisms or pathways (Bernauer, 1995; Haas et al., 1993). Measurement of output concerns the choice of metric, and various contributions assess outputs, for instance in terms of justice, ecological sustainability or economic efficiency. My own focus is on the extent to which the developing countries (DCs) have had a breakthrough for their interests in the negotiation output, and whether and how institutional factors may account for this breakthrough. I then go on to discuss the role of institutional factors in the implementation phase. The first part of this endeavour is a relatively simple task in terms of measurement, the latter two involve identification of causal mechanisms through process tracing (see Chapter 2).

As far as the *independent variables* are concerned, a great many variables have been proposed, more or less linked to coherent analytical frameworks: Key variables include distribution of power, state interests, institutional factors and mechanisms, knowledge and learning processes (Strange, 1983; Wettestad & Andresen, 1991; Young, 1989; Haas, 1992). The space and time constraints of this study do not permit a thorough examination of each study; moreover, such work has already been undertaken in several excellent reviews (Bernauer, 1995; Hansenclever et al., 1996; Levy et al., 1995; Wettestad, 1999). I shall briefly refer to some of the main contributions in the field, to present the "projects" of these scholars and indicate how different questions are associated with different approaches. Do we need to pose different questions when addressing implementation in developing countries? Do we need a different approach to global resource management than to the analysis of regional pollution regimes?

Studies vary regarding whether they take "institutions matter" (how, when) as a point of departure or leave this an open question (whether). The majority of these studies focus on institutional factors, particularly collaboration between OECD countries.[40] I will argue that an analytical approach to implementation in developing countries needs to take overall structural power relations into consideration to a greater extent than is the case with OECD studies. The approach to assessing cost/benefit calculations for the negotiating parties in a global regime must take account of the large asymmetries between developed and developing countries, so more attention must be given to the negotiation process itself: If institutional factors matter in implementation in a global context, they must first be found to have had an impact at the negotiation stage. Studies within this field tend to focus either on regime formation, regime implementation effectiveness, or regime dynamics and robustness.[41] My approach necessitates a focus on both formation and implementation.

Next, there is variation regarding whether to apply a comparative approach and study variance in implementation between states or variance in implementation between regimes, or to limit the study to account for implementation in a single case (Underdal, 1998b; Brown Weiss & Jacobson, 1994). My analytical framework seeks to address DC implementation of IEAs as its scope of applicability – it should thus allow for comparison between the same IEA in different DCs and between different IEAs in the same DC.

Some studies present a combination of rational choice and a more normative approach to study regime effects on implementation (Levy & Young, 1994; Haas et al., 1993). Both may have valuable contributions, but the difference between these approaches needs to be clarified and separated. Within the framework of this study there is moreover a need for careful deliberation on the utility of an ideational perspective, as we are dealing with a global issue-area.

There are also sizeable differences with regard to specification level: Some move at a grand theory level (Bernauer, 1995), while others have a more empirical orientation, describing in detail causal pathways through which institutional factors may affect implementation (Levy & Young, 1994; Haas et al., 1993; Brown Weiss & Jacobson, 1994). For input to the analytical framework, the latter will obviously have a more direct utility, while the former have been found useful in sorting out the different focal points of the various approaches.

---

40 The exception to this rule is Adil Najam (1995). Najam's focus is on IEA implementation in both developing and industrialised countries.

41 The distinction can be found in Oran R. Young and Arild Underdal, (forthcoming) "*Institutional Dimensions of Global Change: A Full Scooping Report*", International Human Dimension Programme.

Furthermore, I shall concentrate on those factors which seem relevant in a developing country perspective. As I believe that most of the factors which may affect a developed or industrialised country may also affect a developing country, this will not go very far as a delimitation. I do, however, believe that the direction and strength of impact may vary between the two, and thus the discussion of relative impact will still be important.

As it is impossible to address all potential factors in the analysis, stringency in the analytical framework must take precedence over exhaustiveness. In order to come up with meaningful inferences about consequences of e.g. institutional factors, we have to focus on a small number of institutional variables. These must be embedded in a clear theoretical argument in order to be parsimonious and assist in theory development.[42]

Most related studies limit their empirical scope to the process within the regime in question. I will emphasise relating the story of what took place also in other, related fora of negotiations. This dimension is seldom found within the analytical frameworks established to study the formation and effect of international institutions, nor do I initially make it part of my own.

In concluding, we need to take a brief look at the pros and cons of applying a deductive approach, as I do, by embedding my analytical framework in established models. The advantage is the possibility of contributing cumulatively to theory building, while at the same time avoiding the risk of "re-inventing the wheel". The drawback may be less flexibility and innovation in approaching a complex and little-studied issue-area. I hope to avoid what might become tedious repetition, by maintaining a critical view of the propositions inherent in the theoretical perspectives as they are confronted with my empirical material.

3.2 International-level Factors: Analytical Framework

The main question in this study concerns implementation of international environmental agreements (IEA): How can institutional factors at the international level boost domestic implementation activities in a developing country? The study of domestic implementation of international environmental agreements necessitates an understanding of how the content of the IEA came about. Let us begin by seeing how I intend to apply the three analytical perspectives. I then present a simple overview of the main assumptions tied to the three perspectives, and briefly discuss the impact of exogenous factors, as this is a necessary starting point in all three perspectives.

i) General comments on the use of the theoretical models

Main questions to ask in this study are:

---

42 See Bernauer, 1995:374.

1. whether the CBD is a result primarily of the overall power and interest structure in international relations;

2a. whether it is accounted for by interest constellations and issue-specific power;

2b. whether institutional factors were instrumental in aligning or altering actor interests, thus tapping the integrative potential, e.g. through package deals including DC interests;

3. whether negotiations induced a learning process resulting in a greater acceptance of common international norms.

Finding support for either of the two latter explanations (2b, 3) would imply a higher likelihood that "institutions matter", than if the first two questions (1, 2a) were answered in the affirmative.

Within each of the three perspectives I shall first concentrate on factors specific to the negotiation process, as these may have an indirect impact on the implementation process, – for instance with regard to how interests have been aggregated and expressed in the *international output*. Second, I shall add factors that may directly affect implementation in terms of *national output*. Third, I highlight factors with specific relevance in a *developing country* perspective, and make assumptions about the direction and strength of these factors in DCs. Fourth, I fit the variables into the three theoretical perspectives, in order to follow Bernauer's advice about being "embedded in clear theoretical argumentation" (1995:374). Within each perspective I start with a general view and go on to make specific *propositions*. This also means focusing on a limited number of factors within each perspective, rather than trying to address all potential factors in the analysis.

On the basis of each of the three perspectives, I shall present propositions regarding:

Pa) expectations as to the *type/content of international output* (regime formation),[43]

Pb) implications for the *scope for domestic implementation* (regime effects), and

Pc) implications for the issue-area: *content of* domestic implementation activities.

The three perspectives highlight different sets of explanatory factors. These will be presented as a guide to the analysis of the formation phase, with subsequent implications for the implementation phase. Pa's will thus refer to the predictions made concerning possible breakthrough of DC interests in the *international output*. Pb's will refer to the predicted *scope* for *national implementation* activities (output).

---

[43] Recall the discussion about output content in Chapter 1: "Existing international structural power relations would hardly predict a breakthrough for DCs interests in IEAs. If, however, regimes matter this could mean that institutional factors have facilitated such a breakthrough: This could be implied for instance by the regime having facilitated learning and by the regime containing incentive mechanisms. In order to be able to offer predictions as to the kind of influence the international agreement may have brought to bear on domestic implementation, I will need to determine whether and how it has come to coincide with DC interests."

On the basis of each perspective, it must also be possible to infer (Pc) implications in terms of the kind of national output to be expected (content). The separation of Pb's (scope) and Pc's (content) is necessary in order to distinguish between the three perspectives: The perspectives may not necessarily predict very different *scopes* for domestic implementation, as they presume that different causal mechanisms may enhance or induce implementation. These different causal mechanisms may, however, focus on different aspects of the CBD objectives, yielding different emphases within domestic implementation activities. Hence, the perspectives may give rise to different assumptions regarding the *content* of domestic implementation (see Table 3.2). The aim is not to refute or strengthen one particular perspective in general, but rather to examine at what stage or phase one perspective or another may have greater explanatory force.

ii) Main assumptions

In the following sections, I present the three theoretical perspectives, their main foci, causal mechanisms and predictions about international output. This is summarised in Table 3.1.

Table 3.1:
Theoretical perspectives: main foci and causal pathways – international output

Main foci: main factors	Causal pathway	Effect on output
1. Overall power and interest: · hegemon	· structural leadership	DC breakthrough unlikely (business as usual)
2a. Interests and issue-specific power: · interest constellations	· basic game power · negotiation power	DC breakthrough possible (package deals)
2b. Interests and institutions: [1)] · number and type of participants · official purpose · agenda setting · information, monitoring and incentive mechanisms	· aligned cost/benefit perceptions · entrepreneurial leadership · reduce uncertainty · trust mutual compliance	DC breakthrough possible (employ integrative potential)
3. Institutions, ideals and norms: [1)] · legitimacy and obligations · epistemic communities	· learning/teaching · intellectual leadership · coalitions	DC breakthrough possible (accepted common norms)

[1)] The factors in 2b and 3 may work through several of the causal pathways identified in the corresponding boxes.

iii) Exogenous factors: A general starting point

In the study of the effects of institutional (*endogenous*) factors related to IEAs on implementation success or failure, important control variables are the inherent characteristics of the issue-area itself (*exogenous factors*). These characteristics may affect the possibility of having relevant parties participate in the first place, as well as influencing the scope for reaching agreement on commitments. Regardless of which of the three perspectives we employ, we will have to check for these exogenous factors, for two main reasons:

- *Measurement of institutional effect*: We need to control for third variables in order to reject the null hypothesis.
- *Evaluation/measurement of implementation*: By examining exogenous characteristics we may come closer in determining our expectations regarding implementation scope. A low degree of implementation in a difficult issue-area may "count" for more, or at least equally much, as a high degree of implementation in an easily resolved area.

What kind of characteristics should we be looking for in order to explore whether one issue-area is more or less easy to handle compared to other areas? In general, studies concerned with the formation and implementation of IEAs seem to concur that the *political* and *intellectual characteristics* of the issue go a long way in determining its "benign/malign" nature, by indicating aspects pertaining to the distribution of costs and benefits among the parties (Andresen & Wettestad, 1995; Miles et al., 1994).

⇨ *Political characteristics* concern the basic structure of property rights, as the users and owners of the resources associated with the issue-area will to a large extent define the interests and capabilities of the actors involved in negotiations. When a resource is a collective or public good, no actor or actors may be excluded from its utilisation, whether they pay for it or not, and thus the problem of free riders may ensue.

> *Solutions to collective action problems may lead to a* redistribution *of costs and benefits, or to enhanced benefits to all parties through a more* rational management *(Underdal, 1980). It is generally assumed that issues predominantly involving* (re)distribution *are inherently more difficult to resolve than issues where there may be greater scope for* integrative *solutions.*

The latter may be solved by co-ordination and *improved management* procedures, producing more goods to be shared by all parties involved (Underdal, 1980; Young, 1982b).[44] In such cases, a compatible interest structure is much more

[44] A related distinction has later been made between *incongruity* and *co-ordination* (Underdal, 1997, draft). For resources management the original distinction between redistribution and improved management seems sufficient to capture the characteristics of the issue-area.

likely. The malignant nature of *distributive* issues may be intensified by asymmetric relations between the parties involved (the parties' interests are negatively correlated), and by cumulative cleavage lines.[45] In that case, former free riders may be asked to internalise (i.e. pay the abatement costs for) externalities from their actions (e.g. air and ocean pollution).[46] It is assumed that when the distribution of winners and losers is asymmetrical and predictably so, aggregation deadlock will be more likely to afflict negotiations.

⇨ The *intellectual characteristics* of an issue-area will also go some way towards determining whether it is easily resolved. The scientific status or knowledge base of the issue is thus likely to have an independent impact on the expected scope for regime formation (Andresen, 1989; Young, 1982b):

> *Relatively easy to resolve will be issue-areas in which there is scientific consensus about cause and effects, the significance of the problem and relatively simple solutions.*

It is assumed that it is easier to *postpone* political action in an area on which there is a large degree of scientific uncertainty or discord, compared to a situation of consensus. The political and scientific characteristics are analysed in Chapter 4. We now turn to the independent factors, starting with the power-based perspective.

3.2.1 Power-based explanations

Issue-areas characterised by a high degree of asymmetry and a general lack of scientific consensus would, from a *realist perspective*, presumably have comparatively low chances for agreed co-operation in the first place. If we find that the biodiversity issue ranks high on these aspects, a first-cut realist regime model would give rise to the *proposition* that states would hardly be able to reach agreement on co-operation in this area, let alone implement any agreed measures.

Now it is a fact that the CBD has been agreed to, and has entered into force. This could simply imply that biodiversity is not perceived as a high salience, controversial issue-area among the parties. On the other hand – and this would apply to the majority of international environmental agreements – why then bother to participate in the negotiations? A realist-model explanation would subsequently point to the presence of a *hegemon* to account for actual international agreement on co-operation. Co-operation on trade and security issues among OECD countries in particular has to some extent been geared towards limiting the tendencies for one hegemon to gain strength among them (e.g. WTO, EU), but these lines of explanation do not figure widely in current IEA studies. In the absence of a clear-cut

45 These characteristics have been adapted from Arild Underdal's article "Patterns of Effectiveness: Examining Evidence from 13 International Regimes" (1997b).

46 Still, if the party with the highest degree of externalities also turns out to be most affected by the same activity, while the one that "pollutes" least is only marginally affected by the overall "pollution", this would be a case of a *benign* and more easily solved problem.

hegemon, the realists indicate the possibility that overall structural power relationships (hegemonic relations) still may have an impact, as interdependence in international relations is not symmetrical. This may well be the case in the relationship between developed and developing countries.

In a realist perspective, the overall structure of international power and interests largely determines the output of international collaboration efforts. The result is not likely to diverge greatly from the preference of the dominant party or parties. This assumption rests on three conditions: that there is a dominant party or parties, that the issue-area is one in which that party has a clear interest (high salience) tied to output, and hence, that the party is willing to employ positive or negative incentives to modify the behaviour of others.

To analyse the case in light of a realist framework, we need to address the following questions:

1. Why would a hegemon act: Is it possible to identify a basis for hegemonic relations within the biodiversity issue-area?
2. To what effect would a hegemon act: To what extent were one or several developed countries able to impose their preferences on the output of the bio-negotiations? If power decides, one would expect to find the interests of dominant state(s) reflected in the final result.
3. Through what causal mechanisms would a hegemon act: How can we trace hegemonic behaviour?

The first two questions of *why* and to *what effect* a hegemon would act are closely linked. Here we may profitably draw a distinction between *coercive* and *benevolent* hegemonic behaviour (Snidal, 1985):[47]

In connection with biodiversity, *coercive hegemonic behaviour* would roughly mean that the North would have access free of charge to the biological resources of the South, while the South would be pressured into putting restrictions on the exploitation of their own natural resources – in order that these may be preserved for future generations.[48] Such pressure might be exerted through threats of trade or loan restrictions.[49] This assumption seems to square well with another notion concerning implementation in developing countries: that external actors play a much more significant role here than in developed countries (Najam, 1994). The notion of coercive hegemony has, however, been little in evidence in realist literature on international relations, ever since the general observation of a decline in US global

---

[47] Snidal (1985) uses the terms "benevolent" and "coercive" leadership, linking the former to interests, motivations and collective action, and the latter to power capabilities to pursue narrow self-interests.

[48] This categorisation of "North" and "South" is clearly simplified, as will become clear in subsequent chapters.

[49] An instance where such policy proved to be a failure is, however, the USA's effort to curb Mexican tuna fishing through measures which were judged to be incompatible with GATT regulations.

hegemony and the end of USSR hegemony in Eastern Europe. Non-realists would add that coercive hegemonic behaviour would be incompatible with accepted conduct in multilateral fora.

Benevolent hegemons on the other hand, would be more inclined to act in the interest of the common good, paying a significant share to obtain this goal, provided that they perceived the issue to be of high salience. In this scenario, the developed countries would still be expected to achieve unrestricted access to the South's resources, but they would be willing to refund a substantial share of the kind of biodiversity conservation they deemed necessary. In realist terms, preservation of ecosystems for the future may be seen to embody an important aspect of the national *security* – which is central in realist thinking. Preoccupation with global environmental concerns clearly originates with the North, which has put the issue on the international agenda. In this light, it may be argued that environmental issues may be moving up the policy agenda to become higher-salience issues, for which hegemons may favour some degree of international co-operation (Haas, 1990).

Thirdly, there is the question of how: *through what causal mechanisms* hegemons attain their goals. The concept of the benevolent hegemon necessitates a clarification with regard to *leadership* in institutional bargaining and, in particular, Young's concept of *structural leadership* (Young, 1991). There is one important aspect linking hegemonic activities to structural leadership, as Young argues that the structural leader employs bargaining power to influence the behaviour of other parties:

1. First, bargaining power may stem from having less to lose than others from no agreement. In this case, weaker states may find some scope for leverage.
2. The second kind of bargaining power envisaged by Young stems from having more to gain from successful agreement – the structural leader would then offer side payment and try to manipulate the other parties. This can hardly be performed by weak states. Hence, both benevolent hegemony and structural leadership presuppose asymmetrical international relations, as well as self-interested actors, acting in line with their own interests.

Thus, we may regard structural leadership as a causal mechanisms through which benevolent hegemons attain their goals.[50]

Whether or not structural leadership is conductive to regime formation will depend on whether the hegemon is acting as a "pusher" or a "laggard" with a view to establishing regulations aimed at solving the problem at hand – in this case, the

---

50 Contrary to the hegemon-notion of the realists, however, Young does not consider structural leadership a necessary condition for the establishment of international institutional arrangements, and secondly he expects such leadership to frequently consist of several parties. Thirdly, Young puts greater emphasis on an actor-level approach in the study of structural leadership, as the talents and characteristics of the agent, acting on behalf of the state, are seen as important factors (Young, 1991).

establishment of a new regime.[51] In the analysis in Chapter 6 (6.2.1), I shall conduct a small case study of the role played by the USA in the negotiations and post-negotiation phase: First, because the United States is still the main contributor to IEAs in financial terms, and to some extent also by way of contributing to environmental problems. Second, because the USA has been acting both as a main pusher and a main laggard at various stages in the negotiation process, making it interesting to examine the US negotiation position more closely.

The realist perspective gives rise to the following theoretical propositions regarding international collaboration:

1. International outputs in high-salient issue-areas will be "dictated" by economically and technologically stronger states.
2. Global environmental issues are becoming increasingly salient with governments, thus opening for (benevolent) hegemonic behaviour to satisfy national interests.

With a view to the empirical case, the following propositions can be constructed regarding international output and the implementation phase:

P1a) expectations for the *type/content of international output* (regime formation):

The international output will reflect the interests of the dominant (structurally powerful) regime members.[52] A breakthrough for DC interests is highly unlikely here, unless they happen to coincide with the interests of dominant parties.[53]

P1b) implications for the *scope for domestic implementation* (regime effects):

Dominant regime members tend to induce compliance in weaker states. This is achieved through structural leadership, as they have the capacity to carry most of the burden themselves or to use sticks and carrots to achieve their goals. Hence, compliance activities in DCs may be expected to be induced by dominant Northern states. In line with the process-tracing method, I shall return to more explicit inferences drawn from this perspective after the analysis of the international output (Chapter 6).

---

51 Arild Underdal, "One question, two answers" in Ed Miles et al., "Explaining regime effectiveness: Confronting theory with evidence" (working title, forthcoming).

52 This proposition obviously has less relevance in an issue-area with small degree of conflicting interests (benign).

53 As indicated in Chapter 1, this classification of breakthrough / not breakthrough is not intended as a dichotomy. Rather it is a question of degrees, including normative breakthrough and / or a breakthrough for material interests. My reason for introducing the concept of breakthrough is to have some kind of standard by which to measure the CBD output, but it should obviously be handled with caution.

P1c) implications for the issue-area – *content of* domestic implementation activities:

> *As the output of negotiations will reflect the interests of economically and technologically dominant states, we might anticipate the development of conservation policies to be based on a notion of global benefits – aimed at preserving the largest number of species and ecosystems. The North would have access free of charge to the biological resources of the South, while the South would be obliged to put restrictions on exploitation of their own natural resources – in order that these may be preserved for future generations.*

If the international output is considered to be a compromise between the North and the South, or indeed represent a real breakthrough for DCs interests, a hegemonic explanation is probably less applicable. This does not necessarily imply that the realist model will be fruitless as an explanatory framework in the implementation phase. As the implementation process at the international level moves from a normative to a factual level, it will become clearer to what extent norms and principles are followed by operative (funding) mechanisms (Hanf & Underdal, 1998). If no mechanisms are devised to substantiate the compromise (or breakthrough), the asymmetry in the distribution of power between the North and the South will appear to be reflected in the implementation phase. If funding mechanisms are fully in the hands of the developed world, this would be an indication of benevolent hegemonic behaviour.

The power-based perspective is essentially concerned with how the interplay between *overall structural power and interests* determines an international negotiation process and its output. There may, however, be other sources of power within a given issue-area beside structural (economic and technological) power. A central assumption in the "after hegemony" perspective of complex interdependence is the notion that power capabilities may vary across issue-areas (Keohane, 1984). Leadership may be based on different sources of power, and capabilities and leader(s) may differ from one issue-area and arena to the next. This element links in with the next perspective, and thus brings us to the interest-based explanations: the interactions between interests, issue-specific power and institutions. There is little dispute that power without interests is meaningless, and interests without power are insignificant. Here I shall again emphasise the main difference between the two perspectives: While the power-based perspective focuses on overall power and interests, the interest-based perspective is concerned with the interaction between interests, issue-specific power and (issue-specific) institutions.

3.2.2 Interest-based explanations

Most studies concerned with the formation and implementation of international environmental agreements are found in the vicinity of the interest-based model, i.e.

models focusing on the interplay of interests, issue-specific power and institutions. There may be various reasons for this:

- First, most studies are concerned with collaboration among OECD countries, and here hegemonic explanations seem largely outdated. The predominance of studies on OECD country co-operation, rather than on issues including North–South relations, may lead to a potentially false conclusion: that the debate about whether regimes matter or simply reflect overall structural power has been resolved once and for all.
- Second, this may reflect an implicit or explicit normative tendency to regard more and better regimes, and greater international order, as goals in their own right.[54]
- Third, the focus on institutional factors is attractive because these, unlike the overall structure of power and interests, may lend themselves to modification and conscious design.

This section will focus on how various institutional factors and mechanisms are assumed to affect negotiation output and implementation. Nevertheless, a thorough examination of the interests and interest constellations involved is necessary. Before seeing how institutions may have an impact in terms of aligning or altering these interests, we need to look into the actors' potential for *issue-specific power*. In a situation of bargaining, where one party controls something in which the other party has an interest and vice versa, the exchange aspect of power may come into play (Hernes, 1975:40). Arild Underdal has made a distinction between issue-specific power which is "purely" interest-based, and issue-specific power which is related to the interplay between institutions and interests (Underdal, 1997a:17):

- The first type has been termed *basic game power*, and refers to the parties' control over the resources in question as well as their economic and technological capacity to make use of the resources.
- The second type is conceived of as *negotiation power*, referring to capabilities based on strength in numbers, coalitions, and leadership. The aspects constituting this second type of power are discussed further below, as I narrow the focus to the interplay between interests and institutions.

Underdal's distinction is important as it highlights how *basic game power* may be of continued use also in the implementation phase, while the utility of *negotiation power* will largely remain limited to the formation phase. The sources of issue-specific power will depend greatly on the characteristics of the issue-area. This will constitute the first step of the analysis in the interest-based perspective (section 5.5.2).

---

54 This potential bias, along with other criticism of regime studies, is pointed out by Susan Strange (1983).

Above I noted that most IEA studies are less occupied with hegemonic relations. This is not to say that most scholars within the interest-based tradition tend to disregard non-institutional (exogenous) factors. On the contrary, exogenous factors are important control variables in most studies of regime formation and impact:

- First, as pointed out above, some problems can be assumed to be harder to solve than others. Many studies have been devoted to analysing the political and intellectual characteristics of the issue-area, which are believed to identify the scope, if any, for both formation and problem-solving capacity of international institutions.
- As regards potential regime impact, compliance or problem-solving may come about as a result of economic or technological changes. Similarly, compliance may be achieved as a by-product of policies in other sectors altogether,[55] rather than as a result of any implementation activities encouraged by the regime.
- Third, external actors other than regime members may influence policies relevant to regime compliance: This point is particularly important here, as developing countries with weak domestic political systems are assumed to be more sensitive to external pressure.

Scholars associated with an institutionalist or interest-based tradition frequently seek to specify institutional factors which may encourage actors to overcome their reluctance to co-operate and follow up their obligations under IEAs.[56] The former features are associated with the *regime formation phase* and the latter with that of *implementation*. In order to see whether and how weaker countries may have had a breakthrough for their preferences, we need to start out with factors that may have had an impact during the formation phase. In the following, I make an analytical distinction between those institutional factors primarily associated with the formation phase and those associated with the operative phase. Obviously, the factors presented for the negotiation phase may be active also in the next phase of implementation, just as some of the implementation-phase factors may have had an impact at the formative stage as well. This is also reflected by some overlap between the two dimensions, as when scientific input is treated twice, both as part of the institutional procedural set-up and as part of the information mechanism provided by institutions. We will also return to the relationship between scientific facts and norms in my third main perspective, in section 3.2.3.

It should be made clear that I describe those values of independent variables which are anticipated to strengthen the scope of regime formation as positive effects, while negative connotations are ascribed to values that have adverse effects on regime formation. This is primarily a semantic and traditional choice rather than a normative one. On the other hand, the normative bias inherent in this choice may be

---

55 As shown in Laugen (1995).

56 See for instance Elinor Ostrom (1990), Arild Underdal (1980) and Oran Young (1989b).

questioned, as no study of IEAs take it as their point of departure to find out how international co-operation may best be *avoided.*

3.2.2.1 Formation phase – main institutional factors:

Several factors have been envisaged through which a regime may enhance agreement among parties in the negotiation phase. Factors associated with the *institution-as-arena* are primarily linked to how the formal *rules and procedures* may affect the parties' rational choices in negotiation games (Underdal, 1997b). In addition, and in line with the definition of regimes, this also includes how underlying *norms and principles* may influence actors' perceptions. Another set of factors relates to the *organisation-as-actor*, and concerns how various actors may affect agenda setting, given a *flexible agenda.* These observations are linked to the assumption that institutional arrangements vary with regard to the interplay of interests and institutions.

Institution-as-arena:

* *Decision-making rules* and *participation:*[57] When objectives of resource management are subject to international negotiation, the political feasibility of finding a solution depends on the inclusion of *all relevant parties* and on *agreement* among them (Underdal, 1980). When such a situation arises, coupled with consensus or unanimity as the decision-making rule, this may lead to *the law of the least ambitious programme* (1980:36).[58] *Participation* represents an ambiguous factor in global problem areas: All states are potentially *relevant parties* to the collaboration, so all must participate to increase both effectiveness and legitimacy. As the number increases, however, the scope for successful co-operation and implementation is likely to be reduced, since the risk of other parties free-riding increases, together with complexity and transaction costs. More information will be needed to make coherent decisions about management of natural resources as a regime's functional and geographical scope expands. Finally, we must note the difference between formal and actual participation; moreover, it is important to examine changes in participation over time (Andresen, 1989).

---

57 *Decision-making rules* have been cited as an important factor, with a distinction between majority voting and consensus rules: The former is believed to be better suited in terms of making efficient decisions, and the decision rule is hence an important factor in accounting for output. As IEA negotiations in general operate on the basis of consensus, however, this factor shows little variation in the universe of cases to which generalisation would be relevant. Hence, it cannot alone explain why various IEAs have had differing results.

58 The law says that "Where international management can be established only through agreement among all significant parties involved, and where such a regulation is considered only on its own merits, collective action will be limited to those measures acceptable to the least enthusiastic party" (Underdal, 1980).

* *Science-policy organisation:* Similarly, there exist various strategies for avoiding or mitigating this *law of the least ambitious programme*. Such strategies include selective incentives and comprehensiveness, for instance through package deals. The possibility of achieving integrated package deals in the agreement may depend on the *legitimacy of scientific input* and thus on *procedures for the organisation of the science–policy relationship*. Legitimacy can be assumed to depend on representation and participation in relevant science-policy fora. Opposed to the comprehensiveness assumption is Young's idea of *salient (focal) solutions* as a factor enhancing the scope for agreement. However, we may assume that no simple focal solution could have been conceived in the biodiversity negotiations.

* *Official purpose:* A related, albeit more elusive, aspect of the institution-as-arena may also affect the extent and content of issue-linking, and hence the scope for agreement. This concerns the general normative orientation of the negotiation forum itself: It is assumed that the *official purpose* of a negotiation forum may be *framing* the issue-area (Midgaard et al., 1973). Such underlying aims of institutions may further increase governmental willingness to co-operate:

- by magnifying public pressure (Haas et al., 1993),
- by enhancing the authority of the regime, and
- by attracting special *types* of participants to the negotiation table (March & Olsson, 1989).

As March and Olsson (1989) point out, there may be certain expectations tied to individual members of political institutions, through the logic of appropriateness.[59] While the sheer *number* of participants may go both ways in affecting regime formation, there is growing agreement that the *type* of actors may matter: for instance, that the participation of non-state actors and epistemic communities may improve the formation and effectiveness of environmental regimes (Levy et al., 1995).

Organisation-as-actor:

The organisation-as-actor dimension gives rise to a proposition with particular relevance to implementation in developing countries: A *flexible, open agenda* and an *active secretariat* may aid weaker countries in getting their interests included, for instance through package deals. This factor is thus partly dependent on whether the international institution is backed up by an active, autonomous secretariat. Such institutional factors may be instrumental in aligning or altering the interests (the cost/benefit calculus) of the parties.

* A flexible agenda is hardly a causal mechanism in itself; in order to visualise the link between institutional design and agenda setting, it may be useful to employ

---

59 There are several similar contributions found on the borderline between interest-based and ideational perspective. See for instance Stokke & Vidas (1996:23) and Franck (1990). Franck points to how the norms and rules of an institution may have a compliance pull of their own.

the concept of *entrepreneurial leadership*. While Young's *structural leadership* can be linked to power, entrepreneurial actions can be linked to tapping the integrative potential among interests in a given issue-area, with entrepreneurs acting as brokers of innovative policy options (Young, 1991:294). In effect, rather than ending up with *the law of the least ambitious programme* or even less than that, entrepreneurs may succeed in constructing creative package deals at a higher feasibility level. The scope for entrepreneurs would thus also seem to link up to the science–policy relationship.

* Against this backdrop, it is also apparent that quite different *types of actors* may be wearing the *entrepreneurial leadership* hat. These actors may include individual members of active and relatively autonomous secretariats, non-governmental organisations, or members of national delegations. This could easily end up in a conceptual stalemate as to which is a benevolent hegemon, making feasible proposals for the common good, and which is an entrepreneurial leader, doing much the same (Young, 1991: 297).[60] Of course, a party that is in a position to be a structural leader may also utilise entrepreneurial skills. The transformation cannot go in the opposite direction, however: entrepreneurial skills may be acquired, while structural power is hardly open to manipulation.[61] This does not go far as a distinction, so I have chosen to associate the entrepreneur primarily with actors who would not have been able to sway decision-making on the basis of overall structural power relations alone.

3.2.2.2 Implementation phase – main institutional factors:

I shall distinguish between three broad types of functions through which regimes may facilitate co-operation: information, monitoring and sanctions. These are directed at fostering and strengthening trust among the parties, and may also affect the parties' perceptions of costs and benefits relating to the regime.

* *Information*: In the "weak" form, the regime may be instrumental in shifting cost/benefit perceptions among the parties by providing relevant and easily accessible *information*. The parties may regard it in their interest to subscribe to an international agreement because this reduces transaction costs and enhances the flow of information (Haas et al., 1993; Levy et al., 1995). To assess whether this factor is relevant in a DC perspective, we must ask what kind of information we are talking about, whether the information is equally relevant to all the parties, and whether it is possible for some parties to control the flow of information:

---

60 Young argues that the motive force of an entrepreneurial leader may be self-interest, e.g. compensation in the form of "prestige, political influence in domestic arenas, or progress toward the achievement of some larger personal goal" (1991: 297). This does not, however, necessarily exclude the structural leader from the definition.

61 See also Raino Malnes (1995). Malnes argues that the use of sticks and carrots, associated with hegemonic power and structural leadership, cannot sustain leadership, properly defined.

- One type of information may include *knowledge about the issue-area and issue-links*, which may in turn give rise to package deals. Here, weak countries may be assumed to have an interest in, and potentially benefit from, increased information flow. This type of information may also be associated with how norms and ideas come to prevail, which is discussed further below.
- A distinction should also be made between *utilisation* of information and *production* of information. Participation in the latter may be a precondition for taking part in the former: In some issue-areas, the South has argued that they had limited access to the fora in which scientific information (later to be utilised in the negotiations) was "produced".[62] While developed countries may have become convinced as to the necessity of co-operating due to a large body of scientific information stressing the urgency of the environmental threat, this may not have had the same impact on developing countries.
- On a similar note, developed countries may welcome an international surveillance system, while the South may fear that this will mean meddling in their internal affairs. This could be the case with surveillance and information about habitat and species loss. Information of this type largely concerns the (other) parties' activities in the area, which again broadly concerns transparency and monitoring.

* *Monitoring*: In the "medium" form, the regime may be instrumental in shifting cost/benefit perceptions among the parties through monitoring compliance and reciprocity. Monitoring may enhance implementation through two mechanisms:

- Monitoring may increase transparency and thus help to *reduce distrust* among the contracting parties (Haas et al., 1993). If trust is enhanced through monitoring mechanisms (e.g. national reporting), the parties are assumed to be more likely to follow up their obligations. This would seem to pertain equally to poor and rich countries. Parts of such surveillance systems could, however, be regarded with distrust in developing countries.
- Monitoring may expose non-compliance (free riders), which might then be subject to *informal sanctions* such as *public pressure* (Haas et al., 1993). Public sanctions against free riders are unlikely to affect developing countries significantly – not only because of the capacity problem associated with DCs, but also because it is the industrialised countries which have had the actual possibility of being free riders in biodiversity matters.

* *Sanctions and incentives*: More "strongly" affecting the parties' interests, a treaty may be equipped with incentive mechanisms such as funding or technology transfers, or various kinds of formal sanctions. Again the aim would be to shift the balance of cost-benefit calculations so as to enhance the benefits of compliance.

---

62 This has been found to have been the case during the early phases of the climate change negotiations (Gupta, 1997). See also Leiv Lunde (1991).

- *Incentive mechanisms*, such as financial and technological transfers, are of greater importance in developing countries than in industrialised ones. Incentive mechanisms may be equally important in medium income (or NICs) countries as in low-income countries (LICs), as such transfers may affect both the *will and ability* to implement. If such mechanisms are provided by the regime, this is assumed to boost implementation in poor countries by *enhancing domestic capacity* for effectuating environmental policies (Haas et al., 1993). To this I would add that these mechanisms must have some degree of accessibility and sufficiency, if they are to have any impact on implementation in developing countries. One of the few and recent examples of such incentive mechanisms can be found in the Montreal Protocol on Substances that Deplete the Ozone Layer, which established a fund to *enable* developing countries to comply with the treaty limitations on chlorofluorocarbons (CFCs). It was clear that (future) major contributors of CFC gases such as China and India would not have been *willing* to ratify the Protocol without such financial incentives, and thus the major donor countries reluctantly agreed to the fund (Chayes & Chayes, 1991).
- Of less importance in a developing country is probably the use, or threat of using, *sanctions*. On the other hand, sanctions may be important with a view to making other (developed) parties in the international community comply with regulations – which may in turn be a prerequisite for DCs to do so. In the absence of a central authority, formal sanctions in international institutions are seldom particularly severe. Even social sanctions, like threats of exposing harmful behaviour, may, however, be felt as sufficiently strict to encourage compliance.

If these factors combine to produce clear, stringent rules and an efficient institution, this will necessarily have important effects also in the later phase of implementation. Equally obvious, however, is the possibility that the particular characteristics of the endogenous (institutional) factors are perfectly correlated with exogenous factors such as actor interests and power capabilities: A regime may have been designed to failure or success by powerful actors, i.e. through the structural leadership of benevolent hegemons. This relates to the scientific-theoretical dispute between those who see the whole as something other or more than the sum of its constituent parts, and those who do not. The former view is a basis for assuming an independent impact from regimes. Once more, this points up the importance of controlling for exogenous variables.

The interest-based perspective gives rise to the following theoretical propositions regarding international collaboration:

1. As power capabilities may vary from one issue-area to the next, the overall structural power relations is only one of many factors which may influence the formation and implementation of IEA. Hence, regimes may be more than mere epiphenomena.

2. Regimes establish some degree of formal rules and converging actor expectations:

- *in the formation phase*, institutional factors may help align actor interests by identifying and boosting the integrative potential among them. This may take place through flexible agenda setting and by providing a forum in which entrepreneurs can forge issue-linking. Rules and procedures, along with the official purpose of the negotiation forum, could also be instrumental in altering or aligning actor interests, and could give rise to mutual expectations among the parties.
- *in the implementation phase*, institutional factors may help reduce uncertainty and insecurity among the parties through information, monitoring and sanctions. These mechanisms may have an impact in establishing trust about mutual compliance among the parties, by reducing transaction costs and by providing a frame of reference for future interactions.

Against this background we may construct the following propositions:

P2a) expectations for the *type / content of international output* (regime formation):

A breakthrough for DC interests is possible. Institutional factors have facilitated the development of agreement between the interests of industrialised countries and DCs.

P2b) implications for the *scope for domestic implementation* (regime effects):

If incentive mechanisms are established as part of international environmental agreements, DCs are more likely to follow up their commitments under the regime. In line with the process-tracing method, I shall return to more explicit inferences drawn from this perspective after the analysis of the international output (Chapter 6).

P2c) implications for the issue-area – *content of* domestic implementation activities:

The negotiation result will reflect a compromise between the interests of the North and the South. In this case we could anticipate the development of DC conservation policies to reflect national priorities; and also anticipate that developed countries will fulfil their obligations in the deal. National priorities in DCs would probably be more concerned with development than with environmental goals.

The establishment of incentive mechanisms as part of an IEA leaves us with some crucial questions regarding explanatory factors:

- Are such incentive mechanisms installed by a benevolent hegemon who consents to pay a larger share for the common good?

- Did incentive mechanisms come about as a result of the regime facilitating a breakthrough for DCs interests?
- Are these incentive mechanisms due to a common international acceptance of the fairness of establishing such mechanisms?

The latter question leads us to the ideational-based perspective, to which we now turn. In order to approach an answer to the questions above, it will be necessary to look closer at the negotiation process leading up to the incentive mechanisms, as well as the practical operation of the incentives.

3.2.3 Ideational explanations

Central to norm- and knowledge-based explanations of state behaviour in international relations are the notions of *interdependence* and *communication and learning.*[63] States are still seen as central actors in international relations, but domestic and transnational actors must also be taken into account. According to the ideational perspective, cost-benefit analysis is not the only basis for predicting state interests and action. Two other broad types of mechanisms, also linked to the international institution, may be identified as central to this model:

- First, environmentally oriented coalitions between domestic and international actors associated with a regime may be developing within *epistemic communities*; strengthening the development of consensus within an issue-area.
- Second, norms may acquire a sense of *legitimacy* or *obligation (pacta sunt servanta)*. The idea here is that convergent expectations and recognisable norms are generated through repeated interaction and *learning*, and common acceptance of international norms and rules developing within specific issue-areas may hence affect state behaviour. Finally, it must be noted that norm adoption may have a trace of *authoritarian* persuasion.

The main point of this approach is not to deny the importance of power and interests in international relations. Rather it is to cultivate a greater focus on learning processes within international institutions, and on the additional effects of the development and acceptance of common norms or ideas. This opens up for an examination of learning processes taking place between non-state actors at different levels, e.g. between international and sub-national actors in this case. As we will see in the presentation of the analytical framework on domestic factors (Chapter 7), this also opens up for regime effects which may be de-linked from the general assumption about a breakthrough for national interests as a prerequisite for domestic implementation.

63 For an overview, see for instance Krasner (1982).

3.2.3.1 Epistemic communities

Peter Haas has developed this perspective further with the concept of epistemic communities: "networks of knowledge-based communities with an authoritative claim to policy-relevant knowledge within their domain of expertise" (Haas, 1993:179). This is defined as a professional group that believes in the same cause-and-effect relationships, uses the same truth tests to assess them, and shares common values. The members also share *consensual knowledge* from which they convert facts to policy-relevant conclusions. The power resource of such a group lies in its authoritative claim to knowledge (Haas, 1990:55). Policy-makers confer with these experts in order to reduce uncertainty and thus boost the legitimacy of decisions at the domestic level. While I associated Young's *structural leadership* with the power-based perspective and his *entrepreneurial leadership* with tapping the integrative potential within the interest-based perspective, Young's idea of *intellectual leadership* builds on similar mechanisms as that of the notion of epistemic communities. While Young's concept is oriented towards individual actors and Haas' epistemic communities involve a macro-orientation in terms of networks of actors, a similar causal mechanism is envisaged: Intellectual leaders relies on the power of ideas to shape the way in which participants in institutional bargaining understand the issues at stake and to orient their thinking about options available (Young, 1991:300).

Members of epistemic communities seems to be at the same time international and national actors: members are drawn from the national arena, they report back to national policy-makers, and may still have loyalties here. At the same time, through interactions in an international environment, they come to identify themselves as belonging to the international level to a greater extent than ordinary national delegation members.

Haas indicates several "channels", or what may correspond to causal pathways, through which epistemic communities may influence regime formation and implementation:

- Mainly formation phase: The group's members may be present in the international secretariats and thus have an impact on shaping the agenda.
- Formation and implementation: They may be able to create transnational and transgovernmental coalitions that hold similar views, and can forge consensual solutions.
- Mainly implementation phase: The group's members may hold administrative positions in government, and thus encourage common policies in their respective governments and push for their enforcement. This would seem to imply that as we open the black box of the URA model, and examine implementation in the domestic arena, the concept of epistemic community would also allow for a study of the role of individuals (members) vis-à-vis domestic governments.

Thus Haas assumes that the countries in which the epistemic community is best represented will be the most environmentally concerned. How sound is this proposition? and what are the implications for the formulation and enforcement of biodiversity policies in Ethiopia? A parallel notion may be found in Levy et al. (1995): Among the pathways through which regimes may affect outcome they mention two: regimes as *bestowers of authority* – by creating respect for legitimately constituted authority among states; and regimes as *learning facilitators* – by changing information and values, and thus altering incentives. A problem with the former view is the seemingly inherent idea that the regime will be more "green" than the states which created it.[64] A related problem is denoting one sort of leadership as structural, i.e. power-based and self-centred hegemonic, and other kinds as entrepreneurial or intellectual in spirit, i.e. with an ultimately common good in mind. This brings up the complex relationship between power and knowledge/norms. All three kinds of leaders seek to make other parties do what they otherwise would not have done – but the two latter are seen as more legitimate, as the changes in other parties' positions are brought about through reason or sentiment. These concepts will have to be employed with caution, although, as already contended, I shall refrain from a discursive attempt to delve into the motivations behind them.

In this section, the main focus has been on the actors involved in "ideational" activities, and the channels through which these actors may promote consensus. The next section provides a more general discussion about the nature of such consensus and of the learning processes involved.

3.2.3.2 Legitimacy and obligations

There are conflicting assumptions as to whether an ideational-based model may contribute to understanding and explaining agreement to and implementation of IEAs in less developed countries. On the one hand, developing countries may be expected to harbour a different set of priorities than the North regarding the global environment agenda (Najam, 1994). This is exemplified by the South's emphasis on development ahead of environment, while the North can afford to be more concerned about preserving a healthy environment for future generations. While the general picture of these differences is hardly false, it may be exaggerated, for two reasons:

First, when we look at *real* politics in developed countries, there is little doubt that here also the overriding priority is still economic growth, rather than environmental protection. Thus, the priorities of developing and developed countries may not really be inherently so different. This is most clearly seen in the concept of *sustainable development,* which has had a definite appeal to politicians in developing and developed countries alike. We could also cite examples of non-compliance on the part of Norway concerning the whaling moratorium of the International Whaling Commission – decisions which certainly have more to do

---

64 For an illustration of this point, see Laugen (1995).

with willingness than ability. Likewise, during the UNCED negotiations, the United States opposed all references to lifestyle changes and sought to minimise the link between consumption and environmental destruction; while Malaysia, Brazil and Indonesia opposed regulations aimed at restricting their right to cut their forests at the current, unsustainable rate (Wapner, 1995).

Secondly, there are several indicators that environmental concerns are entering the agendas of the South as well. From 1972 to 1980, the number of developing countries with environmental agencies grew from 11 to 102 (Williams, 1993). From 1972 to 1995 the growth was from 14 to 30 in developed countries (Wapner, 1995). There has also been a rapid growth of local green movements in the South (Williams, 1993). These two examples might indicate that there is a narrowing gap between Northern and Southern environmental implementation also in terms of ability, not least because politicians in the South also face the choice of making priorities about how to spend the financial resources they do control – but also this notion should probably be handled with care.

If we, against this backdrop, accept the possible utility of a ideational model in understanding DC implementation, at least to the same extent as in developed countries, this may still take many forms. Norm adoption may come about for various reasons, and with varying results in terms of successful implementation:

Norms as learning and teaching:

* First we must ask whether the developed world has been more susceptible than the developing world in accepting and adopting the norms generated within environmental negotiation arenas. This might be put down to the North being more closely associated with epistemic communities within these issue-areas, through e.g. the scientific communities. In that case, norms would play a very limited role in DC implementation. This clearly seems to have been the case in the climate change negotiations (Lunde, 1991). The situation may, however, be different in the biodiversity area, where the South has been both active and well represented in several expert fora (Rosendal, 1991).

> *Even if norm-adoption may not play a direct role in DC implementation, it may have an indirect effect: If norms play a role in enhancing implementation in the North, this may boost implementation in developing countries as the fear of other parties cheating is reduced.*

* Second, the flip side of the learning coin is *teaching*: It may well be that the South has had a breakthrough for their own norms and ideas – which they have been able to generate within the international environmental arenas. The South usually prefers that negotiations take place within the UN system, where they tend to have greater clout. Along with the more general problems of how to identify priorities and distribute costs and benefits among participating states, the implementation of global environmental agreements runs into the added difficulties of dealing with sovereignty among unequal states as well as the related problem of weak domestic

systems and weak state capabilities (Hurrell & Kingsbury, 1992). This highlights how international regimes may in some cases represent an important political instrument for developing countries. This hardly includes fora of economy and trade – although several UN organisations, like the FAO, do provide a wider scope for influence by the developing countries. UN development agencies in particular are thus providing developing countries with relatively more power than would otherwise emanate from their domestically based capabilities (Krasner, 1981).

To the extent that they actively participate in the international environmental negotiation process, DCs may hence be expected to be able to promote their own norms and ideas, as well as being more susceptible to adopting emerging common norms and ideas. This means that international co-operation may lead to converging governmental policies. An added proposition is thus that the higher the level of active participation, the greater will be the likelihood of implementation.

Norms as Authoritative persuasion:

* Another way of viewing norm acceptance and norm adaptation in international relations is in light of the power-based model – as a variety of cultural imperialism (Galtung, 1971). According to the realist view, moral and ethics in the international system are basically a product of power and politics, reflecting the interests of dominant powers. As Elster (1987) points out, some norms bolster self-interest, but even when they do, the fact that norms are followed gives them an independent impact. Furthermore, the one who invokes the norm, whether for manipulative or altruistic purposes, is also constrained by it to act consistently (Elster, 1989:117). For instance, when a country seeks to win international acclaim by displaying a green image, this will be credible only if applied indiscriminately, self-interest or no self-interest. In this perspective, developing country representatives participating in international fora and becoming part of international epistemic communities may concede to Northern ideals and bring them back home. In this capacity these representatives may be more like external actors themselves, and they may or may not find themselves in conflict with local levels of government (Najam, 1994). This is an important point, but I would argue that caution is needed when anticipating a kind of "false consciousness" on the part of actors, especially if such predictions are made for representatives from the South alone. Even though external actors, like international NGOs and multilateral development institutions, may have been effective in shifting the South's attitude toward environmental issues, there may be no more reason to brand this as this *cultural imperialism*, than when governments in developed countries are influenced by the environmental movement.

* A related point, however, is the possible mismatch between the environmental agenda in the North and the South. Even though environmental issues are gaining ground in developing countries, the main environmental challenges are not necessarily the same as those in the North. This means that there is a possibility of a

mismatch between what the elite in the South (learn to) perceive as environmental problems and what local people perceive as environmental problems. If this is the case, there may be perfect compliance and implementation of the international environmental obligations – and still be far from actual problem solving.[65] As problem solving is beyond the scope of this study, this will remain an hypothetical possibility, but one which should be kept in mind.

This highlights the difficulty in assessing the development of conservation policies in DCs: Is there really a basis for anticipating more concern for locally perceived environmental problems in national conservation policies if the international agreement reflects the norms and interests of DC governments? This is an important question, but for the moment, I shall again point out that what we are concerned with here is *implementation* of international agreements and not with environmental impact as such. On the other hand, if we fail to consider this question, it rapidly becomes apparent that it is difficult to distinguish between various factors affecting national legislation outputs; regardless of explanatory model we may end up with rather similar results. Hence, we must first identify the national interests and priorities of DCs, and assess whether these have been reflected in the international output; and secondly, we must look into sub-national interests, and evaluate the implementation process in this light. We return to this in the development of a domestic model on implementation (Chapter 7).

The ideational-based perspective gives rise to the following theoretical propositions regarding international collaboration:

1. Through repeated interaction, consensual knowledge and social learning give rise to commonly accepted international norms and ideas, which to some extent become formalised in international regimes.
2. The direction and scope of such learning processes will depend partly on who "captures" the epistemic communities / provides intellectual leadership, and on the process of reaching consensual knowledge.

Thus, we end up with two diverging case-specific propositions:

P3a) expectations for the *type / content of international output* (regime formation):

Through learning processes, DCs may adopt the prevailing norms and ideas of the developed countries, acquiring a deeper concern for global environmental problems.

P3b) implications for the *scope for domestic implementation* (regime effects):

This could give DCs an increased willingness to implement IEAs.

---

[65] A related debate is whether a central world state (able to coerce state action) or conversely, a decentralisation of the state (able to recognise and handle local environmental problems) present the best answer to solve global environmental problems (See Wapner, 1995).

P3c) implications for the issue-area – *content of* domestic implementation activities:

Much the same as for P1b: Dominant regime members will tend to induce implementation, in accord with their own interests, in weaker states. This is achieved through structural leadership, as they have the capacity to use sticks and carrots to achieve their goals.

Alternatively:

P4a) expectations for the *type / content of international output* (regime formation):

A breakthrough for DC interests is possible. Through active participation, DCs may themselves be teachers, *and succeed in promoting and integrating their own norms and ideas on environment and development into the international agenda and output.*

P4b) implications for the *scope for domestic implementation* (regime effects):

This could give DCs an increased willingness to implement IEAs.

P4c) implications for the issue-area – *content of* domestic implementation activities:

Same implementation score as in P3b, but different content. P4a would indicate that the international community has commonly accepted central ideas and principles advocated by DCs. There would presumably be less focus on wildlife conservation and a greater focus on equity concerning use of and access to biodiversity

However, it may be difficult to distinguish the power of ideas from bargaining power: When the parties agree to include financial mechanisms as part of the IEA, is this a result of issue-specific power (hence: P2)? – or a general acceptance among the parties that this is a fair solution (hence: P4)? In line with the process-tracing method, I shall return to more specific inferences based on this perspective after the analysis of the international negotiations and output (Chapter 6).

3.2.4 Summary

In this chapter, I started out with an overview of the broad theoretical approaches to international relations, and proceeded to order these into three main perspectives with different propositions about whether and how institutions matter. These three perspectives give rise to different assumptions concerning the formation of international regimes, with subsequent implications concerning their operation. In analysing the biodiversity issue I shall accordingly start with the most parsimonious perspective, looking at the explanatory force of structural relations of power and interests among the negotiating parties in determining positions and negotiating strategies. Next, I will see how well the modified regime model squares with the empirical material. This implies looking at the institutional factors and discussing their impact in terms of aligning or altering actor interests. The third perspective directs attention to learning processes and to the development of common international norms and ideas as factors accounting for regime formation and operation.

By way of summation I have already presented a simple table (Table 3.1) on the main foci of these perspectives in terms of how the various factors are assumed to affect international *collaboration and output*. On the same note, I shall briefly outline the relationship between the main foci, main causal pathways and assumed impact on *domestic implementation scope and content*, also taking the content of the international output (from Table 3.1) as a point of departure.

Table 3.2:
Theoretical perspectives: main foci and causal pathways – implementation

Main foci: **main factors**	**Causal pathway**	**Effect on implementation**
1. Overall power and interest: · hegemon	· (structural leadership)[1] · sticks and carrots · unilateral action	high scope: induced compliance in DC "business as usual"-content: wildlife conservation focus in DC and free IC access to (DC) biodiversity
2a. Interests and issue-specific power: · interest constellations	· basic game power · (negotiation power)	medium / low scope: as DC have less basic game power than negotiation power "package deal"- content: mostly symbolic, "play for grandstand" (little reason to trust mutual compliance)
2b. Interests and institutions: · (number and type of participants) · (official purpose) · (agenda setting) · information, monitoring & incentive mechanisms	· (aligned cost/benefit perceptions) · (entrepreneurial leadership) · reduce uncertainty · trust mutual compliance	high scope: implementation in DC encouraged by incentive mechanism and mutual compliance in IC "integrative"-content: higher level of "national interest" in biodiversity policy (development rather than environment)
3. Institutions, ideals and norms: · legitimacy & obligations · epistemic communities	· learning/teaching · (intellectual leadership) · coalitions with similar views	high scope: DC internalise common international norms (or authoritative persuasion) "common IC norms"-content: wildlife conservation focus in DC and free IC access to (DC) biodiversity alt.: high scope: DC obligated to follow own norms and ideas alt.: "common DC norms"-content: equity combined with conservation focus

[1] Factors and pathways in brackets indicate that these are primarily associated with the regime formation (international output) phase. As I believe that the impact in this first phase is also important for understanding the next phase of domestic implementation, I have included them in the table.

BIOLOGICAL DIVERSITY – THE ISSUE-AREA

4.1 Introduction

In the following three chapters I shall describe and analyse the formation and operation of the Convention on Biological Diversity (CBD) in light of the three theoretical perspectives set out in the previous chapter. Chapter 4 will focus on the characteristics of the issue-area, and how the issue was put on the international agenda: Phase 1 (1982–1987).[66] Here I offer a preliminary analysis of the agenda-setting phase in light of the political and intellectual characteristics of the issue-area, in order to analyse possible determinants for international collaboration outputs and implementation. I then go on to describe how aspects (related items) of the issue-area have been dealt with in other international fora. Special attention is given to the central actors and interests involved, and the principal conflict lines.[67] An essential controversy revolves around wildlife and habitat preservation versus equitable sharing of benefits from utilisation of biological diversity – a controversy inherently linked to the question of property rights to genetic resources. I will start by examining this conflict, and then follow the debate on property rights through relevant international fora, briefly reviewing the debates in the UN Food and Agricultural Organization, the GATT Uruguay Round and the Keystone Dialogue. These fora have no legal links to the CBD, but it is important to include them, since they deal with directly related aspects of the problem area under study here.

In Chapter 5, I go on to the presentation and analysis of the formal negotiations and output in view of the three regime models: Phase 2 (1989–1992). Important questions here are whether and how developing countries were able to achieve any breakthrough for their interests. I look at factors related to the institutional design, to see whether they may help to explain such a breakthrough, while first controlling for factors exogenous to the institution which could have affected the output.

---

66 The year 1981 has been chosen as a starting point, because this was the first time the issue of access to, control over, and conservation of plant genetic resources was put on the agenda of an international organisation, FAO.

67 I would like to thank Veit Koester, Head of Ecological Division of the Danish Ministry of the Environment and Chair of several working groups during the CBD negotiations, and Erik Steineger, special advisor to the Norwegian Ministry of the Environment, for their valuable comments at various stages in my work with this section. The full responsibility for any shortcomings resides of course with the author.

In Chapter 6, I draw a new set of inferences for future implementation activities, on the basis of the three analytical perspectives. These general implications for future implementation of the CBD are then discussed in light of the principles embedded in the output and the unresolved conflicts surrounding the issue-area: Phase 3 (1993–1997). Here I look into further implementation measures at the international level.

4.2 Biodiversity: Getting the Issue to the International Agenda

The conservation of biological diversity constitutes one of today's greatest challenges, for two main reasons:

First, the concern with biodiversity stems largely from our increasing awareness that the current rate of species extinction is extremely high, viewed against the natural average ("background") rate. For example, the average species lifetime of mammals in fossil records is in the order of one million years, which would mean roughly 0.5 extinctions per 100 years for the present mammalian fauna of nearly 5000 species. In fact, however, the current rate of extinction of mammals is roughly 100 times higher than this background rate. In other taxa the discrepancy may be even greater (Heywood, 1995:232). Estimates of the number of existing species in the world vary from about 5 to 100 million (Wilson, 1992), of which only some 1.9 million have been described scientifically.

Second, as the new biotechnologies make it possible to utilise the full potential of the world's genetic resources, economic incentives to conserve biological diversity increase.[68] Hence, the interest in genetic material arises from environmental concerns, as well as being based on technological developments and financial gains. By the year 2000, farm-level sales of products of agricultural biotechnology are expected to have reached some US$ 100 billion. As of 1986, the value of global trade in plant-based pharmaceuticals was estimated to US$ 20 billion.[69] Three years later, in 1989, the market value in the North of medicinal plants from the South was estimated to US$ 43 billion (Principe, 1989). Now, such a rapid increase – from 20 to 43 billion dollars – is probably due more to uncertainty and increased attention than to any real increase in value. Nevertheless, these figures give an indication of the economic interests at stake. Above and beyond the ethical and aesthetic value of species diversity, mankind depends on a diversity of biological resources for food, medicines and for raw materials in the chemical industries.

Let us take a closer look at the driving forces involved in the loss of biodiversity. Biodiversity is threatened by the destruction of species and habitats world-wide,

---

68 While the "old" biotechnology includes traditional activities such as brewing beer and baking bread, the concept of "new biotechnologies" refers to activities like tissue culture and recombinant-DNA (r-DNA) techniques.

69 Report of Panel II, UNEP/Bio.Div/Panels/Inf.2, Nairobi, 28 April 1993.

through land-use changes, deforestation, desertification, industrial development and pollution. The destruction and fragmentation of habitats is indirectly endangering or causing the extinction of many species of plants and animals. The introduction of exotic species and the increasing uniformity in agricultural crops – the widespread use of high-yielding varieties of crops – are replacing local varieties in the fields, thus leading to genetic erosion. The dwindling of local crop varieties in the fields leads to genetic erosion within species, leaving the remainder less resistant to pests and diseases. What these disappearing species, plant varieties and the genetic diversity have in common is that as yet, no direct economic value is connected to them. These losses may thus be regarded as the externalities (unintended and unaccounted for consequences) of a variety of actions such as road and dam construction, industrial logging and agricultural development. Other species (among them, elephants, certain whales and fishes) are directly threatened through overexploitation. These species are valued, also in economic terms, but overexploitation proceeds because it yields great profit for traders and manufacturers who can externalise the costs. They themselves have no incentive to moderate their exploitation to preserve species for the future. From an individual, short-term perspective, elephant hunting, the cultivation and use of uniform, high-yielding crop varieties and the exploitation of forests may be rational actions, but in sum, these activities lead to a dwindling of biological diversity.

Gradually, the international debate on biodiversity has come to concern not only conservation, but equally much the distribution of benefits derived from using biological resources. The main bulk of the world's remaining species biodiversity is found in the South, but it is the developed countries that possess the (bio)technology to reap the economic benefits from these resources on a large scale. This potential conflict was realised by the World Commission on Environment and Development (1987), which urged: "Industrialised nations seeking to reap some of the economic benefits of genetic resources should support the efforts of the Third World nations to conserve species" and "Developing countries must be ensured an equitable share of the economic profit from the use of genes for commercial purposes" (WCED, 1987).

The CBD is not the first international treaty to address species or habitat conservation, but it is the first to address conservation of all biological diversity world-wide, and the first to include sustainable utilisation of these resources. Along with the Convention on Climate Change, the CBD is the first global treaty that directly embraces the socio-economic dimensions and recognises that environmental issues cannot be separated from economic development. There exist a great many agreements on international co-operation for the conservation of various species of plants and animals, and their habitats. The Ramsar Convention on Wetlands is one of the most important global measures concerned with habitat protection (Ramsar, 1971). For the Arctic area, there is the Convention on the Conservation of Antarctic Marine Living Resources (CCMLR, Canberra, 1980). Whales (ICRW, Washington, 1946) and tuna (ICCAT, Rio de Janeiro, 1966) have their own Conventions. Then there is the Convention on International Trade in Endangered Species of Wild Fauna and Flora (CITES, Washington DC, 1973). While all these treaties concentrate on conservation, the FAO International Undertaking (1983) on plant genetic resources also includes sustainable use of the resources. Coinciding with these processes, the issue of property rights to, *inter alia,* biological resources created great tensions within an inherently different type of international institution: the GATT Uruguay Round (now World Trade Organization, WTO). In the late 1980s, the Keystone Dialogues hosted a science-policy debate on biodiversity and property rights. Discussions in these latter three fora clearly had an impact on the CBD negotiations and output.

Until the early 1980s, conservation work – whether national or international – was still focused on wild species of plants and animals. An important shift came when questions of access to and control over plant genetic resources were raised by governments of the developing world. The UN Food and Agricultural Organisation became the forum for this heated debate, which resulted in the FAO International Undertaking on plant genetic resources – at the time, the most comprehensive agreement in terms of linking genetic resources conservation to distributive concerns.[70] An Undertaking, however, is not legally binding, as a convention would be. In 1989, the UN Environment Programme (UNEP) was given the formal mandate of negotiating what was to become the all-encompassing Convention on Biological Diversity. This was adopted in Nairobi in May 1992, signed in June at the 1992 UN Conference on Environment and Development in Rio de Janeiro, and entered into force 29 December 1993.[71]

Box 4.1 Central controversies in the biodiversity negotiations

Two crucial and partially linked questions can be identified at the outset of the CBD negotiations: The first concerned whether preservation or conservation is the best method to avoid biodiversity loss. The second was concerned with whether to include both wild and domesticated species.

The traditional preservation ideology basically believes in protecting species and ecosystems from mankind by establishing nature reserves. The idea of conservation includes both preservation and sustainable use of the resources. The conservation concept hence admits for the possibility that people may coexist with "nature" without detrimental effects. At the onset of the biodiversity negotiations, the term preservation was going out of use, as the more politically correct concept of conservation entered the mainstream terminology. In this process of wider adoption, however, the concept has tended to become watered-down – in the sense that it can also be used with the old preservation ideas in mind.

---

70 For an extensive presentation of the historical background for the seeds issue as a North–-South conflict area, as well as a thorough examination of the FAO debate, see Cary Fowler (1993).

71 For more information on the treaties and agreements see Bergesen et al. (eds) *Yearbook of International Conventions on Environment and Development* (prior: *Green Globe Yearbook,* FNI/Oxford).

For an extended analysis of the debate leading up to the arrangements in FAO, see G. Kristin Rosendal (1989). For a corresponding analysis of the UNEP debate, see G. Kristin Rosendal (1991).

Once the concept of sustainable use is included, equitable sharing of benefits from use comes into focus. The link between use and benefit sharing is more explicit in the domesticated sector, as the economic returns from agriculture generally exceed revenue from tourism in the wildlife sector. Wildlife may, however, also provide incomes through bioprospecting, for instance within the pharmaceutical sector (see Box 6.1 on bioprospecting).

These main controversies pertain to the deeper question of property rights to genetic resources, a point which needs further clarification here. I will limit the discussion to property rights to *genetic* resources, as it is this dimension which has the widest connotations to the international conflict. Property rights pertaining to other components of biological resources (land tenure etc.) are obviously also important for conservation and use of biodiversity, but these aspects are to a larger extent linked to potential domestic conflicts – between national and local levels – and these will not be considered here.

4.2.1 Genetic resources: properties and property rights

Biological diversity is a broad concept which has been used to embody the variability among all living organisms, including diversity within species, among species and among ecosystems. *Genetic resources* are the hereditary material (genes) in all animals, plants and micro-organisms; the concept refers to genetic material with actual or potential use or value for humanity.[72] Genetic diversity or variability is necessary to sustain vitality in both wild and domesticated plants and animals, and also for the development of new and improved products.

The world's genetic resources are the raw materials of biotechnology. With the advent of the new biotechnologies has come an increasing awareness of the value of genetic resources. In the 1970s the transnational seed and chemical corporations started applying these new technologies in plant breeding and agrochemicals, and in the course of the 1980s the biotechnology industry expanded rapidly. These developments have had a profound impact on the understanding of property rights to genetic resources:

> *The story begins with the principle of a* common heritage of mankind *and ends with* patents *and* national sovereignty.[73]

Collective goods are usually defined by their character of non-rivalry and non-exclusiveness. Non-rivalry implies that it is possible for more than one user to utilise or consume the good without diminishing the amount available to others. Non-exclusiveness indicates that it is hard to exclude others from using or consuming the good. The air we breathe is generally regarded as an example of a non-rival and non-exclusive good. This used to be the case with clean water as well, but its character of

---

[72] As defined in the Convention on Biological Diversity, Rio de Janeiro, 5 June 1992.

[73] For a more extended expose of the developments in patent legislation see G. Kristin Rosendal (1995).

non-rivalry is rapidly declining in many parts of the world. Common property resources will often display the combined properties of non-exclusiveness and rivalry. This may give rise to problems of collective action, unless some kind of management regime can be established to control access to the resource in question.[74]

Basic to the idea of *common resources* is the element of non-exclusiveness or *open access*: the absence of well-defined property rights. This was the case with ocean fisheries in the past century, in the Grotian doctrine of the freedom of the high seas (Grotius, 1609). *Open access* implies merely that no one can be excluded from using the resources, save by lack of economic and technological capacity. The *common heritage of mankind* (CHM) principle, however, is not necessarily identical with the idea of open access as practised under the high seas doctrine (Bilder, 1980). The common heritage principle may imply that all are entitled to benefit from exploitation of the resources.

In international negotiations, the common heritage principle has been used differently by developed and developing countries. The CHM-principle was first introduced at the UN Conference on the Law of the Sea (UNCLOS) in 1967 by the Maltese Ambassador to the UN, Arvid Pardo, as a guiding principle in governing the exploitation of minerals in the deep seabed. Both in the UNCLOS negotiations and later in the Antarctic treaty negotiations, the intention was to secure greater equity between developed and developing countries in the exploitation of a "common" resource. The South's interpretation of the concept implied *restricted or controlled use, for the benefit of all countries*. The majority of industrialised states objected to the principle as being legally diffuse and practically impossible.

All along, however, the North's interpretation of the principle of common heritage did constitute the international regime for exchange of and access to plant genetic resources (seeds): International gene banks were stocked with seeds from the most commonly used food plants. These seeds were primarily collected from the extensive variation found in the South, and the gene banks were based on the North's interpretation of the CHM-principle – *open access, free of charge*. "Technically", the collection of seed samples was considered as a non-rival and non-exclusive activity. Moreover, no-one questioned this practice on moral grounds, as the seeds of our most utilised food plants were seen to be of basic significance to all mankind. A central agent in this collection of seeds world-wide has been the Consultative Group on International Agricultural Research (CGIAR), which has been controlling the International Agricultural Research Centers (IARCs).

---

[74] See for instance Gareth Hardin's now-classic article, "The Tragedy of the Commons", *Science*, 162, 1968, pp. 1243–1248.

Box 4.2: The CGIAR and plant genetic resources

The Consultative Group for International Agricultural Research (CGIAR) was established in 1971 and has since 1975 been involved in the conservation of plant genetic resources. Jointly with the International Plant Genetic Resources Institute (IPGRI) and the UN Food and Agriculture Organisation (FAO) it has built up the world's largest ex-situ collection of gene material through the IARCs. The IARC system comprises the sixteen international gene banks, such as the International Rice Research Institute (IRRI) in the Philippines, the International Centre for Maize and Wheat Improvement (CYMMIT) in Mexico and the International Potato Center (CIP) in Peru. The IARCs have in turn been largely responsible for what has become known as the Green Revolution.[75] The CGIAR has traditionally been dominated by the developed countries, as these along with such institutions as the World Bank and the Rockefeller and Ford Foundations represent the main financial donors. The main donors in terms of germplasm are developing countries where seed collection has taken place. The geographic distribution of the gene banks storing gene resources for breeding and cultivation is thus: IARCs (41.3%), North America (23.8%), Western Europe (15.9%), South America (6.3%), Asia (4.2%), Africa (4.6%) and others (3.8%).[76]

Certain national gene banks may apply other rules for the exchange of gene material, but what is indicated here is that the developed countries, i.a. through the IARCs, enjoy free access to more than 80 % of the gene material deposited in the world's gene banks.

There are, however, several technical barriers to gene-bank conservation: First, gene banks include a limited range of plant varieties (chiefly seeds from some 30 common agricultural plants). Second, some important food plants cannot be stored in gene banks by ordinary methods (cocoa, mango, avocado, among others) (Prescott-Allen & Prescott-Allen, 1988). Third, seeds deposited in gene banks will not, unlike plants growing in natural conditions, be equally adapted to new environmental conditions (Ashton, 1988).

While most (international) gene banks still operate on the basis of open access to genetic resources, the common heritage regime for genetic resources is rapidly becoming a thing of the past. This change may be traced back to the 1930s with the introduction of hybridisation, which also helped to secure *exclusive rights* to superior plant varieties. More recently, the regime change has come about swiftly, primarily as a reaction to the introduction of intellectual property rights for organic material, which allow private ownership to genetic resources through patents or

75 Modern agricultural methods and systematic plant cultivation seek to develop homogeneous, stable plant varieties which are both adapted to mechanised agriculture and which fulfil the quality requirements for registration and marketing of newly bred races. Successful plant breeding therefore often leads to reduced genetic variation. Though the new types have a far higher yield, (hence the term *green revolution*) the homogeneity renders them more vulnerable to disease and pests, and thus more dependent on pestkilling chemicals. Due to the limited genetic basis, these high yield plants are far less robust and hardy than their local genera. Modern wheat varieties have for instance today an average life of seven years, necessitating constant improvement to increase its resistance (Wood, 1993). As a general rule, the best results are achieved by crossing such high-yield characteristics with the locally adjusted varieties in order to preserve the local adjustment ability.

76 Source: Norwegian White paper, No. 46, 1988/89.

plant breeders' rights, primarily utilised in developed countries.[77] At this point, the moral notion that food and medicinals should be excluded from patentability because of their fundamental importance to basic human needs was rapidly losing ground. In tandem with the developments in biotechnology, it became necessary for developed countries to change, or rather reinterpret, their national patent laws.

> *The new biotechnologies made patenting* an economical necessity, legally possible – *and still a* controversial *issue.*

Economical necessity

The economic need for patenting, as seen from the Northern biotechnology sector, is based on the high costs of research in biotechnology (as compared to traditional breeding methods). Competition is fierce, and research is increasingly carried out by the private sector. The biotechnology sector has been arguing strongly for compensation in terms of royalties, along the lines of other fields of technology.

Legally possible

National and international patent legislation drew no a priori distinction between various sectors of technology. Traditionally, the patent system was limited to technologies dealing with non-organic material. Biological products or processes were originally excluded from patentability on the basis that such inventions could not meet all four of the fundamental patent criteria (see Box 4.3). The barriers represented by the *patent criteria* have now been largely overcome by developments in the new biotechnologies – as well as reinterpretations by patent lawyers.

Box 4.3 Patent Criteria

A patent can be seen as a contract between researcher and society; the researcher making her invention public, rather than keeping it a secret, and society offering royalties for using the invention for a limited period of time (usually 20 years). The patent criteria are:
Novelty: the invention must be novel, meaning basically that it has not been published anywhere before.
Inventive step: there is the criterion of non-obviousness – the invention must display an inventive step.
Industrial utility: the third criterion states that the invention must have an industrial application – a practical utility. One function of this utility requirement is to distinguish between basic research, considered to belong to the public domain, and applied technology, which is eligible for patenting.
Reproducibility: the patent application must fulfil the criterion of reproducibility – the application must describe the invention in such detail that other experts may repeat the experiment and arrive at the same results.
In addition to these criteria, patent legislation commonly excludes from patentability inventions whose utilisation would run counter to *public order or morality*.

77 "Intellectual property rights" (IPR) include patents, plant breeders' rights, copyrights and trademarks. The most relevant IPR categories for genetic resources are patents and plant breeders' rights.

Controversies

At the same time, several areas of patenting within biotechnology remain contentious issues:

- Biological issues: Is a product patent on naturally occurring organisms fulfilling the *inventive step*? Biological material was traditionally regarded as natural products rather than industrial products – discoveries rather than inventions (Crespi, 1988). It is still contested whether the process of isolating and describing a micro-organism or a gene may be defined as "inventive enough" to meet this criterion.
- Legal issues: Unlike traditional breeding methods, the new biotechnologies may meet the *reproducibility* criterion. However, many biotechnology patent applications are so complex that they fail to fulfil the reproducibility criterion. Hence, the deal with patents as an alternative to secrecy is forfeited and scientific exchange and development is impeded.
- Moral issues: There are moral concerns regarding exclusive rights to food, medicinals and living material – based on the moral notion that food and medicinals should be excluded from patentability because of their fundamental importance to basic human needs.
- Equity issues: Patenting is a long and costly business that can primarily be employed by large companies. It is hardly a tool for indigenous and local communities, even though these groups often harbour much knowledge about the use of these resources. Biotechnological products often build on local breeding and knowledge about the medicinal traits of biological resources. In that case, a patent will reward only the one that provided the last piece of a long process of breeding or invention. Patenting may thus be incompatible with fair and equitable sharing of the benefits derived from use of genetic resources. Hence, it is argued that patenting here is a form of *biopiracy* that may be incompatible with the CBD objective of equitable sharing. On a more optimistic note, other actors coin this activity *bioprospecting*: The concept *bioprospecting* denotes among others the prospects for making equitable contracts between local communities and transnational corporations (see box 6.1).[78]
- Environmental issues: A contested question concerns whether intellectual property rights represent a direct threat to genetic diversity in agriculture. Supporters of intellectual property rights argue about the need to introduce high-yielding varieties, and have linked to this the use of *plant breeders' rights* (PBR) and patents, as preconditions for food security. Opponents argue that the precondition for food security lies in the conservation and sustainable use of a

---

78 The situation of inequity is believed to be more aggravated within the agricultural sector and for domesticated plant genetic resources, compared to the medicinal sector and concerning the screening of wild germplasm.

variety of (non-systematically bred) farmers' cultivars or land races.[79] They go on to say that the patent and PBR criteria of reproducibility and *uniform, stable and distinct from existing varieties*[80], inevitably leads to monocultures and loss of genetic diversity.

The principal ruling on the patentability of biological material appeared in the German Federal Supreme Court in 1969 (the *Red Dove Case*), which ruled that a breeding process for animals was indeed patentable (Mooney et al., 1988). In the *Chakrabaty Case* of 1980 the US Supreme Court decided, by five against four, to allow industrial patents for naturally occurring living matter, including both asexually and sexually reproduced plants (Bent et al., 1987). A judge from this case was later employed by the EU Commission in drawing up their formulation of a directive on industrial patents in biotechnology. This draft directive subsequently spent ten years in the EU bureaucracy (1988–98), as it was repeatedly turned down by agricultural and environmentalist interests. In 1995, the European Parliament voted by 240 to 188 to abandon the directive, but the Commission proposed a new draft which was adopted in 1998. I shall return to this subject in section 6.2.

This controversy is not confined to the agricultural sector. There is a growing awareness that the largely unexplored components of biodiversity may conceal treasures, for example of great medicinal value. A much-cited case from medicine is the Rosy Periwinkle (*C. roseus*), a native plant of Jamaica and Madagascar. Since this story appeared in *New Scientist* in 1992, it has made a considerable impression on environmental and developmental circles, even though its integrity on some points has since been challenged. Two components (anti-tumour agents *vinblastine* and *vincristine*) from the plant have been patented and turned into a medicine for the treatment of Hodgkin's disease and certain types of leukaemia. The patent holder is a USA-based pharmaceutical firm, Eli Lilly. The company's annual return on the invention is about US$ 100 million, none of which was returned to the country of origin.[81] This discovery of new drugs from plants is one of the few that has evolved from a random-selection broad pharmacological screening program. Madagascar does profit from the plant today, albeit on a small scale, as the Rosy Periwinkle is now cultivated for export. Madagascar exports about 1000 tonnes of *C. roseus* annually.[82]

This situation gave rise to the central conflict line in the biodiversity issue-area:

---

79 Land races are genetically variable populations that exhibit different responses to pests, diseases and fluctuations in environmental conditions. The genetic diversity in these land races was, and remains, a form of insureance for peasant cultivators (Kloppenburg, 1988:46).

80 Union for the Protection of New Varieties of Plants, 1978, § 2

81 F. Pierce (1992), "Brazil, where the ice cream comes from", *New Scientist*, 17 July, p. 47.

82 The commercial value of wild plant resources may also constitute a new environmental threat. For instance, due to indiscriminate collection and lack of cultivation, *Rauwolfia confertiflora* has almost disappeared. This plant contains high levels of *reserpine*, and is used in treating hypertension.

- Intellectual property rights negate the principle of free exchange of breeding material, and their utility is limited to countries of some economic and technological strength.
- There was a growing tension in the South about their genetic material being acquired as a *common heritage of mankind (meaning open access – free of charge)*, at the same time as the North could impose *exclusive property rights* to the varieties developed from this same genetic material.

4.2.2 Issue characteristics and agenda setting

I now proceed to analyse the agenda-setting phase in light of the political and intellectual issue characteristics outlined in section 3.2. These characteristics were assumed to affect the possibility of having relevant parties participate in the first place, as well as affecting the scope for reaching agreement on commitments: Problems characterised by a high degree of asymmetry, as well as a general lack of scientific consensus, will be comparatively harder to solve. While the realist and part of the interest-based perspective emphasise these factors as the central determinants for international collaboration outputs, they are important control factors also in terms of establishing effects of endogenous factors within the other perspectives. Issue characteristics were also found important for determining our expectations regarding implementation scope, as low degree of implementation in a difficult issue-area may account for more or equally much as a high degree of implementation in an easily resolved area.

4.2.2.1 Political characteristics

The political characteristics are determined by the configuration of actor interests, and the extent to which these give rise to compatible or conflicting preferences. The basic structure of property rights to the resources associated with the issue-area will, to a large extent, define the roles, preferences and capabilities of the actors involved in international environmental and resource management negotiations. In examining the possibility to reach consensus on international co-operation, Underdal (1980) draws a distinction between *distributive* and *integrative* aspects of issue-areas. A distributive issue is characterised largely by being zero-sum; it may also involve redistribution, in that some parties will end up paying a larger share than they did prior to negotiations. This will generally lead to conflicting interests among the parties. By contrast, with an integrative issue, co-operation may lead to better management and consequently a larger pie to be shared among the parties. In such a case, a compatible interest structure is much more likely.

Against this backdrop, issue-areas dominated by *distributive* aspects can be seen as malignant, and presumed to be inherently more difficult to resolve, compared to more benign *integrative* issues. It is assumed that a malignant issue-area may be further complicated by asymmetry and cumulative cleavages. Conversely, a symmetrical interest structure and cross-cutting cleavages may mitigate the malignant nature of an issue-area.[83] Renewable resources are more likely to have an integrative nature than non-renewable ones. In a distributive issue, there is expected to be less scope for learning as a mechanism to reach consensus, than in an integrative issue-area.

The issue-area of biological diversity may involve policies of a dual nature in this respect:

- *Integrative aspect*: Through the rapid developments in the new biotechnologies, the economic potential of genetic resources has been increasing rapidly. More controlled management of biological resources will produce positive gains as the vitality of crops and livestock are ensured. The saving of any particular species, or variety within a species, may prove to be of great importance to pharmacy and agriculture. In other words, improving sustainable management of biodiversity carries with it a potential to increase the overall benefits to be reaped, and shared.
- *Distributive aspect*: At the same time, the issue of biological resources is linked to ecosystem preservation and restrictions on land-use in general.[84] Consequently, biodiversity conservation does not only involve improved management, but also the costs of ameliorating the environment and the costs of abstaining from exploiting a resource.

Here we need to look more closely into the distributive aspects of the issue of biodiversity, in light of the complications introduced by the property rights controversy. In addition to the distinction between integrative and distributive aspects, we may distinguish between issue-areas in which *users* or *owners* of a resource dominate. International resource management regimes (such as fisheries regimes) are often initiated by the users of the resource, whereas issues involving pollution control or resources preservation will generally be initiated by the owners of the resource in question. I shall argue that the evolving property rights structure has bolstered the *asymmetry* between the parties and given rise to *cumulative cleavage lines* in the issue-area.

When the item for international negotiation concerns the sharing of a *common* good outside national jurisdiction, will to participate is seldom a barrier. Cases in point are the minerals of the deep seabed and of Antarctica: everybody wants to get in for a share. In the case of the Antarctic, a main controversy has been linked to the

---

83 These characteristics have been adapted from Arild Underdal's article "Patterns of Effectiveness: Examining Evidence from 13 International Regimes", (Underdal, 1997b).

84 Genetic resources are living but not renewable. Once a species, or genetic variation within a species, is lost, it cannot be reconstructed. Thus, even if an endangered species is saved from extinction, it will probably have lost much of its internal diversity (Wilson, 1988). Many grain and vegetable varieties are disappearing rapidly. A variety of hybrid rice called IR–36 now extends over 60% of the rice lands of South-East Asia where, only a few years ago, thousands of varieties were common. Where 30,000 different varieties of rice grew recently, Indian agronomists now expect that merely a dozen will dominate three quarters of the land (El-Gewely, 1990).

question of inclusion. It has been argued that the Antarctic Treaty lacks legitimacy because membership has largely been limited to the claimant states: the 12 developed countries with territorial claims, more recently joined by India and Brazil (Luard, 1984). Similarly, in the case of deep seabed minerals, inclusion has been regarded as crucial, particularly to states in the South. In both cases, representatives from the poorer part of the world have been demanding common heritage for the resources in question. Today these issues have largely been resolved in what may be termed a realist manner: The exclusiveness of the Antarctic claimant states has found general acceptance, with the exception of Malaysia, who has made regular interventions about common heritage. On much the same note, DCs had a breakthrough for their moral principles regarding the deep seabed minerals under the UN Law of the Sea (UNCLOS), but there has been no follow-up in terms of material mechanisms.

Questions concerning the global environment are likely to give rise to inherently different situations, even though these may also deal with *common* resources. There are certainly similarities between the global issues of biodiversity, ozone and climate change. In these cases, the parties are often asked to modify their behaviour in certain activities under national sovereignty, such as reducing energy consumption or logging. The costs of abstaining from resource utilisation obviously make participation and commitments less tempting, and often lead to free-rider problems in the implementation phase. Nevertheless, participation may be secured if the parties realise that the only way to resolve the problem lies in collective action, and that all may suffer if no solution is reached. The costs of biodiversity conservation in particular resemble the costs involved with climate change, in that the burden may be unevenly distributed, put to bear heavily on economically important sectors of society, in order to restrict and change behaviour. As noted in section 1.3, however, an estimated 95 % of the world's terrestrial biodiversity is found within the jurisdiction of sovereign states, predominantly in the South. In itself, this situation is nothing new. What is new is that biodiversity has been made into a highly contested issue. The background to this can be found in three coinciding developments: First, the realisation that biodiversity loss is accelerating at an unprecedented rate, creating an urgent need for remedial solutions. Second, the technological developments that have greatly increased the economic benefits to be reaped from biological resources. Third, these technological developments have also highlighted the need for changes in the property rights system governing control over biological resources.

Recent years have seen change in the approach to property rights to genetic resources – from the non-exclusiveness (based on the CHM-principle) for all genetic material, to exclusive rights (patents) for parts of the genetic material. In effect, the property rights regime immediately preceding the biodiversity negotiations gave rise to asymmetries between actor interests along two dimensions:

- Increased technological and economical ability to exploit biological resources gave the North close to all of the benefits from harvesting these added cash-flows, with little or no benefits accruing to the South.

- The global distribution of biological diversity gave the South close to all the costs of conservation, with little or no costs directly falling on the North.

This situation guides attention to a specific feature distinguishing the biodiversity issue from other international environmental and resource management regimes:

> *the (primary) users of the resource ask the (primary) owners to change their behaviour – to conduct resource conservation and submit to stricter regulations.*

4.2.2.2 Intellectual characteristics

Scientific knowledge may, broadly speaking, enhance the likelihood of international co-operation through two mechanisms:

- *Scientific consensus* that there is a problem, and as to causes and effects in the problem area, and
- *Critical findings* or external shocks, such as "Waltsterben" and holes in the ozone layer, that act to put the issue on the international agenda.

Conversely, scientific uncertainty, or worse disagreement, about causes and effects in a given problem area can be assumed to hamper negotiations on international co-operation (Miles et al., 1998). Intellectual characteristics may also interact with political characteristics. In issue-areas involving redistribution, scientific uncertainty or disagreement may for instance be used by reluctant actors in order to postpone decisions on a matter.

According to Karen Litfin (1994), efforts to find common solutions to combat the depletion of the ozone layer were greatly boosted by the appearance of several critical scientific findings. While ozone depletion is also a gradual process, genetic erosion is taking place in a much less noticeable manner, never achieving the same dramatic headlines. Furthermore, the main threats to genetic diversity are questioned, as seen in the debate about whether intellectual property rights represent a direct threat to genetic diversity – and food security – in agriculture (see p.12). On the other hand, a sense of urgency about wildlife loss was brought about by headlines concerning destruction of tropical rain forests in the late 1980s.

Moreover, cause/effect linkages are highly complex in the biodiversity area. The great number of sources leading to genetic erosion mean that almost every sector of human activity is involved and may be affected. This also makes the entire biodiversity issue one of high complexity. By contrast, the ozone issue is often referred to as a relatively *benign* area. First, because *all* countries are vulnerable to ozone depletion, and second, because the price for phasing out CFCs has proven to be relatively low. As technological developments made CFC substitution an economically viable avenue, the industrial sector became an important ally in the work to find common solutions to the ozone problem.

The uncertainty and complexity regarding the main sources of genetic erosion have made it harder to narrow the focus to one or two crucial relationships (focal points) around which to centre action. Hence, the agenda-setting effect of natural scientific advice is frequently weak in the area of biodiversity, compared to e.g. ozone and acid rain (the LRTAP regime). Some scientists may feel the need to make exaggerated estimates on biodiversity loss in order to draw attention to a problematic area. In turn, others may refrain from correcting such exaggerations, for fear of minimising the problem.

In recent years, however, *scientific uncertainty* has lost much of its currency as an excuse for postponing action – both in general and in this particular issue-area. This is because of greater acceptance for the *precautionary principle* and the notion of irreversibility: The First Regional Conference on the follow up to the report of the World Commission on Environment and Development in Bergen saw a breakthrough for the *precautionary principle:* The world must take action against environmental disasters, such as global warming, without waiting for scientific certainty about their causes or extent.[85] In this light, the accomplishments of the Montreal Protocol may be explained by the relatively simple nature of this problem area, but also by the irreversibility of the damage to the ozone layer. Irreversibility is also a central element to the issue of genetic erosion: once lost, varieties cannot be revived. Genetic varieties thereby can be seen as non-renewable resources.

Regarding the *severity of biodiversity loss* there is in fact a high degree of scientific consensus. The unprecedented pace of species loss is now generally considered an issue in urgent need of solutions. This awareness increased rapidly in the late 1980s as dramatic reports of rainforest destruction reached the headlines. In this light, the scientific character of the biodiversity issue appears more *benign* than that of climate change, in which scientists still disagree on the environmental dangers involved.

The scientific characteristics of the biodiversity issue may be summed up as follows:

- lack of critical findings (external shocks), no one focal point.
- scientific uncertainty and disagreement regarding causes and effects.
- scientific consensus regarding gravity of the problem.

The potential benefit of the scientific status is a *veil of uncertainty* which may obscure the distribution of costs and benefits thereby perhaps facilitating agreement in the negotiation phase. The drawback is a lack of focal points, which may be most detrimental in the implementation phase. Lack of focal points will reduce the scope for formulating specific targets. In turn, implementation activities will be more elusive and less amenable to monitoring.

---

85 "Action for a Common Future". Conference organised by UNECE (Economic Commission for Europe) in Bergen, Norway, 1990. See *New Scientist*, 26 May 1990, p. 19.

In sum, then, many essential features of the biodiversity issue – scientific uncertainty regarding means, complexity, cumulative and asymmetrical affectedness, and resources under state sovereignty – may largely be characterised as *malignant:* This could be expected to work against the possibility of reaching agreement on international co-operation in this area. There are also two features – scientific consensus regarding ends, and the dual nature in terms of integrative and distributive aspects – that could go both ways.

The next section traces the international debate on biodiversity through the international fora of the 1980s and early 1990s. Although my focus in this study will be on the Biodiversity Convention, an understanding of the broader biodiversity regime – the cluster of norms, rules and procedures constituting this issue-area – necessitates the inclusion of other international arenas.

4.3 The Issue in Related Fora

In addition to tracing the formal biodiversity negotiation process in UNEP, I deem it necessary to include references to three other international fora: FAO, GATT (later WTO) and the Keystone Dialogues. The debate leading up to the FAO Undertaking on Plant Genetic Resources constituted the basis for central elements later included in the CBD, such as the concept of *mutually agreed terms* relating to access and use of genetic resources. The FAO debate has a more narrow focus, in that it is limited to domesticated genetic material (agricultural plants), while the CBD includes both domesticated and wild biodiversity, hence involving a wider range of actors. As to GATT, discussions on intellectual property rights in the Uruguay Round have had significant impact on the question of property rights to genetic resources as discussed within the framework of the CBD. GATT/WTO negotiation parties come from different sectors of government agencies than the CBD actors. In the former we find top-level civil servants from the Treasury and trade-related ministries, whereas in the latter senior civil servants from the ministries of foreign affairs and environment predominate. As these different groups discuss items with an important bearing on the same issue-area, this aspect cannot be left out of this process-tracing endeavour. The Keystone Dialogues represent an informal, consensus-oriented negotiation forum for detailed discussion of matters relating to genetic resources and biodiversity, with participants coming from (but not directly representing) a broad range of public and private sectors.

4.3.1 FAO and the Undertaking on Plant Genetic Resources

The United Nations Food and Agricultural Orgainzation (FAO) is a global organisation, founded on an inter-state treaty. FAO was established in 1945 by the UN Conference on Food and Agriculture and is headquartered in Rome. It works through a Conference, a Council and a Secretariat.

When the industrialised world started to rely increasingly on plant breeders' rights (PBR) and patenting, the governments of developing countries began questioning whether the common heritage principle would eventually apply solely to resources from the South: This is because the non-systematically bred landraces hardly meet the PBR criteria of being *uniform, stable and distinct from existing varieties*. During the FAO "gene war" in the 1980s, they argued that the elaborated material of the industrialised countries was based largely on material from the South, and should thus also be seen as part of the common heritage. This view met with strong resistance from the industrialised countries, who argued that such an arrangement would not be compatible with Northern plant breeders' rights and patent legislation.

It was in 1981 that the issue of control and access to genetic material was first put on the FAO agenda by the developing countries, spearheaded by Mexico. DC governments and NGOs started questioning what they perceived as a one-way, free of charge, flow of genetic resources from the South to the North.[86] One result was a non-binding International Undertaking, a Commission and a Fund for Plant Genetic Resources in FAO.

It was seen as a great victory to the developing world when the FAO Undertaking of 1983 established the principle that **all** categories of plant genetic resources should be subject to free exchange for exploration, preservation, evaluation, plant breeding and scientific research. Article 1 of the Undertaking declared that genetic resources were the "heritage of mankind and consequently should be available without restrictions".

Responding to the then-emerging regime of intellectual property rights, however, an *Agreed Interpretation* of the Undertaking was signed in 1989. With regard to elaborated material subject to legal protection under national legislation, this material was to be made available *on mutually agreed terms*, according to the reinterpreted Undertaking. Basically, such *mutually agreed terms* signifies acceptance of the principle of payment for legally protected varieties. Since the *Agreed Interpretation* regarded *intellectual property rights* as compatible with the Undertaking, the DCs abandoned the *common heritage of mankind* strategy.

The main idea of the FAO Fund for Plant Genetic Resources is crystallised in the concept of *farmers' rights*, meant to represent a counterpart to *plant breeders' rights*. The Fund aims to provide compensation for "the enormous contributions generations of farmers have made to the conservation, selection, domestication and development of plant genetic resources. If breeders, who provide the finishing touches to this process, can secure a title and handsome profits for their efforts, then the farmers too should receive compensation..".[87]

Behind the concept of *farmers' rights* are two NGOs who were active during the negotiations on the FAO Undertaking: Promotion of this idea can be credited largely to the International Coalition for Development Action (ICDA), an NGO which

---

86 See Rosendal (1989).

87 FAO, Working Group of the Commission for PGR, CPGR/89/9, 1989.

played a central role in informing developing countries about their common interests and thus facilitated the maintenance of a strong South alliance (Rosendal, 1989).[88] Often in collaboration with the Rural Advancement Fund International (RAFI), a Canadian-based NGO, ICDA made available various reports and documents on the South's genetic capital during the FAO sessions (Mooney et al., 1988). Pat Roy Mooney, RAFI's leader through 20 years, is also credited with coining the concept *biopiracy*, which set the stage for one of the main controversies in the subsequent biodiversity negotiations. Another central RAFI member, Dr. Cary Fowler, was later employed by FAO and given the mandate to prepare and organise the FAO Fourth International Technical Conference on Plant Genetic Resources at Leipzig (17–23 June 1996). This Conference addressed, among many others, the issues of access and benefit sharing within agricultural biodiversity. *Farmers' rights* were once more a controversial item in the discussions.

The fact that the question of *farmers' rights* was put on the agenda and achieved consensus is still regarded as a victory for the South. To date, however, the concept has had little practical effect – first of all because the FAO Undertaking is not legally binding. The lack of effect may also be ascribed to the inability of the FAO Fund to attract funding, which in turn is basically due to widespread donor-country scepticism to FAO. Another explanation may be found in the difficulties inherent in applying the concept in practice. This concerns the problem of tracing the "contributor" to whom compensation should be made. These plant genetic resources have not been subject to systematic breeding, and rarely fulfil such norms as the UPOV criteria of stability and uniformity – on the contrary, their greatest value lies precisely in their broad and evolving diversity. Over the years, seeds have crossed so many borders and been developed in so diverse parts of the world that such a system would be hard to design and administer. As the FAO Global System on Plant Genetic Resources was not cast in a legally binding mould, efforts are now underway to renegotiate the Undertaking with a view to include relevant parts of it in protocols under the CBD. In essence, then, the FAO output may be summed up as a moral victory for the South, but lacking material implications.

4.3.2 Trade-Related Aspects of Intellectual Property Rights (TRIPs) in GATT

While the developing world seemed to gain acceptance for some of its argumentation within the FAO arena, the issue of intellectual property rights was dealt with in a somewhat different manner elsewhere. Although developing country governments may have achieved a certain footing within international development assistance agencies, industrialised countries generally dominate the fora on economy and trade. Most significantly, questions concerning the widening scope of industrial patents were brought up in the Uruguay Round (1988–1994) of negotiations under the General Agreement on Tariffs and Trade, which included provisions on trade-

[88] ICDA has since then changed its name to GRAIN (Genetic Resources Action International) and is still active lobbying in international biodiversity debates.

related aspects on intellectual property rights (TRIPs). The final text of the Agreements established the World Trade Organisation (WTO) including the TRIPs Agreement adopted in Marrakech in April 1994.

The TRIPs discussion soon became one of the fiercest arenas for the North–South patent controversy. The USA, Japan and (less adamantly) the EU advocated the principle that all countries should provide and respect intellectual property protection in all technical fields – including biotechnology. Disregarding this principle would constitute a contravention of GATT regulations, making the offending country liable to economic sanctions.

Intellectual property rights have previously been handled under the UN World Intellectual Property Organisation (WIPO), which was set up to administer the Paris Convention (1968). Under WIPO each country is free to determine its own level of intellectual property protection, so long as no distinction is drawn between nationals and foreigners. The GATT TRIPs agreement, by contrast, constitutes a global system of intellectual property rights.

Governments in the South were strongly opposed to the GATT TRIPs proposals, arguing that patents benefit those states that are already technologically and economically strong (GATT, 1989).[89] This is a point hard to refute, as the South holds no more than one to three per cent of all patents world-wide (WCED, 1987). In the initial rounds, India argued against patenting of plant and animal varieties as well as food and pharmaceutical products, citing concern for basic human needs (GATT, 1989). Many DCs argued that the application of IPR systems would hinder the transfer of technology to the developing world as well as disregarding the very real contributions of generations of farmers to the world's plant genetic resources, the basis of global food security. For a great many developing countries, this is not merely a matter of contesting theoretical principles. With the introduction of industrial patents, access to improved breeding material may be hampered, as prices for seed increase (Keystone, 1990). Some fear that patents will place constraints on technology transfers in general. As the Uruguay Round drew to a close in Geneva in December 1993, most DC delegations had resigned, but grassroots organisations were still mobilised in large numbers: Indian farmers, Latin American pharmaceuticals manufacturers and Guyami Indians demonstrated in Geneva against the patent regime proposed in GATT.

The opposition has had some success in GATT. This was partly due to the mitigating effect of the *European Patent Convention* (Art. 53[b]), which at the time allowed *plant varieties* to be excluded from patentability. The final agreement on Trade-Related Aspects of Intellectual Property Rights (TRIPs) grants the parties the right to exclude from patentability (Art. 27.3.[a]) diagnostic, therapeutic and surgical methods for the treatment of humans and animals, and (Art. 27.3.[b]) plants and

---

89 GATT (1989), *Communication from India: Standards and principles concerning the availability, scope and use of Trade-Related Intellectual Property Rights*, GATT Secretariat, MTN GNG/NG11/W/37, July.

animals other than micro-organisms.[90] The parties are bound, however, to introduce some kind of intellectual property rights for *plant varieties*. TRIPs require members to provide for the protection of *plant varieties*, either by patents, or by establishing an effective *sui generis* system (a system of its own kind). Those WTO member-states that choose the *sui generis* laws must establish these by January 2000 (the least developed countries have until 2005 to fulfil their TRIPs obligations). There is still some uncertainty with regard to whether developing countries will choose, or feel pressured to, adopt the UPOV-system; or if they may choose to create systems better suited to their present needs. In December 1999, the *sui generis* option within TRIPs will be reviewed.

The latter part of the TRIPs agreement has prompted the farmers' movement in India, among others, to propose that such a *sui generis* system should focus on the rights of farmers in protecting and improving plant genetic resources: *Common Intellectual Property Rights* (Shiva, 1994). As to whether the TRIPs regulations on patenting will have harmful effects for farmers in developing countries, the answer is probably no in the short-term perspective – but that issue is still contested. The underlying threat in the expanding patent legislation – that farmers must pay royalties for reusing seeds – is still a long way from being enforceable.[91] A far more harmful effect of the TRIPs patent regulations is that they reinforce a North–South conflict line in an issue-area where common solutions and co-operation are of paramount importance.

The basic ideals on which WTO rests may also come into conflict with the CBD in a more general respect. Measures to enhance world-wide free trade can contradict the need for stricter control mechanisms to avoid the spread of alien organisms into ecosystems. The rise in incidents of exotic species displacing or threatening indigenous or endemic species has been linked to the growing and liberalisation of world trade.[92] This aspect is currently subject to further negotiations under the CBD Conference of the Parties in terms of establishing a Protocol on Biosafety, so I will not treat it in any detail here. The question of international regulations of biotechnology nevertheless illustrates that the current efforts to put environmental concerns on the WTO agenda have not given rise to a clear-cut North–South divide: On several other matters relating to the trade/environment interface, DCs have remained sceptical to this link, seeing it as yet another protectionist move.

Thus, the overlap between the functional scopes of these two international regimes may represent a potentially serious challenge for the operation of what may

---

[90] Multilateral Trade Negotiations, the Uruguay Round, The Negotiations Committee, MTN.TNC/W/124, 13 December 1993, MNT/FA II–Annex 1C. Section 5, article 27 in the agreement on Trade-Related Aspects of Intellectual Property rights.

[91] A first test-case is presently under way in Canada, as one of the world's largest agrochemical companies, Monsanto, is suing farmer Percy Schmeiser for, they claim, growing the company's patented seeds without paying royalties for it.

[92] UNEP and Norwegian Institute for Nature Research (NINA): Conference on biological diversity and exotic species, Trondheim, June 1996. Forthcoming Report.

appear the weaker one, the CBD. The large number of international environmental agreements have been negotiated in the absence of explicit measures to resolve the frequently overlapping and conflicting goals of economic regimes. While the USA retains a stronghold in the latter fora, it has often found itself isolated, an almost single laggard, on environmental issues.[93] The relative strength between environmental and economic regimes, whether in terms of power and interest structure or in terms of normative persuasion, has yet to be put to any ultimate test.

4.3.3 Informal negotiations: The Keystone Dialogues

A central forum for the debate on biodiversity questions has been the Keystone Dialogues. The Keystone Centre, founded in 1975, is a national non-profit organisation located at Keystone, Colorado, in the Rocky Mountains. For some two decades now, the Keystone Center has been active in organising debates, bringing together diverse parties to resolve conflicts over the environment and natural resources.

The Keystone Dialogue on plant genetic resources started in 1988 to "bring together diverse interests to engage in a structured, off-the-record consensus-building dialogue to promote a strong international commitment to plant genetic resources at the community, national, regional and global levels" (Keystone, 1991). The plenary sessions met three times: in Colorado (1988), in Madras, India (1990), and at Sundsvollen, Norway (1991). This dialogue brought together politicians, scientists, corporations, intergovernmental and international organisations, NGOs and administrators from developed and developing countries with the intention to formulate consensus recommendations and strategies regarding availability, use, exchange and protection of plant germplasm. A total of 92 delegates from 30 countries took part. Many of the participants would later represent their governments in the delegations to the CBD negotiations.[94] Among the most active participants were also members from ICDA (now GRAIN) and RAFI, as well as the then director of the Ethiopian Genebank, Dr. Melaku Worede. Also present were high-level officers from FAO, the World Bank and the IBPGR, as well as a Senior Programme Advisor from the UNCED Secretariat. Furthermore, three of the world's largest multinational seeds and pharmaceutical corporations, Monsanto, ICI and Ciba-Geigy, sent high-level management representatives.

Delegates to the Keystone Dialogues met in their personal capacity, not as representatives of governments or corporations. This seemed to encourage a more

---

93 "Despite talk of U.S. leadership on environmental management, the United States was in reality an obstacle in most international environmental negotiations", Konrad von Molkte (1997:256) claims, referring to the Basel Convention, climate change, biodiversity and the Global Environment Facility (GEF). The USA has, however, championed environmental agreements in some issue-areas, such as the Antarctic minerals regime, ozone and the Convention on Long-Range Transboundary Air Pollution (LRTAP).

94 To take an example from Norway: two of the participants, from the Ministry of Foreign Affairs and the Ministry of Agriculture, were later to join that delegation.

forthright approach to the controversial issues at stake. Decisions reached in this forum are obviously not legally binding, but the discussions have managed to set some precedents and created an impetus for further negotiations in other fora. For example, the report issuing from the first meeting (1988) was presented to the US Congress, and it is believed to have been significant for the work on plant genetic resources in FAO.[95] The entire concept of the Keystone Dialogue may be said to embody an "entrepreneur-like" spirit, its members come close to constituting an epistemic community for political and scientific discussion in the issue-area.

One early "output" from the Keystone Dialogue concerned the debate on patent legislation in GATT. The whole group, down to the patent expert from Ciba-Geigy, agreed that an extension of intellectual property rights, as advocated in GATT, would appear to conflict with the basic premise of treating genetic resources as common heritage, and to challenge the basic intention of the FAO Undertaking on Plant Genetic Resources and the proposed legal instruments for biological diversity. A consensus recommendation was reached "that GATT should assess the impact of extending patenting to genetic resources in consultation with FAO, IUCN, UNEP, UNESCO, UPOV, WIPO and UNCTAD". The group also recommended that "the governments involved in the GATT negotiations consult with experts concerning biodiversity in their countries to assess the potential impact of the negotiations on the preservation and exchange of plant genetic resources" (Keystone, 1991). Thus, some of the most controversial issue-links in the biodiversity negotiations were treated outside the formal international fora, and the strengthened understanding that developed within this setting was subsequently fed into the formal sessions under the UNEP biodiversity negotiations.

4.4 Summary

It is clear that in this first phase of *getting to the international agenda* the parties generally recognised that a collective problem existed, and that this called for joint action. There was, however, no consensus as to just what this collective problem consisted of. Very broadly put, the North was concerned with conservation of biodiversity, and the South with equitable sharing of benefits from use of biodiversity.

Two main conclusions may be drawn with regard to the effects of the political and intellectual characteristics of the issue-area in this agenda-setting phase:

- *Cumulative asymmetry*: The international distribution of biodiversity itself is uneven, as are the capabilities to utilise biological resources economically and technologically, and this leads to a double asymmetry: (i) *Economic benefits for users*. The countries with the lowest level of biodiversity frequently have high

[95] As pointed out by Ms Kathleen Merrigan of the US Senate, during the third Keystone meeting in Oslo.

capacity to *benefit* from its use, due to high technology levels, and vice versa. (ii) *Economic costs for owners*. Conservation *costs* will place a heavy burden on owner countries and will often have a short-term adverse effect on sectors such as agriculture and forestry, on which these countries are heavily dependent. This situation led to diverging demands regarding what items should be put on the international agenda.

- *Uncertain means, consensual goals:* The agenda-setting effects of scientific findings have been relatively weak in the biodiversity issue, as compared to other global environmental issues such as ozone-layer depletion and acid rain. While scientific knowledge cannot be said to have acted as the main booster in terms of being a source of solutions, it has clearly had role as pointing out the causes for concern. There is widespread consensus in the scientific community about the gravity of the problem and that something must be done.

These features clearly affected how aspects of the issue-area were dealt with in FAO and GATT. The changes in property rights pertaining to genetic resources in OECD countries originated outside international negotiation fora, but quickly led to two different sets of demands within these negotiation fora:

1. *North's position*: the North demanded that all countries provide and respect *intellectual property protection* in all technical fields, including biotechnology, in GATT – and it was later to succeed in getting acceptance for this demand.
2. *South's position*: the South had successfully advocated the principle that **all** categories of plant genetic resources should be subject to the *common heritage of mankind principle* in the FAO Undertaking of 1983. Since the FAO "Agreed Interpretation" of 1989 establishes intellectual property rights as compatible with the Undertaking, however, the DCs abandoned the common heritage strategy and turned to *national sovereignty*.

The situation immediately prior to the CBD negotiations may be summed up thus:

- Aspects of the issue relating to property rights were clearly crucial in affecting agenda-setting. The transformation of the property rights regime for genetic resources is the central point to which most other agenda items were subsequently linked.
- The UNEP Secretariat inherited several established and unresolved conflict lines when informal negotiations on the CBD were initiated in 1989.

All in all, this would hardly seem to encourage expectations for a large scope for agreement and co-operation in this issue-area. And yet, the CBD has been adopted and signed, and it has entered into force. How can we account for the international output? What does this mean in terms of "expected" scope for implementation? To what extent and how did the countries of the developing world succeed in getting their items onto the agenda – despite their limited control over their own genetic resources? These are the subjects of the next chapter.

CHAPTER 5

THE CBD – NEGOTIATIONS AND OUTPUT

5.1 Introduction

This chapter will be concerned with the first part of the question dealt with in this thesis:

> *How have international institutional factors had an impact on the CBD objectives (international output)?*

In addition to seeking to account for output, I shall also be concerned with assessing the nature of the output in terms of the extent to which it represents a breakthrough or coincides with the interests of developing countries (DCs). Using the three theoretical perspectives as guides, I shall look into the expected patterns regarding international and national output along the following lines: First, there may be *no correspondence* between the CBD and interests of DCs. Second, the CBD objectives may *happen to coincide* with the interests of the South to a greater extent than was the case under pre-regime conditions. If so, there is little reason to assume that institutional factors may have had an impact on output, as this would indicate compatible interest constellations. Third, the CBD output may represent a *real breakthrough* for the norms and interests of the South, in the sense of a hard-won victory. In that case, institutional factors may have been instrumental in achieving this.

Applying the three models as analytical guides, I shall examine what factors have been most central in accounting for the international output.

1. The *realist regime* model predicts that international outputs in high-salience issue-areas will be determined by the interests of dominant state actors. On the basis of this model it was assumed that any "agreement" among the highly asymmetrical actor interests associated with the biodiversity issue-area can only be accounted for by overall structural power through hegemonic activity (structural leadership). Hence, we would expect to see the interests of the economically and technologically dominant states reflected in the CBD text.
2. The *interest-based* regime model points to how institutional factors such as scope of participation, agenda-setting, the role of the secretariat and scientific input may assist in identifying and boosting the integrative potential among the parties. These rules and procedures, along with the official purpose of the negotiation forum, could be instrumental in altering or aligning actor interests,

and could give rise to mutual expectations among the parties. Moreover, according to the interest-based perspective, power may be issue-specific, and thus hypothetically provide DCs with the capabilities to win through with their interests in the international output.

3. The *ideational* model draws attention to the role of epistemic communities and learning processes in developing common norms and ideas among the actors in a given issue-area.

5.2 The CBD – Negotiation Process

Broadly speaking, the North and the South started out with very different environmental agendas for negotiation of the biodiversity issue: Some of the more powerful countries of the North, notably the USA, set out to get an international commitment to conserve wild species and habitats, and to leave their free access to the biological resources of the South *out* of the Convention.[96] The former objective was explicitly advocated and was spurred on by rainforest destruction and other threats to biodiversity world-wide. The second objective may be seen as a "hidden agenda" and reflected the increased economic and biotechnological interest in biological material. By contrast, the South wanted to be compensated by the developed world for the costs involved in conservation activities – the costs of refraining from utilising their own resources. Secondly, many developing countries wanted to include all kinds of biodiversity in the agreement, including domesticated material, as this highlighted the economic benefits accruing from use of genetic material. Third, the South wanted access to the same technologies that enable industrialised countries to profit from the exploitation of biological resources.

In effect, both main groups of actors were drawn to the negotiation table with high concerns, but with quite different "agendas" in mind. This section will show how the demarcation of the issue-area was a highly controversial process.

It was the users that put biodiversity conservation on the international agenda in the first place, although at the time they did not expect to have to pay or change their behaviour (see Box 4.1). The problem from the users' perspective was basically how to get the owner countries to the negotiation table. The problem from the owners' perspective was basically how to get the users to share the responsibility and costs of biodiversity conservation. In the international biodiversity negotiations, the redistribution of these costs has been a focal point of contention, as the owners have increasingly responded by asking the users to contribute to the solution of the problem.

---

[96] As this presentation will reveal the Nordic countries were of a different opinion.

Preparations for negotiating the CBD were initiated by an *Ad Hoc Working Group of Experts on Biological Diversity*, established by UNEP in 1988. Once formal negotiations got underway, the group changed its name to the *Intergovernmental Negotiating Committee* (INC).[97] The first meeting in the ad hoc group drew experts from 25 countries, as well as some NGO observers, including the World Conservation Union (IUCN) and World Wide Fund for Nature (WWF), and IGOs, such as FAO. The number of delegates expanded rapidly; during the seven formal working sessions of the Intergovernmental Negotiating Committee there were representatives from some 80 nations. Negotiations started informally in November 1989, formal multilateral negotiations began in 1991, and the CBD was signed by 153 countries and the European Community at the UN Conference on Environment and Development (UNCED) in Rio de Janeiro 5 June 1992. The Convention entered into force on 29 December 1993, three months after achieving the prescribed 30 ratifications. As of October 1999, the Biodiversity Convention had been ratified by 175 states and the European Community; in addition it has 10 Signatories who have not ratified.

When the issue of biodiversity was first put on the agenda by UNEP, several parties suspected that this was in fact an attempt to de-link the politicised plant genetic resources debate in FAO from the more traditional values of wildlife conservation in protected areas. What gave rise to this suspicion? The process was led by the US delegation, who submitted a draft proposal for a convention made by the IUCN.[98] Judging by this draft proposals, the idea was to retain a focus on *in-situ conservation of wildlife and habitats*, rather than tackle the controversial issue of *equitable sharing of benefits from use of biological resources*, i.e. to include domesticated genetic resources (Rosendal, 1991). The link to domestication and hence systematic or non-systematic breeding would inevitably lead back to the inclusion of the whole controversial issue of property rights and compensation, as the FAO and TRIPs negotiations had already shown very clearly.

The USA feared that linking the packages of conservation and benefit sharing would hamper negotiations on a biodiversity convention (Rosendal, 1991). It was thus hoped that "the move" to UNEP would quench these fires, and the USA refused to include any mention of biotechnology or discuss the economic value of genetic resources. The controversy concerning utilisation of biological resources was intensified by the consequences this would have for distribution of costs and benefits: Firstly, putting a price-tag on biodiversity might disclose how profits in the agricultural and pharmaceutical sectors in the North are extracted from genetic

---

97 Decisions on negotiations: 14/26 (1987), 15/34 (1989) and 16/42 (1991) in the UNEP Board of Directors.

98 IUCN: an INGO established in 1948 and located in Gland, Switzerland. Members include 62 states, 108 government agencies, 407 national NGOs, 35 international NGOs, 35 affiliates. The aims are to promote scientifically-based actions directed at sustainable use of natural resources, conserving variety of species of plants and animals. The IUCN publishes Red Lists over threatened species.

resources from the South. Secondly, the new perspectives draw attention to biodiversity in a much wider sense and could lead to stricter regulations on agricultural and forestry practises in all parts of the world. While the former issue-link primarily concerns redistributive effects *between* countries, the latter may have redistributive effects *within* countries. The North–South conflict line is rather clear-cut regarding the former (cost-benefit redistribution between countries), but much less so with regard to the latter aspect (cost-benefit redistribution within countries). The latter also has a North–South angle, however, as it highlights whether biodiversity conservation is equally important in all parts of the world, as well as whether there should be a common global responsibility for conservation.

Environmental and wildlife management NGOs like the IUCN and the WWF feared that no conservation agreement would be reached, either for wild or domesticated species, if the latter were to be included in the negotiations (Rosendal, 1991:33). Hence, the first IUCN draft convention presented to the participants reflected the Western traditions of nature conservation (or rather preservation: see Box 4.1) in full. The role of the IUCN was also a manifestation of the type of NGOs represented as observers at UNEP negotiations. These were mainly concerned with habitat and species protection, rather than what had been the case in FAO: In FAO in the 1980s, the International Coalition for Development Action (ICDA) and the Rural Advancement Fund International (RAFI) promoted the seeds issue and helped to advocate and co-ordinate the interests of DC (small-scale) farmers. ICDA had at this point changed their name to Genetic Resources Action International (GRAIN) and was primarily concerned with lobbying the TRIPs negotiations in GATT. Among the reasons for RAFI not to join the CBD negotiations, were their doubts that anything would result, and what they experienced as lack of control within the UNCED framework.[99] GRAIN was not to join the UNEP until at a later stage, and then only as observers.

During the early phases of the UNEP biodiversity negotiations, DC governments abandoned the demand for an all-embracing common heritage regime. In response to the developments in FAO and GATT, they turned the argument around:

> *The DCs' new line of argumentation was to claim national sovereignty over their genetic heritage, demanding that it be regarded as a national asset along the lines of other natural resources, such as oil and minerals.*

Genetic resources differ, however, from oil and minerals in being non-rival and largely non-exclusive goods. Nor is species distribution necessarily confined to national borders. These characteristics will obviously hamper state control over genetic resources. Nevertheless, national sovereignty ended up as the only passageway for reaching consensus about property rights between the North and the South in the CBD text. We shall look closer into the contents of this package-deal and how it came about.

---

[99] Personal communication, Pat R. Mooney, Seminar in Oslo, 22 May 1998.

Initially, the UNEP biodiversity agenda was dominated by a focus on wildlife conservation – protecting biological *hot-spots* such as tropical rainforests and other places of high biological diversity. In addition to the IUCN and the United States, this position was held by UK and the Netherlands, among others. The agenda was, however, soon to be characterised by a high degree of flexibility, and participants kept adding on new elements and making new linkages. At the second informal meeting (February 1990) in the UNEP ad hoc Working Group, the UNEP secretariat prepared a document on important and controversial links within the issue-area: i) the link between conservation and development; ii) how to establish the value of biodiversity and find mechanisms for transfer of technical and financial resources; iii) access to genetic resources and hence the need to include domesticated species; iv) the link to biotechnology; and v) the position of indigenous people. As the informal negotiations were drawing to a close (July 1990), the UK, Canada and the Nordic countries (among others) were reported to be putting emphasis on a wide mandate for the CBD, integrating sustainable use of the resources in all sectors. At this point, Germany, the Netherlands and the USA were among those who still advocated a more narrow scope, focusing on classical preservation. In was only in the very last negotiation round that the three-fold objective of the CBD – conservation, sustainable use and equitable sharing – was accepted. The Nordic countries were ardent pushers for this triple objective, fearing that the, by then, watered-out concept of conservation alone would not suffice (see Box 4.1).

Secondly, the Nordic countries, together with some others, stressed that biodiversity conservation is essential in all countries, regardless of the number of species. This was an effort to counter the bias towards tropical forests and to emphasise that the CBD should include all biodiversity.[100] Moreover, it was primarily the Nordic delegations that emphasised the development aspects and an improved utilisation of resources as a means for providing incentives for better conservation of natural species or habitats. It was a representative from the Norwegian delegation who formulated the crucial text of article 16.5, saying that IPR-systems should "ensure that such rights are supportive of and do not run counter to its (the Convention's) objectives". These views coincided with those of many DCs, and were rapidly gathering support as the formal negotiations were about to start. At the same time, the South started advocating the principle of national sovereignty over biological resources in response to the ongoing GATT negotiations

[100] The link to climate change may be important in this respect. The tropical rainforests, and not least that of the Amazon, are not only home to a large portion of the world's flora and fauna, they also represent a significant "sink", a buffer against atmospheric CO_2 build-up. This may add to enhance the US interest in Brazil's forests, albeit from a slightly different angle. Estimates of *deforestation* have recently placed countries like Brazil at the top of the list of greenhouse gas emissions (WRI, 1990), a (controversial) "fact" that put many tropical countries on the alert with regard to the parallel negotiations on a climate change convention. From Brazil, Indonesia and the Republic of Congo's viewpoint, it seems that they would have more to win from insisting that forest conservation be approached from a biodiversity perspective (Rosendal, 1995e).

on intellectual property rights. Eventually, the majority of developed countries realised that postponing these issues would mean that the DCs, especially countries such as Brazil and Malaysia, would not join the CBD. As the main bulk of biodiversity is located in the tropics, negotiations could hardly proceed without the South. In response to the *property rights* question, it was agreed that the questions concerning access and control over biological resources could not be treated separately from the ongoing GATT/TRIPs negotiations. The principle of *national sovereignty* was accepted.

These combined efforts introduced the call for a *common responsibility* for biodiversity conservation as well as for *compensation to the countries providing* biological resources. In response to the *common responsibility* and *compensation* question, the parties eventually did agree on the Global Environmental Facility (GEF) of the World Bank, UNEP and UNDP on an interim basis. The GEF was, and is, controversial with the South, for two reasons: First, developing countries fear that the Global Facility, largely dominated by the World Bank, would leave them with little control. Second, they were (and still are) sceptical about whether the Facility will consist of genuinely additional money. At the outset, the GEF was controversial with the UK and USA as well, as they preferred bilateral agreements between owners and users of biological resources, rather than a multilateral fund.

A Fund under the World Bank auspices would leave the developed countries with much more control over the contents of projects. Final decisions in the World Bank are made by weighted voting, a system favouring donor countries. The Bank also carries much more (financial) weight than its two counterparts, the UNDP and UNEP. In addition, the establishment of the Global Environmental Facility took place prior to international agreements on what standards should govern the management of genetic resources. Nevertheless, in what seems like a surprisingly short time, the GEF obtained the necessary support – although it is still a contested issue. A major reason for the support is the amount of funding involved. The Global Environmental Facility in its pilot phase from 1990 already operated with some US$ 1.4 billion for a three-year period. This is of a magnitude far surpassing other proposals for international funding in the area and is thus not to be treated lightly. The question is whether this is *real additional money* or if the money will simply be transferred from traditional development budgets. Another crucial question is whether GEF is a genuine proof of the "greening" of the World Bank – and, more important, whether this means the DCs will gain a more equal say, or whether it is a bribe used by the industrialised world to side-step the massive claims from the South.

The essence of this section concerning the negotiation position of the South may be recapitulated as follows: The South was relatively unified and focused in their argumentation, and their argumentation was partly supported and never disputed by advice from the scientific community. The South's argumentation was, however, marred by limited control over the resources in question. Nevertheless, the CBD was adopted by consensus in Nairobi in May 1992. A few weeks later, at the UNCED

Rio de Janeiro Conference, the USA was the only developed country to hold out against signing the Convention.

I will shortly abandon the chronological description in order to present a picture of the activity level among the negotiating parties. This effort is necessary to a better understanding of which parties are more inclined to seek influence, and this will hence provide a backdrop for the subsequent discussions of leadership and pushers. In section 3.2.3, on ideational-based explanations of regime formation, I put forth the proposition that countries with a high activity level probably stood a better chance of advocating their own norms and principles – as well as enhancing the possibility of a breakthrough for own interests in the output. The basis for this line of thinking is that parties may not only be subject to learning during negotiations, but there is also the possibility of teaching others.

5.2.1 Indications of activity level

To get an approximate indication of activity level I have made a count of the parties' interventions during the three first COP meetings (Bahamas, 1994; Jakarta, 1995; Buenos Aires, 1996). The count is based on transcripts from Earth Negotiations Bulletin's minutes from the Conferences on discussions of central agenda items: financial resources, technical transfers, capacity building, intellectual property rights, forest management, indigenous people, access to genetic resources, biosafety, conservation, monitoring and agricultural biodiversity.[101]

Box 5.1 Activity level: COP1 & COP2

High-level intervention scores from Australia, Austria, Japan, New Zealand, Norway, Sweden, USA and UK among the OECD countries; Brazil and Colombia among the Latin American countries; China, India and Malaysia in Asia; and Kenya, Malawi and Zimbabwe from Africa.

In the *middle range*, there are, Denmark, Finland, France, Germany, Italy, the Netherlands, Spain and Switzerland within OECD;, Argentina, Chile, Costa Rica, Cuba, Peru and Venezuela from Latin America; Bangladesh, Indonesia, Korea, Pakistan and the Philippines among the Asian countries; and Benin, Burundi, Cameroon, Congo, Egypt, Ethiopia, Ghana, Mauritania, Nigeria, South Africa, Tunisia, and Uganda from Africa.

Again the interventions from EU and G–77/China top the list, with 20 and 22 respectively.

In addition, a large number of single interventions on the part of many African countries.

In a similar counting from COP3, I find:

[101] Internet, http://www.unep.ch

Box 5.2 Activity level: COP3

The count for the COP3 meeting in Buenos Aires, 1996, revealed scores ranging from one to 17 interventions for single countries, with 28 and 18 for the EU and G–77/China respectively.[102] There were in total 301 interventions from DCs, 162 from OECD countries and about 25 from the "economies in transition" (Eastern Europe and Russia).[103]

In the high-level activity range (score from 9 to 17 interventions) we find Argentina, Brazil, Colombia and Cuba from Latin America; China, India, Indonesia, Malaysia and the Philippines from Asia; Malawi, South Africa and Tunisia from Africa; and Australia, Canada, New Zealand, Norway and Switzerland from OECD countries.

In the middle-level activity range (4–8 interventions) we find Chile, Bolivia, Mexico, Peru, Uruguay and Venezuela from Latin America; the Republic of Korea from Asia; Cameroon, Ethiopia, Morocco, Tanzania and Zimbabwe from Africa; Austria, Finland, France, Germany, Hungary, Italy, Japan, the Netherlands, Sweden, Spain, the UK and the USA from OECD; and Russia.

Within the low-level activity range (1–3 interventions) there are 20 African countries, nine from Latin America, seven from Asia and the Pacific respectively, five additional OECD countries and six from Eastern Europe.

In addition, there were a few co-ordinated interventions on behalf of the Andean countries, the Nordic countries and the African Group.

It should also be noted that while the G–77/China frequently sought to harmonise the positions of its 132 member countries, this was not so for OECD as a group. Within the 28 OECD countries, the parties can be divided separated into several groups, mainly the EU members and the JUSSCANNZ group (non-EU industrialised countries).[104]

In sum, we find that Asia is dominated by five high-level activity parties, Africa has a broader range of participation, as well as three high-level and some co-ordinated interventions. Latin America resembles Africa in all these respects, and Western Europe, with EU in the lead, dominates (not surprisingly, in view of sheer numbers) among the OECD countries. The high-level parties tend to be "all-rounders" making interventions on all agenda items, while the middle-level parties generally concentrate on a few items.

Looking at activity level in all three sessions, we find that the following countries show a persistently high level:

- OECD: Australia, New Zealand and Norway

---

[102] G–77: the Group of 77 and China was established at UNCED in 1967, and its membership has increased to 132 developing countries. G–77 seeks to harmonise the negotiating positions of its members.

[103] OECD includes Australia, Austria, Belgium, Canada, Czech Republic, Denmark, Finland, France, Germany, Greece, Hungary, Iceland, Ireland, Italy, Japan, Republic of Korea, Luxemburg, Mexico, the Netherlands, New Zealand, Norway, Portugal, Spain, Sweden, Switzerland, Turkey, United Kingdom and United States of America. I have chosen to count Mexico as part of the G–77/China rather than part of OECD, on account of Mexico not being member of the OECD during the negotiations, and also due to Mexico's important role in bolstering the South's position during the negotiations.

[104] JUSSCANNZ includes Japan, USA, Switzerland, Canada, Australia, Norway, and New Zealand, as well as Iceland, Mexico and Turkey.

- Africa: Malawi
- Latin America: Brazil and Colombia
- Asia: China, India and Malaysia

Several countries display a high/medium or medium/medium activity level throughout the three COP meetings:

- OECD: Italy, Japan, Germany, the Netherlands, Sweden, Switzerland, UK and the USA
- Africa: Ethiopia, South Africa, Tunisia and Zimbabwe
- Latin America: Argentina, Chile, and Peru
- Asia: Indonesia, Philippines, and Republic of Korea

This summary indicates a high degree of consistency among the high-profile countries, as only Canada and Kenya among the once-high activity-level parties fail to be registered in this summary. In general there are thus great similarities between the interventions in the three COPs, although the COP1 and 2 counting reveal a higher activity level on the part of several OECD countries, including Japan, Sweden, the USA and UK. In addition, some countries from the South which were very active in COP3 (Bolivia, Mexico, Tanzania and Uruguay) were less profiled in COP1 and 2.

What do these figures tell us? The developing countries seem to be more co-ordinated as a group compared to the developed countries. This fits well with the presentation of the negotiation process, where it was clear that the majority of developed countries changed their position. The exceptions are the USA on one side, and on the other side the Nordic countries and Canada, all of whom basically stuck to their original positions.

Apart from the obvious methodological problem of omissions, which is however likely to even out among the parties, there is a high chance of under-representation by parties with small delegations, primarily among the DCs. Not only does this imply lack of relevant experts on the various agenda items, but there are often parallel sessions going on in these international conferences, leaving countries who have only one delegate at a disadvantage.

Moreover, this visual degree of activity level may fail to register both the concern and influence of parties, for various reasons. Interests and positions may be pursued through informal channels, and co-ordinated interventions by groups of parties may disguise the main architects behind a proposal. One example is the high activity level registered for Norway and the Nordic countries, with Sweden coming out with medium-level activity. When Denmark does not figure on the list at all, despite its being an important and active entrepreneur behind the Nordic position, this reveals one shortcoming of this approach. There are clearly additional channels and means by which negotiations may be conducted – a point to which we return in section 5.2.2. It should also be kept in mind that co-ordination within the two major groups, EU and G-77, is disguising the activities of many countries.

Another problem concerns chronological order: Ideally, the same scrutiny should have been applied to the negotiation process prior to the CBD. However, we lack such data, so the minutes from the COP meetings may serve as a reliable substitute. The indications provided by the COP examinations square well with my general impressions from having followed the biodiversity issue since 1988. I shall return to a more thorough analysis of these data in section 5.5.2.

5.3 The CBD – Negotiation Output

In Krasner's definition, international regimes are "principles, norms, rules and decision-making procedures around which actor expectations converge in a given issue-area". I shall rely on the first part of this definition in the following description of the CBD; – and later return to the second part, when discussing whether *actor expectations* do or do not *converge* in this issue-area (Chapter 6).

5.3.1 Principles and norms

The main objective of the Convention is threefold:

- the conservation of biological diversity,
- the sustainable use of its components, and
- the fair and equitable sharing of the benefits arising out of utilisation of genetic resources, including by appropriate access to genetic resources and by appropriate transfer of relevant technologies, and by appropriate funding.

The developing countries were concerned that the common heritage principle (in the North's interpretation) would still be applied to genetic resources. This led, for the first time in the operative text of a treaty, to the incorporation of the Stockholm Declaration Principle 21, which provides that "States have... the sovereign right to exploit their own resources pursuant to their own environmental policies, and the responsibility to ensure that activities within their national jurisdiction or control do not cause damage to the environment of other States or of areas beyond the limits of national jurisdiction".[105] The CBD states that each country has the sovereign authority to determine access to its genetic resources.

[105] Decided at UNCHE, the United Nations Conference on Human Environment, in Stockholm, 1972.

5.3.2 Procedures

Box 5.3 The CBD organisational set-up: [1]

- The Conference of the Parties (**COP**) makes all the formal decisions and meets annually. COP's mandate is further to review scientific, technological and technical advice; consider reports by parties on their implementation of CBD; consider and adopt protocols, amendments and annexes; establish necessary subsidiary bodies. Decision-making is by consensus, or as a last resort by two-thirds majority of the parties present and voting. In the case of a decision on the Financial Mechanism, no agreement has been reached on the voting procedure.
- The **Secretariat**, an independent body, is administered by UNEP and located in Montreal, Canada. The staff of the Secretariat includes 21 professionals and 13 support staff by February 1998 (Bergesen et al., 1998). This is down from 30 professionals and 19 support staff as of January 1997 (Bergesen et al., 1997).
- Global Environment Facility (**GEF**), the interim financial mechanism, was established in 1990 and is run by the World Bank, UNEP and UNDP.
- On the basis of Article 25 of the CBD, the first COP established a Subsidiary Body on Scientific, Technical and Technological Advice (**SBSTTA**) which is open to all parties and which also has annual meetings. SBSTTA is composed of experts representing governments. Considerable effort has gone into discussing the relationship between COP and SBSTTA, and between SBSTTA and the Scientific and Technical Advisory Panel (STAP) of GEF.
- There is also the **Bureau**; which is meant to strengthen the scientific networks and mobilise the scientific communities. Its 10 members are drawn from national delegations, representing the five regional groups. In addition to the COP, SBSTTA and the Open-ended Ad Hoc Working Group on Biosafety have their own bureaux.
- COP I agreed to a **Clearing House Mechanism** (implementing the provision in Art. 18, para 3 of the CBD); this is to promote international scientific and technical co-operation, and disseminate information on the lessons learned during implementation.

[1] This presentation of the organisational set-up has been brought up to date – providing information of what institutions have been established by 1998, while also indicating the CBD paragraphs on which their establishment rests.

Moreover, the Convention entails several control and monitoring mechanisms for national implementation.

Box 5.4 Control and monitoring mechanisms of the CBD:

- Art. 23.4: the COP shall keep under review the implementation of the Convention and shall for this purpose undertake any action that may be required for the achievement of the purposes of the Convention;
- Art. 26: Contracting Parties shall, at intervals to be determined by the COP, present reports on measures which they have taken for the implementation of the Convention;
- Art. 23.5: NGOs will be admitted as observers to the COPs unless at least one third of the parties present object; and
- Art. 24: a Secretariat is provided for.
- In the event of a dispute, the parties shall seek solution by negotiation or seek mediation by a third party (procedures set out in Part 1 of Annex II). Failing this, they may submit the dispute to the International Court of Justice.

The administrative core budget for the Trust Fund was US$ 4,787,000 in 1995, $ 6,896,000 in 1996, $ 6,561,041 in 1997, $ 6,765,208 in 1998, and $8,303,000 in 1999. The administration and secretariat of the CBD is funded by the Trust Fund which was operative from 1995. Outstanding contributions for 1998 and prior years as of 31 August 1998 totalled $ 2,317,373. In addition there are a Special Trust Fund for additional voluntary contributions to the core budget and a Special Voluntary Trust Fund for facilitating participation of developing country parties (Bergesen et al., 1999/2000).

5.3.3 Rules

The Convention sets out obligations and objectives for nations to combat the destruction of plant and animal species and ecosystems. The Contracting Parties shall develop national strategies, plans and programmes for conservation and sustainable use, and integrate conservation and sustainable use of biological diversity into relevant sectoral plans and policies, and develop systems of protected areas (Art. 6). The parties shall identify components of biological diversity important for its conservation and sustainable use; monitor the components through sampling and other techniques; identify activities which have or are likely to have significant adverse impacts on conservation and sustainable use of biodiversity and introduce environmental impact assessments with a view to avoid or minimise such effects; establish a system of protected areas (in-situ conservation) and adopt measures for ex-situ conservation, preferably in the country of origin; integrate consideration of conservation and sustainable use into national decision-making; adopt economically and socially sound measures to act as incentives for conservation and sustainable use; establish programmes for scientific and technical education and training for identification, conservation, etc. and provide support for such training in developing countries.

Each Contracting Party is to present *national reports* on the measures it has taken towards implementing the provisions of the Convention and how effective these have been in meeting the objectives. It is left to each party to decide on which measures are most effective to conserve biodiversity. The international community is given responsibility for conserving biodiversity in developing countries, including the most environmentally vulnerable – such as those with arid and semi-arid zones, coastal and mountainous areas. The CBD provides that developing country parties must only implement their conservation and sustainable-use obligations to the extent that developed country parties meet their commitments related to financial resources and transfer of technology. Furthermore, the Contracting Parties agree to respect, preserve and maintain knowledge and practices of indigenous and local communities, and encourage the equitable sharing of the benefits arising from utilisation of such knowledge and practices (art. 8j).

The CBD also confirms that access to genetic resources requires *prior informed consent* (art. 15.5) and must be on *mutually agreed terms* (art. 15.4), and that a country providing genetic resources is entitled to benefit from the commercial use of its resources (art. 15.7). The Convention envisages three basic mechanisms by which a country may benefit from use of its genetic resources: i) participation in the research using the resources, ii) receiving technology which embodies or utilises the resources, and iii) sharing the financial benefits realised from commercial exploitation of the genetic material or resource. These references reflect the developing countries' concern to link conservation and development goals and to avoid exploitation of their natural resources by the North on unfavourable terms. The CBD goes on to give practical effect to these equity concerns:

Box 5.5 CBD equity provisions:

- Provision of new and additional financial assistance to developing countries, over and above ODA, to cover "agreed full incremental costs" of implementing the Convention and enabling compliance (art. 20.2).
- Transfer of environmentally safe technology, including biotechnology and technologies covered by intellectual property rights, on "fair and most favourable terms", facilitated by the financial mechanism if necessary (art. 20).
- Obligations for developed and developing countries to share equitably benefits arising from utilisation of the knowledge, innovations and practices of indigenous and local communities with the countries concerned (art. 8, and 12th preambular).
- Obligations to advance priority access to developing-country parties and to share equitably the benefits and the results of research and development arising from the commercial or other utilisation of genetic resources, particularly with developing-country parties providing access to genetic resources (art. 15)

- Obligations to advance priority access to developing countries and to share equitably the results and benefits arising from biotechnologies, based on genetic resources, particularly with developing-country parties providing access to genetic resources (art. 19).
- Acknowledgement that developing-country parties' abilities to comply are conditional upon the "effective implementation" of developed-country parties' financial co-operation and technology transfer and that "economic and social development and the eradication of poverty are the first and overriding concerns of the developing countries" (art. 20.4).
- In funding and technology transfer, special consideration for least-developed countries, and consideration to special conditions resulting from the dependence on distribution and location of biodiversity within developing-country parties, especially small island states and those that are environmentally vulnerable (art. 20.5, 20.6, 20.7).

5.4 Assessment of Output

5.4.1 Principles and norms

Very generally speaking, "the North" primarily wanted the CBD:

- to focus on biodiversity conservation.
- to keep treating non-systematically bred genetic material[106] as part of the common heritage of mankind (CHM) – hence, in their liberalist interpretation, freely available for users.
- to avoid any language representing restrictions on biotechnology and the acquirement of intellectual property rights (IPR systems) on biotechnology products.

This position may roughly be described as preserving the status quo regarding access to and transactions with biological resources, while adding a higher national awareness and responsibility for conserving biodiversity.

"The South's" main priorities included:

- a focus on sustainable utilisation and hence – equitable sharing besides conservation.

[106] Non-systematically bred material will primarily comprise the landraces of the South that have not been subject to systematic breeding – and that are hence, not *uniform, stable and distinct* from other varieties. These would not be elegible to plant breeders' rights or patents.

- that all genetic resources (including systematically bred material) should be part of the CHM, that is, for the benefit of all humankind. *Alternatively:* that genetic resources should be regarded as commodities under national sovereignty for which payment must be made.
- to restrict or regulate IPR systems, or counter them with national sovereignty and improved access to (bio)technology.

This position was intended to alter the status quo, by making users of biological material pay a higher share of the costs associated with conservation, as well as paying for access to biological resources.

The CBD text, as presented in section 5.3, leaves little doubt that it is largely the position of the South that is reflected in the output.

5.4.2 Rules

The CBD text includes many of the ethical arguments presented by developing countries in the negotiations. As pointed out by Farhana Yamin (Yamin, 1995) and Veit Koester (Koester, 1994), among others, there is still a high degree of uncertainty concerning whether the Convention will actually lead to more equity in the area, and concerning whether the CBD objectives will indeed be implemented. This uncertainty is due largely to the legal nature of the rules set out in the Convention:

- There is a rather large scope for interpretations of many of the provisions, due to usual legal formulations such as *as far as possible* and *as appropriate*. The rules may generally be described as vague and there are no set standards to which the parties are required to commit themselves.
- The concept of *new and additional funding* as well as *full incremental costs* has resulted in relatively modest assistance so far. In article 39, the Global Environment Facility (GEF) of the United Nations Environment Programme, the United Nations Development Programme and the World Bank is accepted as the interim financial mechanism of the Convention. The control and further development of GEF remains a controversial issue. Basically, developing countries' obligations are limited by article 20.4 with the acknowledgement that developing-country parties' abilities to comply are conditional upon the "effective implementation of developed-country parties' financial co-operation and technology transfer" and that "economic and social development and the eradication of poverty are the first and overriding concerns of the developing countries".
- Some of the more controversial provisions, not least concerning access to genetic resources and biotechnology, are also ambiguously drafted. The relationship to GATT and TRIPs is particularly unclear. As far as patenting is concerned, the Convention stipulates that "technology transfers shall be

provided on terms which recognise, and are consistent with, the adequate and effective protection of intellectual property rights". Seeking to reconcile the conflicting interests in the patent issue, the CBD (art. 16.5) states that "the Contracting Parties shall co-operate to ensure that intellectual property rights *do not run counter to its objectives*" (emphasis added) – a formulation proposed by the Norwegian delegation.[107] This sentence was one of the main reasons why US President George Bush refused to sign the CBD in Rio. Moreover, this has remained a major concern with the current Clinton Administration, as the decision to sign (but not yet ratify) the Convention was followed by an interpretative statement addressing intellectual property rights as well as the provisions for financial mechanisms. As regards links to biotechnology, the final version of the Convention also sought to smoke out another controversial issue. As the United States was strongly opposed to international regulations on "living modified organisms resulting from biotechnology", it was left to the parties to consider the "need for and modalities of" a protocol on biosafety in the future (art. 19.3).

- The provisions dealing with access to genetic resources still need a lot of hammering out to provide specific regulations for equitable sharing both between and within countries. Furthermore, national sovereignty over genetic resources does not include genetic material in international gene banks which was collected prior to the Convention entering into force.
- The "all-embracing" functional and geographical scope of the Convention makes it difficult to assess implementation in a traditional sense. Scientific uncertainty regarding what exactly constitutes conservation and sustainable use of biodiversity makes it difficult to assess the problem-solving ability of the CBD.

Nevertheless, the CBD has already sparked a lot of activity, at the local, national and international levels. Chapter 6 will look into some of the most important sectors to investigate further Veit Koester's conclusion that: "The Convention and the process leading to its conclusion has probably already caused behavioural change" (Koester, 1994:112).

5.4.3 Procedures

It is up to the Conference of the Parties to decide on policy, programme priorities and eligibility criteria relating to access to the financial mechanism (GEF), among other points. As the COP meets annually and operates on a one-state, one-vote system, this may be assumed to endow the bioregime with a flexible and efficient approach in the event of external changes. This assumption is strengthened by the

[107] Personal communication with Jan P. Borring, Advisor to the Norwegian Ministry of Environment; 4 October, 1998.

active role played by the secretariat thus far. The relative strength between the COP and the World Bank in controlling the GEF remains to be seen.

The main compliance and monitoring mechanism of the CBD lies in the national reporting system, which was up for its first round preceding the fourth COP (1997/98). The parties are to present reports of measures they have taken towards implementation of the CBD objectives, and their effectiveness in meeting these objectives. So far, around 100 national reports were submitted at the fourth COP in Bratislava, May 1998.

5.4.4 Summary

Like most international regimes, the CBD can be described in terms of its *principles*, *norms*, *rules* and *procedures* (see section 1.2.1). In addition to the conclusion that the CBD norms and principles to a large extent represent a breakthrough for DC norms and interests, it is necessary to look into its potential as a successful international institution. The latter question may be addressed by examining the rules and procedures and comparing these against those of similar international institutions.

A main feature of the CBD is its complexity, the combination of a set of different dimensions: On the one hand, there is the commitment to make national plans and programmes for conservation and sustainable use of biological diversity. On the other, and complementing this, there is a commitment to share the costs of conservation and to share the benefits deriving from use of biological resources. The consensus that these are necessary means by which to achieve the main goals of the convention can be seen as the basic *principles* of the CBD. A central *norm* of the CBD is the establishment of a new property rights regime for genetic resources, abandoning the system of common heritage and compromising on the combination of intellectual property rights (IPR) and national sovereignty. The principles and norms set up differentiated roles among the participants, and this is considered a strength, mainly associated with the new generation of international environmental agreements. While the *rules and regulations* associated with the Bioregime are admittedly weak, the *procedural* aspects may be suited to bolster the institution and give added strength to future decision and protocols under the regime. These latter aspects may stand out more clearly in contrast to other institutions.

By contrasting the CBD to similar global regimes, we may get a better grasp of its strengths and weaknesses. Here I shall briefly go through some of the main aspects, comparing the CBD primarily to the UN Framework Convention on Climate Change. This necessitates stepping out of the chronological order, looking ahead past 1992, in order to present a brief view of the relative achievements in the CBD regarding rules and procedures (monitoring mechanisms, reporting systems, protocols and ratification).

A straightforward manner for assessing the *monitoring mechanisms* provided for implementation of the CBD is to compare it with similar international environmental agreements; here the most relevant seems to be the UN Framework Convention on Climate Change (UNFCCC). With regard to *procedures*, the UNFCCC has already spurred three rounds of synthesis reports, based on 15 (1994), 34 (1995) and 32 (1996) national communications out of the 37 required in Annex I. In the first round, developing countries were exempt from reporting, and by the first deadline for their submission of reports, only two were presented. By comparison, the first round of national reporting for the CBD were submitted by the parties prior to COP4 in May 1998 (Bergesen et al., 1997). During this first presentation, about one hundred reports were submitted – clearly a very high return.

Initially, the administrative bodies and capacities for dealing with the activities of the two conventions were not very different; the *secretariats* and *budgets* started at about the same size, and the subsidiary bodies and ad hoc working groups are about the same in number. Recently, however, the gap between the two conventions has been widening in terms of budgets and staff: The UNFCCC budgets are increasing, while the CBD budgets show little change either way. Moreover, the UNFCCC Secretariat has increased from 27 to 35 professionals, while that of the CBD has been reduced from 30 to 21. This reduction can hardly be interpreted as a response to the greatly added workload brought on by the recent round of reports, and it is hard to see how the diminished staff can handle the complex and unstructured national reports on biodiversity policies. On the other hand, the CBD Secretariat has received additional support from the Secretariat of the FAO Commission on Genetic Resources for Food and Agriculture (CGRFA). CGRFA is dealing with issues relating to access on mutually agreed terms to plant genetic resources, and consideration is being given to including the FAO International Undertaking on Plant Genetic Resources as a protocol to the CBD.[108] A closer scrutiny of the CBD Secretariat homepages reveals, however, a staffing list with 19 vacancies out of 51 positions. About half of these vacancies concern high-level professionals.[109] As the reduction is thus not due to direct budget cuts, this may be interpreted as a good sign in the long term, but the short-term effect may clearly be problematic for the efficiency of the CBD secretariat.

With regard to specific *rules and regulations*, the UNFCCC achieved its Kyoto Protocol in December 1997, while the CBD is still in the process of negotiating Protocols on Biosafety and on including a renegotiated version of the FAO International Undertaking on Plant Genetic Resources. On the other hand, the conclusion of the Kyoto Protocol is only a first step, and there are severe misgivings regarding its scope for implementation. Although specific targets are included, the Kyoto Protocol does not contain enforcement mechanisms, nor does it include obligations for reduced emissions in developing countries. The United States acted as one of the main "laggards" throughout the negotiations (not unlike the case with the Biosafety Protocol); indeed, it is held that the US negotiators would not have consented to the Kyoto Protocol without the leadership of Vice President Al Gore. It has also been made clear by the US administration that ratification of the Protocol

---

108 See reference to country reports to this Commission in section 8.3.2.1: Ethiopian country report to FAO/CGRFA.

109 http://www.biodiv.org/press/staff.html

will hardly be possible without some kind of commitments from key developing countries (Agrawala & Andresen, 1999).

The CBD and the UNFCCC have approximately the same number of *ratifications;* only three months separated their entering into force. The rapid entry into force and its world-wide acceptance reflect widespread recognition of the global stakes in the loss of biodiversity. Compared to the UNFCCC and UNCLOS, with 50 and 60 *ratifications* required respectively, the CBD threshold for entry into force was set at the very low level of only 30 ratifications. Nevertheless, unlike the case with UNCLOS, the CBD did not draw ratifications primarily from developing countries; today it has received ratifications from all OECD countries except the USA and Turkey. Like the UNFCCC, the CBD has now received more than 170 ratifications in all. In this light, the low threshold does not seem to have weakened the co-operative spirit of the CBD; rather it reflects the urgency with which the parties regard its entering into force. On the other hand, the CBD is still largely a framework convention, and it remains to be seen how well its principles and objectives will be reflected in practical policies among the contractual parties. Also, rapid ratifications may signal that the parties do not see much need for behaviour change. The latter aspect could explain why the parties were so much less reluctant to ratify the CBD (and the UNFCCC) compared to UNCLOS.

Compared to other international *resource management agreements*, the procedural design of the CBD would be likely to receive a relatively high score. The "first generation" of wildlife management treaties, such as the Western Hemisphere Convention (1940) and the African Convention (1968), are generally seen as having had little impact due to lack of administrative support. The "second generation", which may be exemplified by the Berne Convention (1979) and CITES (1973), are generally rated as successful instruments, partly due to their effective administrative structure, regular meetings of the parties, and reporting systems. In addition, CITES requires that each of its parties establish a national authority meant to enforce the permit system set up by the convention. As far as the administrative set-up is concerned, the CBD has clearly been constructed in the second generation mode, although not going as far as CITES in terms of requiring national administrative bodies. Designing national agencies specifically mandated to follow up international obligations may become part and parcel in the construction of third-generation treaties. In his evaluation of 27 treaties on wildlife management, including those on single species, on regional habitat conservation, and the "big four" (CITES, Bonn, Ramsar and World Heritage), Simon Lyster finds the major differences in effectiveness to be based on institutional mechanisms, rather than on the functional or geographical scope of the treaties: "The treaties which have achieved the greatest level of compliance are, by and large, those who keep their Parties active, have a central administrative body to oversee enforcement and have some means of chastising Parties which do not comply with their treaty obligations" (Lyster, 1985:301).

On the basis of the principles, norms, rules and procedures expressed in the CBD, I would offer the following two preliminary conclusions:

- I have found support for the hypothesis that the CBD objectives coincide with the South's interests to a greater extent than under the pre-regime conditions. Moreover, in view of the quite strong opposition from several developed countries to adopt these *norms and principles* in the CBD text, this seems to qualify as a "real" breakthrough, rather than a mere "happy coincidence". Still, this does not necessarily imply support for my initial hypothesis that the breakthrough has come about due to institutional factors.
- The CBD appears as a relatively "standard" international environmental agreement. While it is somewhat lacking in terms of *rules* on specific standards, the organisational set-up and the *procedures* attached to it do not deviate from the mainstream of contemporary adopted conventions. Recent negative trends in administrative capacity may, however, indicate a reversal of this status, at least in the short term.

In order to come closer to an explanation of the CBD output, I shall henceforth discuss it in light of the three perspectives. Concerning the negotiations leading up to the CBD, the central questions are:

- How did the parties manage to achieve agreement on the framework Convention, in the face of what were apparently insurmountable differences?
- How did the output come to represent what seem like a breakthrough for DCs' interests?
- To what extent and how did institutional factors impact on this process?

5.5 The CBD – Analysis of Negotiations and Output

5.5.1 The nature of the issue-area – scope for hegemonic behaviour

In section 4.2.2, I argued that most of the inherent characteristics of the biodiversity issue-area seemed malignant, and would tend to work against international agreement and collaboration. The biodiversity issue has involved strong and asymmetric interests, both between states and at the sub-national arena. It exhibits a high degree of complexity, involving major societal sectors such as the seeds and pharmaceutical industries, agriculture and forestry. Moreover, it affects a great many policy-areas, from land-use practice to international trade. Due to its complexity, the biodiversity issue lacks *focal points* to guide negotiations; moreover, it is difficult to identify any particular situation or events which may have spurred international action in the area (*external shocks*). Rather, the need for action has become apparent over time, first with the realisation of the loss of species and habitats, and then with the growing economic interest in the wake of developments in the new biotechnologies. Estimates of the value of biological resources to agriculture and pharmacy give further indications that at least some aspects of the biodiversity issue may be characterised as being of *high salience.* There is certainly a high level of scientific consensus that "something must be done" to halt the rapid loss of biodiversity world-wide, but remedial action is likely to involve restrictions on resources use, and is thus prone to transgress on national sovereignty.

Considering the international structural power relations in combination with the nature of the issue-area, it was assumed that Northern interests would prevail, while the South would hardly see their interests reflected in the CBD objectives. This proposition has been largely refuted by the explicit wording of the CBD, where we can see a high degree of correspondence with ideas and interests of the South.

One external, complicating factor here was clearly the ongoing and contrary process in GATT (later the World Trade Organisation, WTO). Derrick Purdue (1995) points to the hegemonic nature of the international patent discussions: "The attempt to establish uniform global intellectual property rights over living material is a hegemonic project driven by the biotechnology industry with its complex articulations of molecular biology, agro-chemical transnational corporations and state intervention". Purdue also argues that: "The construction of a hegemonic project can also be seen as the establishment of a particular discourse, as the general epistemological horizon or limit within which meanings are defined and accepted. Thus, forms of knowledge of precise historical and geographical origins, such as liberal trade theory or industrial patents, combine to constitute the limits of legitimate debate globally". This links up with the notions discussed in 2.2.3 that norms may be seen as a testimony of cultural imperialism – in other words, norms emanating from the overall structural power relations. Even though the biodiversity regime itself does not encompass these norms, and rather tries to contradict them (article 16.5), the trade regime established by the WTO may affect the scope for action also within the biodiversity issue-area.

One interpretation of this situation could be that the North simply stopped worrying about the result of the biodiversity negotiations, confident that their interests would be secured by the TRIPs regulations in the WTO. This interpretation is built on one rather doubtful assumption of the URA-model – a high degree of internal co-ordination within states. Following the same interpretation, it would seem that the issue has been raised to a higher level: It is now subject to potential conflict between international institutions, in addition to between states. This interpretation may go some way in explaining the output, but, on the other hand, it disregards the fact that it was the North which started out as pushers for the establishment of a Convention on Biological Diversity in the first place. The CBD was initially part of the North's global environmental agenda, and while the WTO/TRIPs touches on central aspects of this issue-area, it is certainly not a tool for conservation of biodiversity. Second, the CBD text explicitly seeks to counter the possible detrimental effects from the WTO/TRIPs. It also works against intellectual property rights by introducing national sovereignty, prior informed consent and mutually agreed terms to regulate access to biological resources. Third, this interpretation does not explain why the United States as the only developed country declined to sign the CBD in Rio. Other countries may have been satisfied with the result, but not the USA (see section 6.2.1).

In his general analysis of the new WTO, Helge Hveem (1996:190) points out that realism, institutionalism and cognitive perspectives all provide insight and explanations about the output. DC interests were to some extent overrun by the more powerful industrialised countries as well as by the newly industrialised countries (NICs), but the result was also a result of a compromise in which developing countries achieved increased access to the world market but had to accept restrictions on access to technology. To this Hveem adds the impact of learning, as an increasing number of NICs actually changed their perceptions and strategies during the negotiation process.

Returning to the biodiversity issue, the realist regime models may nevertheless shed light on the early stages of agenda-setting and negotiations: In the early stages, the issue of biodiversity was "moved" from what the developed world perceived as a politicised FAO arena. When the issue was put on the UNEP agenda, it is clear that at least the USA and the IUCN hoped to de-link the conservation aspects from the controversial aspects of utilisation, economic value of, and property rights to biological resources. The "move" to UNEP hence represented an effort to start afresh with new rules, in an arena over which the USA believed itself to have greater control. Had this effort been successful, it would have indicated rather high explanatory power on the part of structural factors.

In view of the nature of the issue-area and the high conflict level among the negotiating parties, there seems to have been little room for reconciliation on the biodiversity issue. This strengthens the notion that institutional factors may have aided consensus making. Before turning to the analysis of institutional factors, I will go more deeply into the structural background of actor interests and capabilities in the issue-area. This should aid in assessing whether the parties actually behaved in accordance with these presumed interests during negotiations, and thus give an indication of the extent to which structural factors may be useful in predicting negotiation output.

5.5.2 Actor preferences and capabilities

In this section, I ask whether it is possible to give a power- and interest-based explanation of the roles played by the various countries in the biodiversity negotiations. The presentation of the biodiversity negotiations revealed a central conflict line between *conservation* on the one hand and *equitable sharing* on the other. These two dimensions will provide a basis for predicting actor interests in the issue-area. Next, I will examine whether these interests confirm with the parties' actual positions in the negotiations. The third focus of this section is on the relationship between parties' interests and capabilities to achieve their goals.

- First, I need to identify the *material* basis for preferences within the issue-area: I shall make predictions about the parties' issue-specific interests based on their being "rich" or "poor" with a view to biological resources and biotechnology respectively. Against this background, the parties may be assumed to act as *pushers, bystanders, intermediates* or *laggards*. I shall examine whether these predictions match actual positions in the negotiations (as indicated also by sections 5.2 and 5.2.1).[110] To the extent that these do not match, *ulterior* interests will be considered.

- Second, I shall comment on the relationship between *power capabilities* and *interests* tied to the issue-area on the one side, and *actual breakthrough* in the CBD output on the other side. This relates to the link between the pusher role and various types of *leadership*.
- To the extent that overall power and breakthrough do not match, I shall discuss the *issue-specific power* base of the participants. Important here is the distinction between *basic game power* and *negotiation power*.

Against this backdrop, it is clear that two levels of power and interests need to be considered in order to clarify the basis for the parties' positions and roles in the negotiations:

- Ulterior interests and issue-specific, material interests.
- Overall structural power and issue-specific power (basic game power and negotiation power).

The basis for establishing issue-specific interests is set out in the following two boxes:

Box 5.6 Patterns in the global distribution of biodiversity

Agriculture – plant genetic resources:
Virtually all parts of the world are mutually dependent with regard to breeding material for the chief agricultural plants. Nevertheless, the variation in the degree of dependence is relatively great when we consider the utilisation and origin of the 20 most-used food plants (Kloppenburg & Kleinman, 1987). The countries of Asia and Latin America are least dependent upon external supplies of gene material, as 56.8 and 52.2 % respectively of the food plants they cultivate are indigenous. According to Kloppenburg's (1988) estimates, the corresponding "subsistence" rate for the industrialised countries is under 7%, which means that no more than about 7 % of their plants in agricultural cultivation have their origin there. Of crops of economic importance, only sunflower, blueberry, cranberry, and Jerusalem artichoke originated in North America. As for Africa, the figure is also low, at 12.3 % "subsistence". Nevertheless, African countries are regarded as being far less dependent than the industrialised countries, since African farmers traditionally use a far wider diversity of food plants than is usual in the industrialised countries of the North.

(*box continued on next page*)

[110] See Detlef Sprinz & Tapani Vaahtoranta (1994).

Relatively speaking, many African countries are poor in genetic diversity compared to current international exchange of genetic resources (Kloppenburg, 1988). Ethiopia may be said to be the exception that proves the rule, being one of the few African countries featuring on the list of Vavilov's variation centres.[111] Ethiopia is seen as the variation centre for such plants as coffee, barley, durum wheat, linseed oil, sorghum, sesame, date and finger millet. In fact, it would probably be correct to include Kenya, Somalia, south-east Sudan and northern Uganda in that variation centre list (Berg, 1995; Rosendal, 1995b). The Mediterranean, Asia Minor and Central Asian countries are listed as variation centres for wheat, among other things. India, Burma and the Philippines are associated with great diversity in rice; Southeast Asia for banana and coconut; China for tea, soy bean and oranges; Peru, Ecuador and Bolivia appear on the list for potato, tomato and tobacco; Brazil and Paraguay for cocoa and rubber; and Central America for maize.

Wildlife:
Disregarding food plants and commercial plants, Africa stands out as even richer in biological diversity. The world's third largest tropical rainforest is found in the Republic of Congo (Zaire), and both Madagascar and Republic of Congo are amongst the twelve countries of the world with greatest biological diversity (so-called mega-diversity countries). The first six of these, in addition to Republic of Congo and Madagascar, are Brazil, Colombia, Mexico and Indonesia. On the lists of countries with the largest areas of closed tropical forests and greatest primate diversity, we find in addition Cameroon, Gabon, Guyana, Nigeria, Central African Republic, Uganda and Angola; Peru, Bolivia, Ecuador, Venezuela, and Surinam; India, Papua New Guinea, Burma, and Malaysia (Mittermeier, 1988). Tropical Africa is one of the richest regions for mammals, though it is poorer in flora than either South America or Southeast Asia. Southern Africa, however, has an especially rich floral diversity (Huntley, 1988). Generally speaking, however, there is little official knowledge about plant genetic resources in Africa compared to Asia and Latin America (Juma, 1989).[112]

---

111 *Nikolai Vavilov* was a Russian biologist and agricultural expert who identified a number of areas around the world where one or more important agricultural plants have an especially wide genetic diversity. He claimed this variation was linked to the fact that it was in this place that the plant was developed into a cultivated (domesticated) plant (Kloppenburg, 1988). Areas where the plants were domesticated are not necessarily identical with the places where the plants originated. Barley and wheat for instance came originally from Southeast Asia, and beans, cassava and sweet potato came originally from the Americas, although these species have provided the progenitor for *secondary variation centres* in Africa. For the food plants sorghum, pearl millet, finger millet and cow beans, Africa is regarded as the primary variation centre. Moreover, Africa is regarded as the primary variation centre for two important international articles of trade: coffee and palm oil. For a complete list of the Vavilov centres, see p. 48 in Kloppenburg (1988).

112 Juma points to problems to do with tropical diseases (malaria, dysentery) as a possible cause of this lack. Fear of contracting these diseases has simply restricted plant collection expeditions.

Box 5.7 Trends in the new biotechnologies

In OECD reports from the time of the onset of the negotiations, we can find indications both of the trends in the new biotechnology sector and of the interests tied to it.[113]

In 1989, OECD published a report on *Biotechnology. Economic and Wider Impacts*. Here it is stated that:

"* New biotechnology is distinguished from other major technologies of the 20th century by the fact that its impacts on the quality of life, its human and social consequences, are arriving earlier and may go deeper than macro-economic impacts measured by productivity, investment or GDP growth.

* New biotechnology is clearly a technology of highly industrialised countries, both with regard to R&D requirements and market potential. Companies will exploit the advances in plant genetics to replace Third World crops, which might be increasingly grown in OECD countries, thus reinforcing the concentration of world trade within the OECD area.

* Among the conditions which promote the diffusion of new technologies, the most important are public acceptance and patent protection.

* At the moment, competition in biotechnology is waged principally among OECD countries, while most trade impacts are experienced by some developing countries. In the future, genetic modifications of cereal grains could lead to trade impacts for cereal-exporting OECD countries as well. Biotechnology is reducing the overall demand for primary products from developing countries.

* For OECD countries, being competitive in biotechnology presently still means essentially preparing for the future, by reinforcing the scientific, technical, manpower, industrial, data-bank and other components of the biotechnology infrastructure. (pp. 9–16)"

It is important to note that in 1989, there was no (explicit) awareness within the OECD about biotechnology needing access to raw material in the form of genetic resources from the South. Patenting is mentioned as a topic of importance to biotechnology.

In a report from 1992, *Biotechnology, Agriculture and Food*, the OECD concludes that

"* the agrofood biotechnology is increasingly dominated by the leading multinational corporations (p. 20), and this may result in a major part of the germplasm production being controlled by relatively few companies (p. 58).

* A major constraint for the technology is the shortage of novel valuable genes for crop improvement, as few genes are well characterised and available (p.19). For the most part the genetic variation comes from mutations that occurred long ago in natural or during cultivation. Regardless of the genetic novelty introduced by the biotechnologist, most of the genetic variation required for crop improvement will remain what has earlier evolved in the species. This is why it is important to keep large representative stocks of the natural genetic variation of key crop species. It cannot be replaced once it is lost. More resources should therefore be put into germplasm collections (p. 38).

* Current assessments of biotechnology impacts on developing countries are less alarmist regarding substitution for tropical crops (p. 22).

(*box continued on next page*)

---

113 While classical biotechnology goes back to traditional baking, brewing and fermentation processes associated with bread, beer and wine, as well as traditional breeding and selection procedures in agriculture, the new biotechnology involves genetic engineering, like recombinant DNA technologies where genes are isolated from an organism, manipulated in the laboratory and injected into another organism.

* To reduce the level of uncertainty, agrofood companies would like to see high priority given to more international harmonisation of safety principles and intellectual property rights (p. 22). In Europe, legal protection for genetically modified plants and animals is not as straightforward as in the United States and Japan, in spite of recent efforts by the EC to raise protection to the levels of those two countries. Opposition to patenting in these areas no longer comes mainly from legal circles, but from animal rights and environmental movements (p. 24). The consequences of patenting to agriculture in developing countries may be a hindrance to use of improved germplasm from developed countries (p. 58)"
In this report, the problem of access to useful genetic material is addressed, but not linked to the countries of origin. Concern about intellectual property rights has risen high on the agenda.

Interests and capabilities may thus be inferred by distinguishing between gene-rich and gene-poor countries. If we add the dimension of technology-rich and technology-poor (indicating the ability to reap economic benefits from use of biodiversity), the following Table 5.1 may provide a basis for indicating *pushers* and *laggards* in the biodiversity issue:

Table 5.1: Material interests in the biodiversity issue-area

	Gene-poor	**Gene-rich**
Techn.poor	1. Middle East, West and North Africa in general	2. includes Ethiopia, Madagascar, D.Rep. of Congo, Uganda; Indonesia, China, Philippines, Burma; Peru, Colombia, Bolivia, Ecuador
Techn.rich	3. OECD countries in general	4. (includes Brazil, India, Mexico, Malaysia) [1]

[1] The brackets in box 4 is meant to indicate that some of these are NIC countries, and all of them may be ranked as technology-rich, compared to the low income developing countries. They will generally be considered technology-poor (and thus belonging to box 2) compared to OECD countries. The relative size of the biotechnology sector within OECD countries should also be taken into account.

Table 5.2 has been developed with pollution issues in mind and may not lend itself with equal clarity to the biodiversity issue. Even though biodiversity loss is widely recognised as a global environmental problem, it is not equally amenable to this kind of calculations as are various forms of cross-border pollution problems. All the same, it may be interesting to examine to what extent the boxes in the two tables overlap each other, in terms of providing correct assumptions regarding the parties' positions in the biodiversity negotiations.

Table 5.2: Environmental Vulnerability and Abatement Costs

		Environmental vulnerability	
		low	high
Abatement costs	low	1. Bystanders	2. Pushers
	high	3. Laggards	4. Intermediates

Combining the two Tables (5.1 and 5.2), we may ask: Is it reasonable to expect the gene-rich/technology-poor countries to be *pushers* in the negotiation process, and the gene-poor/technology-rich countries to be *laggards*? Is this the way the countries actually did divide themselves in the negotiations? It seems less questionable that gene-poor/technology-poor countries would tend to become *bystanders,* but is it a fair assumption that gene-rich/technology-rich countries should end up as *intermediates*? While *intermediates* will presumably seek package deals, the activity of a *bystander* is less easily predicted. The *bystander* will have few *material interests* tied to the issue, and may hence either be indifferent or use the arena to pursue *ulterior interests*.

The idea of *pusher* countries is obviously related to *leadership*, and hence ultimately to the three different kinds of leadership (*structural, entrepreneurial, intellectual*), as discussed in the three theoretical perspectives. The close link to leadership illuminates the two dimensions inherent in the pusher category. The label 'pusher' refers not only to the relationship between a party's *position* and *interest*, but also to the *capabilities* available for pursuing these interests. In the following sub-section, I discuss the type of interests tied to various parties' positions, and then compare this to their leverage on the ensuing result, looking into the type of leadership role linked to the various positions.

i) Interests and positions: Ulterior interests and issue-specific, material interests

The concept of *environmental vulnerability* is generally meant to indicate a country's vulnerability to environmental damage, for instance acid rain and air-borne pollution. If vulnerability to others' actions (pollution import) is higher than own emissions (pollution export), then the damage costs are higher than abatement costs, and hence the country may be assumed likely to become a *pusher*. The CBD does address situations where activities are "likely to significantly affect adversely the biological diversity of other States". Such situations may arise from introduction of exotic species or plant varieties that may distort the natural balance between indigenous species.[114] The CBD, however, primarily directs its recommendations for conservation policies *within* national borders. Obviously, the gene-rich countries may feel threatened by claims for stricter conservation demands from the international community. The *conservation dimension* would thus give rise to the following predictions concerning country positions:

- The gene-rich countries may have cause to emerge as *laggards* on the *conservation dimension*, if they are expected to pay the costs of conservation alone.

[114] The controversial proposal of creating a protocol on Biosafety under the CBD is aimed at handling such situations, particularly as these may arise from biotechnological experiments. It may be assumed that gene-rich/technology-poor countries may feel more vulnerable than others. As predicted, these countries have been *pushers* for a Biosafety protocol.

- The user countries (gene-poor/technology-rich) can be expected to act as *pushers* on this dimension.

Table 5.3: Positions on the Conservation Dimension

		Users	
		low (techn. poor)	high (techn. rich)
Owners	low (gene-poor)	1. Bystanders	2. Pushers
	high (gene-rich)	3. Laggards	4. Intermediates

These predictions accurately capture the positioning regarding *conservation* claims in the early phases of negotiations, as exemplified by a few high-profile countries: The United States was the most active *pusher* for a conservation convention, while Brazil and Malaysia were most outspoken against it. On the conservation dimension, the NIC countries are obviously more likely to act in accordance with their status as gene-rich rather than ostensibly technology-rich. Seen in this light, the US position may be interpreted as one of structural leadership in the early phases of the negotiations.

The distinction between high and low *abatement costs* in the biodiversity area may be indicated by the obligations (i.e. costs) each country is expected to fulfil under the IEA in order to contribute to solving the problem. Abatement costs in this area would be costs associated with conservation activities such as setting up reservation areas, which often implies the costs of abstaining from resource use. Owner countries (gene-rich) might be ranked quite high in this respect, but their abatement costs could still be low if they succeed in using this situation as a lever to extract more financial resources from the (gene-poor) user countries, linking financial and technological transfers to access to and use of biological resources. As we have seen, this strategy was partly successful, as the developing countries' obligations are limited by Article 20.4: This article acknowledges that abilities of developing-country parties to comply are conditional upon the *effective implementation* of financial co-operation and technology transfer from the developed countries. Abatement costs would then turn out to be higher in the gene-poor/technology-rich countries. The *equitable sharing dimension* would thus give rise to the following predictions concerning country positions:

- The user countries may be expected to emerge as *laggards* on the *equitable sharing dimension*, if they are obliged to pay for their use of biological resources.

- The owner countries can be expected to appear as *pushers* on this dimension.

Table 5.4: Positions on the Equitable Sharing Dimension

		Owners	
		low (gene-poor)	high (gene-rich)
Users	low (techn.-poor)	1. Bystanders	2. Pushers
	high (techn.-rich)	3. Laggards	4. Intermediates

On this basis we could expect to find the gene-rich developing countries in box (2): *pushers*. These would be countries such as Indonesia, Colombia, the Philippines and Ethiopia. This is very close to reality, as these countries did indeed display a profiled position in the negotiations – but so did Brazil, India, Mexico and Malaysia (see section 5.2.1 and Box 5.6).[115] They were the most eager in fighting for financial transfers, threatening to become *laggards* if such mechanisms were not installed.

This does not, however, explain why countries such as the Democratic Republic of Congo (then Zaire) and Madagascar were absent from the pusher role, fitting better into box (1): *bystanders*. Finding Ethiopia and Madagascar (both gene-rich, both LICs) in different boxes, and the NIC countries in the pusher box, indicates that whether a developing country ends up as *pusher* or *bystander* cannot be put down to the difference between gene-rich NICs and gene-rich LICs alone. Neither does this account for the high profile of Tunisia, a relatively gene-poor developing country.

These examples bring to light an inherent ambiguity in the classification of pushers and laggards. The *pusher* label is directed not only at predicting *positions*, but also at *activity level*, and these two features do not necessarily overlap. While positions concern the type and level of preferences, activity level also involves the *capacity* to advocate these preferences:

First, strong concern either way is not necessarily followed by numerous interventions in the negotiations, particularly not in countries represented by small delegations. Furthermore, there are several channels through which parties may express and advocate their views, and official interventions is only one of them. Other channels include lobbying, coalition-building, distribution of non-papers and informal talks in general. It could be suspected that a party which is active through one such channel might in general be active in all of them, but for small delegation-countries this is less likely. The latter might even be overlooked if they were using informal channels, as my registration of activity level during COP meetings captures only the official interventions.

Second, a country may choose to focus on a few selected items of particular importance, rather than making statements on a wide range of topics. This may strengthen that country's credibility with regard to the main arguments it wishes to put across. Hence, even though this strategy leaves a party with a relatively low score on activity level, it may in the end have been a successful pusher for its special cause. On the other hand, my impression from reading the minutes from the negotiations is that the active parties are consistent in arguing for their objectives on the broad array of related items on the agenda.

The interests and positions predicted by the four categories require further scrutiny. Here we should note that the predictions concern *material* (issue-specific) interests, and do not take *ulterior* interests into consideration. Accordingly, there are several anomalies:

---

[115] I base this ranking both on interviews with members of the Norwegian delegation about Phase 2 and on my own summing up of interventions in section 5.2.1

- Gene-rich/technology-rich countries were clearly assuming the pusher role on the *equitable sharing dimension*, and they were also the most outspoken laggards with regard to the *conservation dimension* during the early stages of negotiations.
- The relatively gene-poor Tunisia appears as a pusher with a high activity level.
- Ethiopia was generally much more active in the negotiation phase than was Madagascar, while both (gene-rich/technology-poor) were predicted to have relatively high stakes in the matter (and hence a high activity level).

Against this backdrop, a purely *material interest*-based explanation seems less applicable. Of course, the NIC's status of being gene-rich may have been a more important determinant for their position than their relative strength in terms of technological development. Nevertheless, there may be various *ulterior interests* behind the parties' positions in the negotiations:

- First, several NIC countries can be assumed to pursue a high profile in international negotiations due to their *ulterior interest* in (and their capability for) displaying a leadership role. Hence, the high profile of the gene-rich/technology-rich countries seem to indicate that interest (preferences) is not the only factor affecting whether a country assumes the role of a *pusher* – capability seems to have an effect as well. This indicates that we should take both will and ability into our calculations also in the regime formation stage, contrary to the assumption that ability comes into play mainly in the implementation phase.
- Secondly, both developed and developing countries may have *ulterior interests* in pursuing certain general principles. Developing countries usually seek to link international negotiations to principles concerning new international economic order (NIEO), while developed countries, most notably the United States, will seldom pass up the chance to advocate free trade principles. The biodiversity negotiations are also clearly embedded in this larger agenda on economy and trade.
- Another type of *ulterior interest* could be related to strategic positioning. When a relatively gene-poor but capable country like Tunisia puts up a high profile in the negotiations, this could be a strategic move to get as much as possible out of the deal, even though the odds may seem low.

Finally, *communication and learning* may have played a part, as neither differences in capabilities nor ulterior interests can account for the difference in activity level between Ethiopia and Madagascar. A closer look at Box 5.6 may provide some indications for explaining the "absence" of Madagascar and the Democratic Republic of Congo, and the high profiles of Brazil, Colombia, Ethiopia, India, Malaysia and Peru: Patterns in the distribution of biological diversity seem to indicate that being gene-rich in the agricultural / domesticated sector may urge a country to a higher activity level, compared to those which are gene-rich in wildlife only. Obviously, the objective of equitable sharing may have more direct and explicit consequences in the agricultural sector. Moreover, the arguments linked to this section had already been advocated and developed in the debates on the FAO Undertaking and through the Keystone Dialogues, thereby providing the relevant countries with a qualified platform from which to proceed (more about this in section 5.5.4).

In box (1), *bystanders,* we would expect to find countries with relatively minor problems in relation to biodiversity loss and a low degree of biotechnology industry to defend. The gene-poor LICs largely seem to fit into this category (with the exemption of Tunisia), even though in the long run these countries may end up as the real losers, as the international financial flows may become redirected, e.g. from Mali to Brazil. The *bystander* label might also fit the Nordic countries. Norway especially does not have a large biotechnology sector to defend. Alternatively, the Nordic countries might be predicted to turn out as *laggards,* being gene-poor/technology-rich countries. And yet, the Nordics were in fact ardent *pushers* on the *equitable sharing dimension* throughout the negotiation process. A possible interest-based explanation could be that the Nordic countries were ahead of their time in realising their dependency on access to genetic resources. However, their position could also be interpreted in light of *ulterior* interests:

- It was an inexpensive way of enhancing their green image, as well as their image of bridge-builders between the North and the South.
- At the actor level, the bystander suit can be said to have provided the individual delegates with the opportunity to pursue an entrepreneurial leadership role.

These lines of analysis are not contrary to the assumptions tied to the bystander role, as this category of actors is the least predictable. As these actors have little to lose from negotiations either way, the setting is ideally suited for pursuing ulterior interests. Thus the Nordics combined the pusher role with the traditional bridge-building role in relation to the developing countries, without this apparently conflicting with their national interests:

- For Norway, this was possible thanks to the fact that Norway is a sparsely populated country with a great deal of unspoiled nature, as well as the fact that Norway, unlike many other OECD countries, has no major domestic biotechnology industry to heed.
- The Danish and the Norwegian delegations seemed least hampered by industrial interest groups lobbying at the national level, even though Denmark, like Sweden, has significant domestic interests in biotechnology industry. In Denmark, the biotechnology industry groups have opted for openness as a means of pacifying the general public, rather than fighting for deregulations.[116]

---

[116] During the CBD negotiations, Denmark was not, as Sweden would later become, hampered by its EC/EU membership. Denmark has a long tradition of non-conformity to the EU Commission. See Jon Birger Skjærseth & G. Kristin Rosendal (1995).

It should be noted that this *bystander-turned-pusher,* having little to lose from negotiations either way, and hence apt to pursue ulterior interests, is also less predictable in the implementation phase. While such "idealistic pushers" may choose to advocate fair and green solutions during international negotiations, partly in order to increase their green reputation at home and abroad, there is no guarantee that these countries will remain role models when the issue moves from a normative to an operational level. There may well be less guarantee that actors of this type will follow up their obligations, compared to the purely interest-based pusher, who has a material interest in implementing the regime objectives. This recalls the core dispute between ideational and power-based explanations: "Policy may be norm driven as well as power driven, but the application of norms cannot be taken for granted" (Weale & Williams, 1998:89).

In box (3), *laggards,* we find the USA, much as expected. On the one hand, the United States has a very good record as far as the *conservation dimension* is concerned. In a summary of percentages of protected areas of the world, North America and Australia get the highest score with 12.6% and 12.2% respectively; the Caribbean, Central America and Pacific also score high at 9.5%, 9.0% and 8.4%; among the middle-range scores we find Sub-Sahara Africa (5.7%), South and Southeast Asia (6.0%), South America (6.3%) and East Asia (5.8%); while in the low-score range we find North Eurasia (3.1%), North Africa / Middle East (2.8%) and Europe with 0.9% of their area protected (Heywood, 1995:983). The USA would thus seem to have less to fear with regard to abatement costs. In the early phases of negotiations, while *conservation* dominated the agenda, the United States was clearly among the *pushers.* However, the USA made no secret of its concern for its domestic biotechnology industry, should genetic resources become a commodity for which the provider-countries could require compensation in terms of financial or technological transfers. When *equitable sharing* nevertheless did become a central dimension in the negotiations, the USA's position was transformed into that of a *laggard.*

How and why was the USA left as the almost sole laggard, rather than persisting as a victorious pusher for the conservation dimension? What seems puzzling from an issue-specific perspective is that no other biotechnology-rich countries (most OECD countries) persisted as laggards and stood out as openly as the USA in this matter. European countries could have been expected to have equally high concerns as the United States in protecting their biotechnology industries. Three lines of explanations may shed light on this matter:

- First, Europe does differ from the USA with regard to material interests. This is partly concerned with the generally low European score on protected areas: They may have had less to win from a *conservation* focus only. Looking at their slightly more neutral position in the TRIPs negotiation, European countries seemed at the time to be less adamant in their claims for extended patent protection, as well.

- Second, as to *ulterior* interests, it is not uncommon for the other developed countries to rely on the United States to act as a laggard in environmental issues. I shall return to a closer examination of this aspect in section 6.2.1.[117]

- Third, whatever the issue and preferences involved, the United States would hardly evince a low profile in international negotiations. The status and economic capacity of the USA give to it a natural leadership role, although it may be hard to distinguish the type of leadership wielded by the USA.

This third element points to the ambiguity inherent in the different types of *leadership* concepts. Put bluntly, a structural leader is by definition a selfish utility-maximiser, employing overall structural power to achieve his/her goals; while the concepts of entrepreneurial and intellectual leadership imply the pursuit of common welfare by way of creative and legitimate argumentation. This raises the question, however, of whether or not a structural leader (like a benevolent hegemon) may be concerned with common welfare, and whether or not entrepreneurial and intellectual leaders may advance their selfish interests through creative and legitimate argumentation. The choice of label is closely linked to our ideas about which roads lead to common welfare maximisation, and which do not. Likewise, the concepts of *laggards* and *pushers* are tied to the "preferred" output. In other words, while the USA is a *laggard* with regard to regulating access to genetic resources, it would be correct to term the USA a *pusher* for liberalisation in this area.

The following, simplified table depicts this potential ambiguity in how the different types of leadership can be classified according to my own perceptions of the issue-area. Employing these labels on the various actors in the negotiations shows how the labels are not value-free.

Table 5.5: Controversial classification of positions and leadership types

		Conservation	
		Pushers	**Laggards**
Equitable sharing	**Laggards**	*structural leadership (early phases):* USA pushing for status quo (: free access to resources of the South)	
	Pushers	*entrepreneurial leadership*: the Nordics, Canada, among others, pushing for creative package deals (: conservation & sustainable use; IPR & national sovereignty)	*intellectual leadership:* DCs pushing for greater acceptance of the legitimacy of norms (: equitable sharing and common responsibility for resource conservation)

[117] This is based on an article analysing the background for and implications of the US refusal to sign in Rio (Rosendal, 1995c).

Turning to the relationship between interests and capability to win through with these interests, it is clear the *overall structural power* does not correctly predict the CBD output. Hence, we must look into the distribution and type of *issue-specific power*.

ii) Interests and capabilities: basic game power and negotiation power

It may be argued that the South had an unusually strong bargaining position in the international negotiations leading up to the CBD. A crucial question is whether this issue-specific power consists of *basic game power* or *negotiation power*. By looking at the two dimensions of *conservation* and *equitable sharing* and the interests tied to these two dimensions, we may discern a tendency towards a more symmetrical relationship between the North and the South in the issue-area. The linking of these dimensions signified the development from an asymmetrical interest structure to a semblance of symmetry between actors' preferences. As the North and the South came out as *pushers* on the respective dimensions, a certain complementarity in interests seemed possible. This development could indicate increased *basic game power* for the gene-rich South. It could be argued that the North's interest in securing access to genetic material has given the South increased *basic game power*. On the other hand, the North may in principle remain free-riders in this respect: First, because they nevertheless retain free access to genetic resources in the international gene banks. Second, because of the difficulties in controlling the flow of genetic resources.

Here there is a certain parallel to the development of oil and mineral cartels, where some DCs have also been able to set the international terms of trade. Nevertheless (as argued in section 4.2.1), genetic resources are not quite comparable to natural resources such as oil and minerals, as access to genetic resources is much more difficult to control. Another relevant parallel may be the fishery regimes set up by UNCLOS; although the coastal-state DCs achieved their 200-miles exclusive economic zones, it has proved difficult to control foreign fleets fishing in their zones.

The analysis provides several indications of the scope for the South's *negotiation power*: Throughout the negotiation phase it was of great importance for a country like Ethiopia that the developing countries joined forces in order to get their interests onto the agenda. The high activity level by the G–77/China group points in this direction. The large number of high-profile developing countries in the biodiversity negotiations could also be seen as an indication of relatively high *negotiation power*. It might also underscore the proposition (Chapter 3, P4a) that a high activity level may have had a "teaching" effect, in the sense that the South has succeeded in promoting and integrating its own norms and ideas into the agenda and the result.

In the implementation phase, however, the gene-rich DCs may be assumed to acquire diverging interests: For a gene-rich but otherwise poor country like Ethiopia, the "Blue Sky" alternative might be a situation where the donor countries fulfilled their obligations under the CBD, while other gene-rich DCs failed to do so – leaving

Ethiopia to trade its genetic material at the highest price, with as little competition as possible. This assumption is linked to the potential bias towards bilateral arrangements in the CBD. On the other hand, strength in numbers may be equally important in the international implementation phase, i.e. the further negotiations of protocols to strengthen the objectives of the CBD. This may indicate a continued benefit in pursuing a unified DC alliance: in order to prolong the effect of *negotiation power*. For instance, the South be anticipated to have a common interest in fighting bilateralism, and rather aim for increased influence in multilateral donor organisations. The countries of the South may have a strong coalition in this respect, as their richness in biodiversity provides them with some, albeit limited, control over something the developed world has a great deal of interest in: hence, still providing them with some degree of *basic game power*.

Summing up, this analysis gives rise to three preliminary conclusions. I will argue that:

- The *material interest* structure does go a long way in predicting the parties' negotiation positions as laggards and pushers and to some extent bystanders. We must also consider *ulterior interests* in order to predict the positions of intermediates and bystanders.
- The *overall structural division of power* does not go very far in predicting the breakthrough for these interests. This is illustrated by the transformation of the USA from a pusher for conservation to a laggard concerning equitable sharing.
- *Issue-specific power* takes us one step further: *Negotiation power* seems to go some way in accounting for the breakthrough for DCs' interests, while there is only a limited scope for *basic game power*. As negotiation power will be worth less than basic game power in the international implementation phase, this distinction should be kept in mind, and it will be further examined in the next section (5.5.3).

Hence, we see that interest constellations and issue-specific power cannot fully explain how the CBD came to represent a breakthrough for the interests of the South. As demonstrated in section 5.5.1, the realist regime model did shed light on the early stages of agenda-setting and negotiations. This was because the move to UNEP could be seen as an effort by structural leaders to de-link the conservation aspects from the controversial aspects of utilisation, economic value of, and property rights to the biological resources. I interpreted the move to UNEP as an attempt to start with new rules, in an arena where the USA believed itself to have greater control. This effort was however not successful, and it thus indicates rather low explanatory power on the part of overall structural factors in the process so far.

For a closer explanation of the CBD output, let us now turn to the next set of explanatory factors. I shall look at the interplay between interests and institutional factors to examine whether this perspective provides better or differently "fitting" explanations. In this section I shall elaborate further on the negotiation situation and issue-specific power within this particular institutional setting.

5.5.3 Institutional factors and the formation phase

As will be recalled, the interest-based perspective gave rise to a number of theoretical propositions regarding how international institutional factors may affect collaboration among states in specific issue-areas. In this study, the broad question concerns whether and how regimes matter. The more specific question concerns whether and how institutional factors can explain the particular output associated with the CBD – the breakthrough for DC interests.

Narrowing in on the interplay between interests and institutions, I treated institutional factors in three broad categories:

First, looking at the *institution as an arena* for repeated interaction, we assumed that procedures for participation and decision-making act to pattern state expectations and affect behaviour in specific issue-areas. These rules, along with the official purpose of the negotiation forum, could be instrumental in altering or aligning actor interests, and could give rise to mutual expectations among the parties.

Secondly, I identified the potential effects of the *organisation as an actor*, indicating how various actors could provide entrepreneurial leadership. Through such leadership the Secretariat or other actors could help tap the integrative potential of the issue in question. In the formation phase, institutional factors may thus help align actor interests by identifying and boosting the integrative potential among them, through flexible agenda-setting and by providing a forum for the science–policy dialogue, as well as for entrepreneurs forging issue-linking.

Thirdly, in the implementation phase, institutional factors may help reduce uncertainty and insecurity among the parties through *information, monitoring and sanctions*. These mechanisms may have an impact by stabilising mutual expectations among the parties, by reducing transaction costs and by providing a frame of reference for future interactions. In this phase in the analysis I am not yet concerned with how these mechanisms affect implementation, but merely to what extent and why they have been established as part of the regime: The main point here is that the inclusion of incentive mechanisms may have been a precondition for some of the parties to agree to the IEA in the first place.

Once more, we should bear in mind that the way institutional mechanisms such as these are constructed may in itself be contingent on power and interests. In line with the process-tracing method, I shall return to more explicit inferences to be drawn from this perspective after the analysis of the international output in this chapter.

5.5.3.1 The institution-as-arena

Basically, the institutional nature of the international system is anarchical, so anarchy also characterises international negotiations. This implies lack of a central authority to guide and enforce implementation of concerted action. Global environmental problems represent an area where attempts are made to diminish this trait by the establishment of regimes, aimed at constructing solutions to problems which cannot be handled by single states. According to current theories in political science, some of these shortcomings of the international system may be mitigated by designing well-functioning fora for negotiations, geared towards achieving agreement from relevant parties. If the parties do not regard rules and regulations as legitimate, they cannot be counted on to comply with them. Nevertheless, free-rider problems remain an important challenge to the possibility for concerted action and the anarchic character of the international system puts restrictions on how far agreement can reach in the first place. The politically feasible output of a bargaining process tends to be subject to *the law of the least ambitious programme*, i.e. collective action will be limited to those measures that are acceptable to the least enthusiastic party (Underdal, 1980). This is closely linked to the dominant decision-making rule in negotiations between sovereign states – which is that of *consensus.*

In order to assess where we have come (success or failure), we need to take stock of where we started from. Two features are central to understanding the situation under the old regime:

- About 95 % of the world's terrestrial species diversity is found within the *jurisdiction of sovereign states*. This aspect distinguishes the CBD from other IEAs such as ozone, the global climate and the Law of the Sea, which are concerned with common resources like the atmosphere and the oceans.
- Genetic resources were traditionally seen as part of the common heritage of mankind, except the systematically modified genetic resources of the OECD countries. These were increasingly being made subject to patents and plant breeders' rights.

The CBD establishes a new type of property rights regime, where national sovereignty is introduced to counterbalance intellectual property rights. In turn, this means that the negotiations have been directed at regime change. This consensus-based decision was bolstered by building on the Stockholm Conference (UNCHE) principle of national sovereignty over natural resources. Apart from that, however, it had very little material basis, due to the difficulties known to be involved in enforcing the new property rights regime, i.e. the "elusive" character of genetic resources.

Section 5.5.2 has demonstrated how the two dimensions of *conservation* and *equitable sharing* apparently combined to *reduce asymmetries* between the main conflicting parties in the negotiations. The combination of these dimensions seemed to give rise to a situation of exchange in terms of issue-specific power capabilities. The exchange aspect of power has been defined in terms of a relationship in which party A controls something which party B has a great deal of interest in, and vice versa (Hernes, 1975:40). The exchange aspect can be described thus: (1.) The developed world controls biotechnology, to which the South is interested in gaining access. (2.) The main bulk of biological resources, in which the North has a great deal of interest, are situated in the South.

At the same time, the actual limited control over these resources in the South – both with regard to domesticated and wild material – was pointed out as a factor that pulled in the opposite direction. We noted for instance that the North had access to large quantities of genetic resources through the international gene banks. On this basis, I made the preliminary conclusion that the *overall structural power* and interest constellations did not go very far toward explaining the relatively high degree of breakthrough for the interests of the developing countries in the CBD objectives. On the same note, it was concluded that the countries of the South had limited *basic game power,* as they have limited control over the resources. Finally, however, the analysis showed that the South's position was strengthened by *negotiation power.*

These observations indicate the need for a further discussion of the particular institutional setting for this negotiation game:

First, as argued in section 4.3.1, it seems easier for the developing countries to win through with their arguments in UN-related fora such as FAO and UNEP, compared to international fora on trade and economy. This links up to the interplay between power and institutions, as the latter type of institutions obviously may have a greater overall impact, since they are the ones who stipulate the economic and trade-related framework that a large array of policies in other sectors must abide by. In addition, international fora on economy and trade have stronger ramifications to principles linked to high-salience issues, and these principles will tend to frame the issues differently than in e.g. UN bodies.

Similarly, we could surmise that in this particular arena, an undercurrent of normative persuasion became linked to the recognition that these resources did indeed belong to the South. In this light, the effect of the institution in terms of its *official goal framing the issue* should not be neglected. While such sentiments would probably have less impact in fora on economy and trade, the goals associated with the biodiversity negotiations in UNEP may have been more receptive to accept the legitimacy of granting the countries of the South their "rights".

Secondly, and strengthening this trend, these negotiations culminated with the signing of the CBD at the UN Conference on Environment and Development. The underlying norms in the UNCED setting were even more clearly geared towards appeasing the South. Even though a recurring criticism from the South was that UNCED was concerned with "the environmental agenda of the North", developmental issues were nevertheless more overt in this forum than in the preceding UN Conference on the Human Environment (Stockholm, 1972). On the other hand, choosing the UN forum to advocate their environmental agenda may be seen as a strategic move on the part of the North, as they could hardly expect to achieve political acceptance for environmental change in developing countries from regulations originating from any other forum.

Timing may also have played a positive role. As the UNCED Earth Summit approached, and mounting global public attention with it, it became necessary for top politicians to be seen to achieve specific measures during this meeting. Public expectations and pressure for politicians to finally "do something" about the environmental threat may never have been greater. Both the speed and the contents of the Climate Change Convention and the CBD may have been given an added boost by this situation. An important aspect of the institution-as-arena was thus the advantage for developing countries in negotiating within UN-related fora like UNEP and UNCED, rather than e.g. in the WTO. Hence, some of the credit for the success may be due to the negotiation arena itself.

The next aspect concerns *participation*. Global environmental conventions (such as the CBD) represent a novelty and a new challenge in international relations: When the number of parties increases, the scope for agreement on an IEA will often be correspondingly reduced, as it will be harder to identify package deals with "something in it for everybody" and subsequently harder to discourage free-riders by means of efficient control and verification mechanisms. The costs of making collective decisions are apt to rise as the membership in a group increases. Generally, in order to tap the integrative potential, institutional arrangements must succeed in striking a balance between legitimacy on the one side and effectiveness and efficiency on the other. In a truly global issue-area, this factor may seem less relevant. Its relevance may, however, pertain to the efforts to gain consensus from an increasing number of parties, and to the dynamic process by which countries are perceived as relevant parties.

For example, when catch quotas were being negotiated in the management of the Northeast Atlantic fisheries, all parties actively involved in exploiting the resource should ideally be included. Similarly, an agreement on dumping reductions in the North Sea would lose most of its significance if the main contributors of pollutants did not participate. As to our case, the "most" relevant parties would be those harbouring substantial biological diversity and also those with the greatest ability to contribute to the conservation of this material. The latter may include both economically and technologically affluent states, as well as skilled engineers. As no region or single nation is, however, independent in its use and need for genetic resources, the conservation of biodiversity is a global concern in which *all* states are in fact relevant parties.

Participation, however, points to both the *number* and *type* of actors, and was anticipated to represent an ambiguous factor in global problem areas: All states are potentially relevant parties to the collaboration, but as the number increases, the scope for successful co-operation and implementation is presumably reduced as the risk of other parties free-riding increases. In the biodiversity area, one solution might have been to concentrate on participants representing the largest donors (as conservation will be costly) and the so-called mega-diversity countries (with high levels of biodiversity endemism). The political danger in this approach was that this would have put the main responsibility for conservation on the mega-diversity countries of the South. Hence, it was deemed not politically feasible to limit the CBD to issues of wildlife conservation: The much more disputed aspects related to utilisation of biological resources were added to the agenda – hence, compensation to the South became a topic. This tendency was probably strengthened as the *actual* participation in the CBD negotiations increased dramatically during the first years –

from 25 in 1989, to 153 signatories at UNCED 1992. Moreover, this all-encompassing approach increased the scope for issue-linking, thus enhancing the legitimacy of the agreement with all parties; developed and developing alike. In sum, the *number* of participants seems to have broadened the scope for achieving integrated solutions.

Participation also involves a *dynamic* element, in cases where the same actor participates over time: As pointed out by Koester (1997:179), the chief negotiators knew each other from previous meetings, and this may have created a team spirit and an added will to reach a solution, in spite of their differences. Participants that have taken part throughout a long negotiation process may also be more inclined to fight for an output. Koester argues that many negotiators may have been involved in preparing their own instructions and taken care to suggest them being as flexible as possible, as they would avoid a situation where they or their country would be blamed for having caused failure.

Another facet of participation concerns the *type* of actors present in the negotiations: The inclusion and participation of the NGO community has occasionally been found to carry some weight in reaching consensus: This seems to have been the case in the preceding FAO negotiations on plant genetic resources, but not so much in the biodiversity negotiations. Even though the environment and development NGO community was eventually granted observer status in the formal negotiations, it seems that they may have been more active "backstage", e.g. through the Keystone Dialogue meetings and in other fora. We return to this aspect in the next section. Moreover, national delegations in these fora (UNEP and FAO) were clearly differently composed than delegations to the WTO. This brings to light another aspect of the participation factor, as it points to the sectors of the national arena from which delegation members are drawn. Ministries of the Environment as well as the environmental and development NGO community were duly represented; the latter to some extent in UNEP and largely during UNCED. UNCED also included a larger number of top-level bureaucrats, but this seems to have pulled in both directions.

At this point we should step outside the URA model in order to visualise how the "national interest" may be represented differently by different sectors of the administration. In Allison's well-known words, *where you stand depends on where you sit*, and bureaucrats from environmental ministries may thus be expected to be more environmentally inclined than their colleagues in e.g. the Treasury. A telling example of this situation from the biodiversity negotiations themselves concerns the US delegation: While the US delegation was reported to be optimistic about the output at its adaptation in Nairobi, the differently composed US delegation in Rio refused to sign the CBD. In terms of participation effects this seems, however, to be the exception to the rule, as the other delegations at UNCED did not turn hostile towards the CBD.

Clearly, different fora make for different compositions of both governmental and non-governmental actors, and this may have implications for the extent and contents of the agreements reached. Realist explanations may nevertheless come into play at later stages, as at some point the Treasury may have its say. In the preceding section I made the comparison between coastal-state LICs' lack of control over their fisheries resources under UNCLOS and lack of control over genetic resources under the CBD, to indicate their limited basic game power. It is nevertheless true that the South did gain acceptance for some of their interests also in the UNCLOS negotiations: This may have come about due to negotiation power. As presumed, negotiation power was less useful to DCs in the implementation phase: their lack of control made clear their hollow victory in the Law of the Sea. In other words, while it was relatively easy to make oral concessions to grant the South their "rights" in the fisheries regime, this has hardly been followed by practical policies in the North to enforce these rights in material terms. It remains to be seen whether this will also become the fate of the South's victory in the CBD negotiations.

5.5.3.2 The organisation-as-actor

A proposition with particular relevance to DC interests was that a *flexible agenda* and an *active secretariat* may aid weaker countries in getting their interests through. It was presumed that the secretariat or other actors could act as entrepreneurial leaders, bringing about *creative proposals* suited to get negotiations going, through package deals. As the bio-negotiations moved from a narrow to a more comprehensive scope – from wildlife preservation to conservation and use, including the issue of biotechnology – these factors may have been instrumental in aligning the parties, and particularly in meeting the demands of the South.

The role of the UNEP secretariat in directing the negotiations is interesting with regard to agenda formation:[118] First, as the negotiations proceeded, the UNEP secretariat was able to integrate controversial items, one after another, into the agenda. Thus the conflict level was raised, but in order to secure agreement from all relevant parties (especially the mega-diversity tropical countries) this was gradually accepted as a necessary move. This active issue-linking was thus, as perceived by the Northern laggards, predominantly malign in nature. Nevertheless, the issue-linking between biodiversity and property rights seem to have enhanced the scope for consensus. The *scientific* input in the process was furthermore adamant in maintaining that biodiversity conservation is important in all types of ecosystems, and as biodiversity loss is irreversible, the dangers of neglecting seemingly insignificant parts may be great, albeit uncertain. As these views became accepted among the parties, this may also have increased the scope for agreement. Opposed to the comprehensiveness assumption I cited Young's idea of salient (focal) solutions. As the description of the negotiation process has made clear, no simple focal solution could have been conceived in the biodiversity negotiations.

Second, as June 1992 and UNCED were approaching, the Chairman of the Intergovernmental Negotiating Committee and the Executive Director of UNEP, Dr. Tolba, set up small working groups focused at reaching consensus on the issues still

[118] The negotiation process is discussed in Rosendal (1995d).

remaining (Svensson, 1993). Applying two useful concepts developed by Haas, Keohane and Levy (1993), we may say that this can also have boosted the overall *governmental concern* as well as the *contractual environment:* These working groups provided several leadership roles for the proponents of environmental action, roles to which they would not have had access in an anarchic international system. Moreover, the establishment of these fora added to entrench regular interaction between the parties, again a factor believed to reduce distrust concerning mutual compliance.

In order to visualise the link between institutional design and agenda-setting, we may see the activities of the UNEP secretariat as *entrepreneurial leadership*. Such leadership was presumably important for identifying the integrative potential in the issue-area. An international secretariat is not, however, the only actor which may assume this leadership role. Indications of such leadership seem to be found also in the Nordic countries, along with some of the more active NICs and LICs, such as Malaysia, Brazil, India and Ethiopia. The Nordic countries maintained the trust of the developing countries, and helped to present several compromise formulations including both dimensions of *conservation* and *equitable sharing*. This seems to have been important in conciliating the parties.

5.5.3.3 Institutional mechanisms: Information, monitoring and incentives

First, it was assumed that parties may see it as being in their interest to subscribe to an international agreement because this will reduce transaction costs and enhance the flow of *information*. Several types of information can be envisaged, and we assumed that information about issue-links might be more important than scientific information, in terms of aiding weaker countries in getting a breakthrough for their interests in the output. While issue-linking has primarily been associated with the formation phase, successful issue-linking may clearly be important for the implementation phase as well. I will argue that the issue-linking which took place during the CBD negotiations, not least through the work of the UNEP secretariat, has largely been accepted as a legitimate part of the output with all parties involved. The issue-linking was to a large extent supported by scientific “facts”, but they were just as much advocated on a normative basis. We return to the normative aspect in discussing who provided the information in the next section on the ideational perspective.

Second, the regime may be instrumental in shifting cost/benefit perceptions among the parties through *monitoring* compliance. Monitoring was assumed to enhance implementation through two mechanisms: increased transparency, which generates trust; and exposure in case of non-compliance. The CBD includes a *national reporting* system as its main monitoring mechanism – neither a particularly harsh nor a particularly lenient measure as international agreements go. This may be seen as a guarantee for the contracting parties that the secretariat of a convention intends to monitor compliance. While it could be speculated that parties might be reluctant to join an agreement lacking this kind of mechanism, it is nevertheless so much part and parcel of these agreements, that we would find it difficult to point out an independent effect of monitoring, whether in the formation phase or in the implementation phase. As to the latter, it also remains to be seen how the national reporting system works out, and whether this kind of mechanism will provide incentives or disincentives for the parties to comply.

Another type of monitoring mechanism associated with the CBD are the principles concerning *prior informed consent* and *mutually agreed terms* regarding access to biological resources. The establishment of such mechanisms (if only in principle) may have been effective in increasing trust among the parties in the formation phase, but the effect in the implementation phase may be less straightforward. In the regime regulating prior informed consent in pesticide trade, this mechanism has been criticised for actually boosting trade in pesticides: By serving as an effective forum in which the pesticide industry may trade their products to developing countries, this has been cited as the main reason why the industry turned from fighting to backing this international instrument. Similarly, there is still uncertainty as to who will in the long run benefit from these mechanisms in the biodiversity issue, and what kind of environmental effects they may produce in the end. On the other hand, the main stumbling block with regard to the principle of prior informed consent in the bioregime would seem to be how to translate it into an operative mechanism.

Third, and presumably even more strongly affecting the parties' interests, a treaty may be equipped with *incentive* mechanisms (e.g. funding, technology transfers), or various kinds of formal *sanctions*. Again, the aim would be to shift the balance of cost-benefit calculations in order to enhance the benefits of compliance. Active implementation from developing countries has seldom been an issue in international environmental agreements (Najam, 1995). When the participation of LICs is needed, there has been a tendency to include "sweeteners" in the IEA in the form of financial transfer mechanisms. Such compensation is not unprecedented in IEAs, as seen in the Montreal Protocol on the Ozone Layer. However, the situation is complicated, as the very nature of genetic resources makes it difficult to control access to their utilisation.

It is the GEF that is the main incentive mechanism of the CBD – and this highlights how the same factor may have different effects in the formation phase and then later in the implementation phase. While the development of the GEF was somewhat of a stumbling block throughout the negotiations, and still is to some extent, it may nevertheless be regarded as one of the main mechanisms for fostering implementation. The CBD text includes a number of symbolic (moral) statements to the South's liking, but it is the GEF that serves as the main material measure for securing the co-operation of the developing world. We may assume that the CBD would not have received the assent of the South without the establishment of a financial mechanism.

As to the effect on implementation, however, it remains to be seen to what extent GEF is regarded as being accessible and functional by donor and recipient countries alike. The establishment of financial and technological transfer mechanisms has been highlighted in regime literature as an important pathway for international institutions to enhance capacity-building and hence, environmental policies in developing countries (Haas et al., 1993). Few studies have, however, examined whether such mechanisms ever attain sufficient scope for any political change to take place. We return to this issue in section 6.3 and in the second main part of the analysis in Chapters 8 and 9.

As far as *sanctions* are concerned, we assumed that their importance would lie mainly in making other parties comply, and hence, like monitoring, enhance trust among the contracting parties. In section 6.2, we look into the application of social sanctions in relation to CBD principles and norms.

The final Convention text represents a delicately balanced package deal with many relatively vague formulations. The real test of its success will come in the implementation phase. Only then will it be possible to say whether the new bioregime may succeed in appeasing the underlying conflicts, or whether the CBD will prove useless as an instrument for affecting state behaviour. This does not mean that evaluating the implementation phase will be a straightforward matter: The objectives of the Convention do not include any tangible standards for measuring activity. Hence, any conclusions regarding improvement of conservation efforts and improved equity in sharing will have to be approached in a qualitative manner. Chapter 6 will look at some of the major leads which may affect future developments.

5.5.3.4 Summary

At the conclusion of this section, I shall draw attention to two features revealed by the analysis of institutional factors:

- The evident differences in the way related aspects of the issue is treated in various international fora strengthen the presumption that institutions matter.
- On the same note, the *official purpose* of the negotiation forum seems to have had a greater impact on the formation process and output than was originally assumed in the analytical framework.

At this point, we may conclude with some confidence that we have come closer to explaining the CBD output. Although this implies an added indication that institutions matter, this effect may still have come about through other or additional causal mechanisms beside those described here. Let us now turn to the next set of explanatory factors: those associated with the ideational perspective. This means looking at the interplay between institutions, ideals and norms, and seeing whether this perspective provides alternative explanations which may increase our understanding of the CBD output.

5.5.4 Norms and learning

The ideational perspective indicates learning processes and the development of common norms as an additional basis beside cost-benefit analysis for predicting state interests and action in international collaboration. This in turn raises two questions:

- Through what mechanisms are these learning processes taking place?
- Who is doing the learning, who is doing the teaching and what is being taught?

The *first* question has been addressed in some detail. Two broad types of mechanisms, also linked to the international institution, may be identified as central to this model:

First, environmentally oriented coalitions between the IEA and domestic groups may be developing within *epistemic communities,* strengthening the development of consensus within an issue-area.

Second, through *repeated interaction*, consensual knowledge and social learning give rise to commonly accepted international norms, which to some extent become formalised in international regimes. These norms may acquire a sense of *legitimacy* or *obligation (pacta sunt servanta),* and common acceptance of international norms and rules developing within specific issue-areas may hence affect state behaviour.

The assumptions tied to the *second* question are much more vague:

The direction and scope of such learning processes would seem to depend on who "captures" the epistemic communities, who provides intellectual leadership, and on the process of reaching consensual knowledge. The discussion in section 3.2.3 gave rise to two diverging propositions: Either the developing countries may adopt the prevailing norms and ideas of the developed countries, acquiring stronger concerns for global environmental problems, through learning processes. Or, through active participation, DCs may succeed in promoting and integrating their own norms on the environment and development into the international agenda and output. Finally, we should note that norm adoption may have a trace of authoritative persuasion.

I will start by examining the presence and activities of an epistemic community in the issue-area. Next, I will consider the general and specific conditions for learning processes and norms of the kind developed within the biodiversity issue-area.

5.5.4.1 Epistemic communities

Haas (1993) has employed the concept of *epistemic communities* to operationalise and identify the possible effects of norms and learning in international collaborations. One may also, with Young, point to the role of *intellectual leadership* in effectuating learning processes and advocating common norms (Young, 1991). Haas defined epistemic communities as "networks of knowledge-based communities with an authoritative claim to policy-relevant knowledge within their domain of expertise" (1993:179). Recall also the three pathways through which Haas sees the possibility of epistemic community influence:

- Members of the group may be present in the international secretariats and thus have an impact in shaping the agenda.
- They may be able to create transnational and transgovernmental coalitions holding similar views, and that can forge consensual solutions.
- Members of the group may hold administrative positions in government, and thus encourage common policies in their respective governments and push for their enforcement.

If *one* epistemic community were to be pointed out within the biodiversity issue-area, I would argue that the Keystone Dialogues comes close to this description. First, and directly related to Haas' first pathway, some of the Keystone representatives belonged to international organisations (FAO, Preparatory Committee to UNCED), and the UNEP secretariat was also found to have had an impact later in shaping the agendas. As the biodiversity issue can best be compared to that of climate change, both being fairly novel, highly controversial and complex global environmental problem areas, let us have a look at some parallels between the Keystone Dialogues and the IPCC, the scientific panel of the Climate Change negotiations. Unlike the IPCC, Keystone was not set up as a scientific forum, nor was it given a scientific mandate by governments or any intergovernmental organisation. It was more of a hybrid between a scientific community and an environmental / development non-governmental movement. Discussions in Keystone were generally more concerned with ethics and justice rather than "pure" scientific "facts" – although a great number of the participants were renowned experts in their fields. Multidisciplinarity was a prominent trait of the forum, which brought together legal experts (patent lawyers), agricultural experts, sociologists and biologists, among others. The stated goal of Keystone was to reach consensus among experts from these varied disciplines, through informal debates. This relates directly to Haas' second pathway.

Moreover, Keystone made possible informal and consensus-aimed debates between two sets of initially highly conflicting interest groups, as per Haas' second pathway: The first conflict was found between the environmental NGOs themselves, with wildlife conservation versus domesticated biological resources: Even though the latter side was dominant among the participants, there were high-ranking officers from the World Bank and the UNCED Preparatory Committee (PrepCom.) present, and these may have brought novel ideas back to their agencies. The second conflict involved the interests of biodiversity conservation, use and equitable sharing, pitted against the interests of the biotechnology industries: Keystone's main achievement may have been to bring together people from grassroots organisations and major transnational corporations – Monsanto, Ciba-Geigy and ICI Seeds. It was clearly important for the development of consensus on issue-linkages within the CBD to reach agreement between the wildlife-oriented and the development and agriculture-oriented NGOs. In addition, the links to biotechnology and property rights to genetic material are not only contentious, but also highly complex. The Keystone group managed to proceed towards a broader common understanding about these controversial and complicated linkages.

Compared to the IPCC, there was also a higher, indeed equal, degree of developing country representatives present, with both government, scientific and NGO affiliations. Developing countries were outspoken in their criticism of IPCC for its initial lack of representation, and they also had severe doubts about the independent standing of IPCC scientific research (Gupta, 1997: 88, 150–156). Keystone seems to have been more successful in achieving legitimacy in North and South alike. This aspect of the Keystone Dialogues is interesting for two related reasons:

- The first point relates to Haas' assumption that the countries in which an epistemic community is best represented will be most environmentally concerned: The participants' nationality largely overlap with those countries most active in the biodiversity negotiations. From the South we find Brazil, Chile, China, Ethiopia, India, Indonesia, Kenya, Mexico, the Philippines and Tanzania. From the North there were Belgium, Canada, France, the Netherlands, Norway, Switzerland, Sweden, the UK, the USA and the (then) USSR. With a view to Haas' third pathway, all may not have equally easy access to their national constituencies or administrations: Among those with a high likelihood of national influence were Dr. Melaku Worede, the Director of the Ethiopian Genebank, and Dr. M.S. Swaminathan, the Director of the Centre for Research on Sustainable and Agricultural Development in India, both internationally renowned experts in their fields. Likewise, from the Netherlands came Dr. Jaap Hardon, the Director of the Center for Genetic Resources.
- Secondly, this aspect points up an important difference between the CBD and the UNFCCC in terms of DC *negotiation power*. According to Joyeeta Gupta (1997), DCs failed to find a concerted negotiation strategy in the climate change negotiations, as they had to rely on foreign research which did not take their domestic interests into consideration. Part of the explanation for the relatively high negotiation power of DCs in the biodiversity issue may be the unifying effect that the Keystone Dialogues helped to induce.

On the other hand, the Keystone Dialogues were less visible and achieved much less attention in the media compared to the IPCC. Their conclusions never reached the front pages with headlines about increased scientific consensus. Nevertheless, the Keystone consensus reports were often distributed to all delegations to the biodiversity negotiations, sometimes directly by the participants themselves. The third of Haas' pathways seems to have come into action here. This was reported to have been the case by the participant from the US Senate, as the Keystone recommendations were presented to the US Congress.

Nonetheless, this example may highlight the danger of operating with the idea of only *one* epistemic community within an issue-area. It may lead to an exaggerated notion of consensus, while in reality there may be several epistemic communities, some more visible than others, and with contentious views. Some have argued that the IUCN and the WWF constituted epistemic communities in the negotiations on biodiversity (Raustiala, 1997). As we have seen here, these international NGOs initially belonged to and advocated the Northern conservation coalition, rather than acting as a consensus-building, cross-cutting epistemic community. Sabatier and Jenkins-Smith (1993), with their central contributions within domestic implementation literature, may have a more realistic conception of this situation in their idea of *advocacy coalitions*. These are knowledge-based coalitions advocating different values within a specific policy-area and a decision-making process. In this perspective, the IUCN and the WWF can be said to represent a specific advocacy coalition, while the Keystone forum would fit the label of *broker* between the various interests or values. In this light, it becomes clear that the concept of epistemic community may easily be wrongly applied to, and confused with, *interest* coalitions.

Similarly, the fact that I have had access to and information about Keystone, but not to any potential epistemic community within the multinational biotechnology corporations, does of course not mean that the latter is non-existent nor non-influential. While the biotechnology sector is most likely to carry substantial weight in lobbying their respective governments, it can, however, hardly be defined as an epistemic community. Still, governments may in the end be more susceptible to "scientific advice" from these groups, as their advice is likely to coincide with domestic economic interests.

5.5.4.2 Legitimacy and common norms

When the CBD was adopted and signed, it can safely be said that environmental issues loomed larger on most political agendas than (ever recorded) before. Only five years later, at the UN General Assembly special session (UNGASS, 1997) convened to discuss governments' performance after Rio, attention was low. In a relatively short-term perspective, we could talk about the cyclical nature of environmental issues, as indicated by the high levels of attention during the UNCHE in Stockholm in 1972 and at UNCED in Rio in 1992. During the two decades that led up to the Rio Conference, the majority of states, both developed and developing, had acquired some form of environmental agencies. Moreover, there has been an unprecedented growth of non-governmental environmental organisations, both at the domestic and international level, and these NGOs have increasingly been granted observer status in international environmental negotiations.

Typical of the growing environmental movement is also a use of scientific arguments. Science has in addition frequently been granted a central place in the negotiations through the establishment of specific research institutes and advisory bodies – linked to the negotiation forum itself (for instance the IPCC). The focus on environmental issues both from public opinion and from politicians, coupled with the steady growth in international agreements, has provided these issues with increased salience and momentum. States have become increasingly aware of their environmental *reputation*, and few have been indifferent as to whether they appeared as laggards in international environmental co-operation.

While this general "climate" for international environmental activities is definitely an important factor, it is nevertheless difficult to show any one-way causality between the development of international norms and the establishment of IEAs. The discussion rapidly ends up in a "chicken or egg" debate, and the safest conclusion may be that the relationship between norms and regimes is one of mutual reinforcement. The causal relationship between scientific "facts" and the development of norms is also interesting and little explored in the environmental issues. The critical scientific findings regarding ozone holes, "Waldsterben" and loss of biodiversity through rainforest destruction – all in the nature of transborder environmental problems – have led to increasing public awareness that "something must be done", and the realisation that concerted international action is necessary.

I shall now address the notions of norms, ideas and learning processes within the biodiversity negotiations in more specific terms. The central feature on which learning processes and development of common norms hinge in the negotiations is clearly the third objective of the CBD: the concept of *fair and equitable sharing of benefits arising from utilisation of genetic resources*. Main proponents for this concept have been parts of the NGO community and the South, with some "like minded-countries" (notably the Nordic countries) as bridge-builders. Opponents are also found among the NGO community, as well as among OECD countries. Three aspects are important to keep in mind:

- *Controversy within the NGO community*: NGOs such as ICDA/GRAIN and RAFI were advocating sustainable use & equitable sharing (environment + development aspects) on the basis of the value added to biodiversity by small-scale farmers in the South. The WWF and the IUCN were advocating essentially conservation/preservation aspects mainly in terms of the environmental agenda of the North.
- *A high degree of mobilisation of NGOs across international fora and sectors*: Within the agricultural sector the NGOs focused on Keystone, FAO, and the CGIAR. UNEP was initially dominated by the conservation/preservation NGOs. The cross-cutting issues of biotechnology and patenting mobilised NGO-lobbying also in the GATT Uruguay Round. The campaign in GATT exhausted the capacity of GRAIN for a long time; or, judged from a different point of view: GRAIN chose to focus its energies on lobbying in this forum.
- *The controversy among NGOs largely follows the same conflict lines as that among states*: This was most clearly seen in the *conservation / equitable sharing* contradiction, with the USA and the IUCN pushing for the former. The interesting message behind these two aspects is that the conflict is both of a

normative character and it has clear *material* effects in terms of who should get what. GRAIN's argumentation tends to come across as more "value-laden" than that of the IUCN. The IUCN argumentation seems more pragmatic, but this may be simply because the IUCN argued from the point of status quo.

This brings up the difficult question of how to distinguish between values/norm-based argumentation and material/interest-based argumentation in the negotiations. In some issue-areas, this distinction may be seen more clearly – for instance the negotiations within the International Whaling Commission. The parties arguing for a moratorium on whaling can hardly be said to have strong material interests tied to this position. In the biodiversity issue, however, the advocacy of equitable sharing of benefits is of a dual nature. References to compensation for farmers' and local peoples' contributions to seed selection and knowledge related to pharmaceutical uses of plants are largely of a symbolic nature. If the principle should achieve broad acceptance in practical politics, however, it would obviously involve high material gains as well.

Our analysis seems to indicate that the biodiversity negotiations and the final result were not in the end dominated by the global environmental agenda of the North, at least not to the same extent as can be said for the climate change issue. This is so even though biodiversity itself clearly ranks among the issues on the Northern agenda along with climate, ozone and marine pollution. If the South has a common environmental agenda, it is believed to be more "local" in character and involve issues such as soil erosion and depletion of local water resources, as well as a higher emphasis on the development aspects.[119]

Against this background we must contend that the concept of *equitable sharing* has both a moral and a material interest side. The question thus becomes: Does the inclusion of this concept in the CBD represent a breakthrough both for DCs norms *and* interests? When advocating material interests, this may or may not be framed in terms of values, but most interests will inherently rest on a normative view of why that particular interest should win through, i.e. why it should carry legitimacy. The CBD text is largely framed in symbolic terms, as it proved difficult to set any technical standards pertaining to any of the three objectives (conservation, sustainable use, equitable sharing). At this point in the analysis, we can answer in the affirmative for the normative part only. From this it follows that we are more likely to be dealing with "norms as learning", rather than "norms as authoritative persuasion" (i.e. as a form of cultural imperialism), as the winning coalition came from the South. In others words, an interplay of norms and structural power is not very likely to have taken place in this process. As the agreement and inclusion of the relevant agenda items took place over a rather long period of time, it is appropriate to describe this as a learning process. We will need to follow the implementation of

[119] Who has identified this as the South's real environmental agenda? As contended in section 3.1, this type of analysis is clearly not suited to reveal the "sources" of the norms advocated by Southern governments, environmental and developmental NGO community, and i.a. the Nordic countries.

the CBD as it moves from the normative to the factual level, in order to assess the interest-based part of the concept. This is the main focus in the discussion in Chapter 6.

While the international "environment" dominant during UNCED and the conclusion of global environmental conventions such as the CBD and Climate Change may have led to an added inclination for states to adopt IEAs, the same may not necessarily be true of the implementation phase. Environmental concerns are seen to be falling in importance on the public opinion agendas, and hence on the political agendas. In the following, I discuss the general outlook for the CBD implementation phase, but here I will start by briefly summing up the findings from the analysis of the regime formation process.

5.6 Summary

Where should we ideally concentrate, when looking for processes and events with a possible bearing on the formation of international agreements? The realist perspective teaches us to look for overall structural power relations among the negotiating parties; but this is not necessarily limited to the Bioregime itself. On the contrary, the realists are not much impressed by "issue-specific power". Hegemons may not even regard the Bioregime as an interesting arena, if the "national interest" can be pursued through other, more influential fora. The WTO/TRIPs has been and still is such an alternative forum, in terms of the North/South controversy over the biodiversity issue. In a sense, the North/South controversy has been raised to a "higher" level, as the conflict has now become institutionalised between the two regimes of free trade and conservation and sustainable use of biodiversity. This may indicate that even though the first-cut analysis of the biodiversity negotiations and output did not seem to leave much explanatory power to the realist perspective, this picture changes if we look outside the biodiversity forum itself. Moreover, in the first phase, getting to the agenda, we may interpret the efforts of the USA and the IUCN to de-link biodiversity from the debate on property rights in FAO in terms of structural leadership, i.e. in realist terms. Still, the analysis in Chapter 5 indicates that we cannot account for the degree of breakthrough for DC norms and ideas in the CBD objectives in terms of overall structural power and interest constellations alone.

The examination of actor preferences and capabilities provided insight in how the role of *bystander* made it possible for individual delegates to act as *entrepreneurial leaders*. This analysis of structural power and interests made it clear, however, that this perspective could not account for the result of the CBD negotiations. This is most clearly pictured by the USA's transformation from a confident pusher in the early phases, to a rather isolated laggard towards the end of negotiations. The examination of the interplay between interests and issue-specific power brought us closer to an explanation of output: It is essentially issue-specific power in terms of *negotiation power* that seems to have strengthened the DC case.

Hence, we needed a closer examination of the interplay between power and institutions.

Turning to institutional factors, there are several pathways for approaching an explanation of the DC breakthrough in the CBD. First, the nature of the arena itself seems to have contributed to a greater acceptance of environment and development concerns among the parties. The main argument concerning the institutional factors is that the normative undertow pervading the UNEP/UNCED arenas were influential in framing the issue in an environment- and development-friendly manner. This general inclination, linked with the particular type of actors drawn to these negotiations, seemed to have enhanced the scope for agreement among the parties. This is partly due to the *official purpose* of the arena, which was geared towards finding common solutions to global problems at a time when world attention was especially focused on these issues. It is hardly surprising that it was easier for the countries of the South on this arena to find sympathy for their argumentation about the contributions of small-scale farmers and indigenous peoples in developing and maintaining biodiversity, compared to arguing for increased market access and higher prices for the same products through the WTO.

This aspect was most likely strengthened by the particular type and persistence of *participants* who were drawn to the UNEP and UNCED negotiation table. We could also speculate on the choice of advocating the environmental agenda of the North in an arena in which legitimising language and financial "sweeteners" came as part and parcel of the effort to boost environmental policy change in developing countries.

Second, it seems that both the Secretariat and some of the (smaller) negotiating parties appeared as *entrepreneurs* in finding and tapping the integrative potential for reaching agreement through creative package-deal proposals and linking the issues of conservation and equitable sharing.

Strengthening this direction of the CBD output is the *learning process* which appears to have taken place during negotiations. This was partly facilitated by what may be identified as the biodiversity *epistemic community* – the Keystone Dialogues. Keystone is likely to have had an impact through high-ranking officers from the international secretariats who may have been instrumental in bringing the consensus back, and later getting it included in agenda-setting. Second, several participants held important positions in national bureaucracies, and they may thus have had access to encourage common policies in their respective governments. This kind of impact may have been limited to rather few countries, however. There are also some data that corroborate an active impact from the third of Haas' pathways, i.e. that the epistemic community was able to create transnational and transgovernmental coalitions holding similar views, thus forging consensual solutions. This is not only creditable to Keystone, but due mainly to the fact that many of the same actors reappeared in a variety of fora relating to the biodiversity negotiations. Such repeated personal contact between actors with different backgrounds should not be neglected as an important factor in enhancing the contractual environment.

The most important contribution of the Keystone Dialogues, however, may be their role in unifying the South's position and helping DCs develop a concerted line of argumentation. This seems to have added to the *negotiation power* of the South. Here we may question the epistemic character of Keystone, and ask whether it assumes the properties of an interest coalition. In that case, the IPCC must also be seen as an interest coalition, advocating the views of Northern governments.

A related conclusion at this stage, which may be of more immediate academic interest, is the importance of the debates underway in other international fora and their significant impact on the formation of the CBD. This conclusion concerns the effects of *exogenous institutional factors*, the issue-linking prompted largely by processes in other international fora. This is an aspect which is relatively poorly elaborated in related regime studies, and it was not included in my original analytical framework – as my framework was built on established theories in the field. In FAO and the WTO, biodiversity is not the main focus, rather it is a sideline to their more general orientations. Nevertheless, the CBD agenda and the issue-linking taking place in the UNEP negotiations greatly reflected the developments in these other fora. This was seen most particularly in the decision to include property rights and hence, the link to domesticated biological resources and the idea of equitable benefit-sharing. As illustrated by GRAIN's choice of lobbying GATT/TRIPs rather than the CBD negotiations, they anticipated that the impact from this organisation (the WTO) on access to, and control over genetic resources, had the potential to be greater than the impact of the CBD itself. This "competition" from an external international forum may even have boosted the activity level in the biodiversity negotiations themselves. This could have been a reaction to the considerable frustrations experienced by the active DCs vis-à-vis the TRIPs debate, which spurred them to intensify their argumentation for a compensating CBD. Following this line of thought, it could even be inferred that endogenous institutional factors had less impact than "*exogenous institutional factors*": it could have been the perceived (and real) injustice of the decision-making rules (weighted voting) within GATT which fostered a stronger alliance between DCs, some bridge-building countries, and parts of the NGO community in the biodiversity negotiations. Of course, the *official purpose* of the UNEP-led negotiation also permitted such an alliance to have a greater impact here, compared to that of GATT.

Nevertheless, the combined effect of these different, albeit related, fora leads to the conclusion that a narrow focus on one (issue-specific) regime can hardly reveal the whole story concerning regime formation. It may even leave the observer ignorant of some of the main explanatory factors. While regimes by definition are issue-specific, there is a high probability that important aspects of the issue itself are likely to be addressed in other arenas as well, specifically in fora with a broader, more encompassing scope. The biodiversity issue is not likely to be the only one influenced by broader trade and economic frameworks set out by GATT and what originated as the Bretton Woods institutions. However, this is not an argument against the explanatory force of regime theory: On the contrary, *it underlines the importance of regimes in framing issue-areas through different official purposes*. A traditional realist perspective would probably have led attention more directly towards the more powerful arenas, but it would hardly have predicted the mitigating effect of the "bioregime".

One analytical implication of this observation is that it points out the inherent tension between the issue-specific definition of regimes and the notion of issue-linking taking place within a regime. This may be due to the historical roots of regime theory – as it originated as an offspring/alternative to neo-functionalist integration theory. Not only may the issue-specific regime be influenced by processes in other fora, it may also adopt related aspects of the issue from such arenas. The opposite situation could also be envisaged, with the regime "shedding" parts of the issue to related fora. This points up the caution needed when identifying the often dynamic and contested functional scope of a regime. A recent trend in regime studies seems to be more inclined to look into such situations, with a focus on "institutional linkages" (Young, 1996), and a reintroduction of the concepts *embedded* regimes (Ruggie, 1982), *nested* regimes, as well as *clusters* of regimes and *overlapping* regimes.

The term *embedded institution* actually characterises all regimes, as every effort at international co-operation takes place within a broader context of existing principles. Secondly, a growing number of international agreements display the character of a *nested institution*. This trait captures the establishment of protocols under wider framework conventions, as well as efforts to reproduce and operationalise, at regional levels, those wider, more diffuse commitments made at the global level. A third ideal type is the *clustered institution*, in which different functional arrangements are combined into comprehensive package deals. Finally, there is the phenomenon of *overlapping institutions*, where the functional scope of one regime protrudes into the functional scope of others.

The relationship between the CBD and the WTO/TRIPs is clearly a case of *overlapping*, which tends to be unintentional and often involves contradictions. According to Young, such unintended effects may be dealt with by establishing procedural arrangements to resolve conflicts, and this is very much in line with how the parties seem to be dealing with these institutional overlaps (see section 6.2).[120] Not only are the CBD and the WTO *overlapping*; the two regimes are also *embedded* in very different principles: The CBD builds on the principles that environmental concerns and equity should take precedence over economic concerns, while the WTO is based on the principles of trade liberalisation. The controversial element in the relationship between the CBD and the WTO would thus presumably be further reinforced (Rosendal, 1999).

Another lesson to be drawn from this observation, also pointed out by the Tübingen team, is that *defining and delineating an issue-area may in itself be a highly political (and controversial) process* (Hansenclever et al., 1996: 192).

[120] See also Oran R. Young (1997) (ed) and Oran R. Young (ed), *International Society: Enhancing the Effectiveness of Environmental Governance Systems*, forthcoming, ch. 9.

Likewise, the choice of forum in which to debate certain aspects of an issue-area may also be a highly political and controversial process.

The analysis has revealed two rather unexpected patterns concerning explanatory factors. Both underscore the importance of international institutions. Nevertheless, neither has figured as central in the literature in the field, and hence did not get much attention in my analytical framework:

- First, a central conclusion concerns the seemingly strong impact of the official purpose of the negotiation forum in framing the output. This underscores the importance of institutions in framing international negotiation outputs. This tendency was probably enhanced by the link to UNCED, a world summit meeting which drew major attention to the negotiating parties. This conclusion – that it will be possible to increase achievements in terms of environment & development objectives during summits like UNCED – can most likely be generalised to other global environmental issue-areas. The more general conclusion concerning the effect of the official purpose may have yet wide applicability, applying to most regime formation processes.
- Another central conclusion points to the importance of other fora, dealing with related aspects of the issue-area, for understanding how the issue was dealt with in the formal CBD negotiations. The manner in which related aspects of the issue has been, and is being dealt with, in overlapping and partly competing fora may still carry structural effects, negating efforts to achieve the objectives of the CBD. Dominant actors may still have their way, as they may have greater impact on choosing the "right" forum for advocating their interests. According to this line of thinking, UNCED was clearly the best forum in which to achieve environmental concessions from the South. This conclusion, which emphasises structural power relations, would seem applicable mainly to other global environmental issues where the North–South cleavage line is likely to dominate.

CHAPTER 6

PROSPECTS FOR IMPLEMENTATION

6.1 Inferences about the Implementation Phase

At the time of writing, seven years have passed since the conclusion of the CBD in Rio. This is not a long time in terms of studying implementation success or failure, and it is probably premature to expect solid conclusions either way. Moreover, the lack of explicit implementation standards makes it difficult to give a complete and coherent picture. As a result, the methodology applied to data collection in Chapter 6 is relatively weak, or at least it is less guided by the analytical framework than the other parts of this study. The time-frame allows for some speculations only. With these shortcomings in mind, I still deem it possible and interesting to investigate some of the trends in the aftermath of the conclusion of the CBD. I will look at some of the implementation activities that have taken place, and discuss whether there are trends supporting or negating my conclusions in Chapter 5. The focus will be on two items which, I shall argue, are central to the implementation process.

As of 1999, there have been four meetings in the Conference of the Parties (COP) to the Convention. COP is currently discussing the development of several protocols under the CBD as well as the relationship between CBD and related international fora. Various items are still pending, among them:

- the financial resources and mechanism (primarily the GEF);
- general measures for conservation and sustainable use;
- conservation and use of agricultural biodiversity and forestry issues;
- knowledge, innovation and practices of indigenous and local communities;
- access to genetic resources;
- issues related to technology transfers;
- intellectual property rights and the relationship to TRIPs;
- biosafety;
- identification, monitoring and assessment mechanisms.[121]

It is not possible within the scope of this study to go into equal detail on all these aspects of the issue-area. Nevertheless, some of these aspects form clusters of issues

[121] This list of items stems from documents to COP3.

which may be treated together, and some aspects are clearly of more direct interest with regard to the thesis question than others. In the following, I concentrate on two such clusters of issues, discussing these in light of the three theoretical perspectives. I will argue that the relationship between IPR systems and the CBD objectives represents a basic challenge for future implementation activities, as do the questions related to the development of the financial mechanism. Both address the principle of equitable sharing as a precondition for conservation and sustainable use of biodiversity. These two items represent a symbolic breakthrough for DC norms, and the question now is whether they will be translated into practical political action. The international approach to and implementation of these items will also have an impact on such related aspects as access to genetic resources, traditional knowledge, technology transfers and conservation and use in general. First, however, let us consider the main inferences based on the three theoretical perspectives.

On the basis of the three perspectives the following inferences may be constructed:

1. In light of the *power-based* perspective, a likely assumption is that the new principles, norms and rules concerning equitable sharing of revenue from utilisation of biological resources will hardly be reflected in practical political action. Norms – and regimes – do count, but it is power that decides.

- the TRIPs regulations, and Northern intellectual property rights (IPR) systems, will take precedence over the objectives in the CBD.
- the GEF (financial mechanism) will be fully in the hands of (benevolent) hegemonies, and will be used to further the global environmental agenda of the North.

2. From the *interest-based* regime model, a likely inference is that the parties will to some extent follow up their international commitments pertaining to the three objectives of the CBD. Implementation activities are likely to take place, due to the monitoring and incentive mechanisms of the CBD. The clause about national reporting will enhance monitoring, transparency and information, as well as make the parties sensitive to social sanctions if they fail to fulfil their obligations.[122] Together these factors will reduce insecurity and uncertainty about other parties' free-riding.

3. From the *ideational* perspective we may infer that even in the absence of strong sanctions being invoked against a breach with the CBD objectives, the parties will follow the new norms as they learn to regard them as legitimate.

In order to assess these inferences, I need examples of activities relevant to the implementation process. It is too early in the implementation process to make any

[122] Enforcement mechanisms are not equally necessary for all kinds of regulations emanating from international environmental agreements. Some rules and regulations may be easy to follow up, involving little costs to the parties concerned, while others may not be heeded despite enforcement mechanisms.

final conclusions, so the following discussion will be based mainly on specific examples of relevant activities.

6.2 The CBD and Intellectual Property Rights

One crucial test for the future implementation and development of the CBD is probably linked to whether industrialised states regard the treaty primarily as a threat or as a possible advantage for their growing biotechnology sector. The United States has signed but not yet ratified the Convention, although it is expected to do so eventually in order to gain more than observer status at the annual meetings of the Conference of the Parties (COP). Compliance from the developed countries with their obligations under the CBD is most likely a prerequisite for implementation in developing countries.

As indicated in section 3.2.1 and as an introduction to the relationship between the CBD and WTO/TRIPs, I shall briefly review the position of the USA with regard to intellectual property rights (IPR) under the CBD. In the next sub-section, I shall make more general comments on the relationship between IPR and the CBD.

6.2.1 The case of the USA

I will look at the behaviour of the United States as a test case and discuss that country's reason for not signing the CBD in the first place, as well as the probability that it will ratify now that it has signed. The US government is still fighting the CBD on several important points. It was, among other things, the wording of Art. 16(5) that IPR systems should *not run counter to the objectives in the CBD*, that made President George Bush refuse to sign the Convention in Rio in 1992.

Moreover, the US government has been strongly opposed to a protocol on biosafety, and was initially alone in this argumentation.[123] It has since been followed by Argentina, Australia, Canada, Chile and Uruguay (constituting the so-called 'Miami Group'), all major agricultural exporters, and these countries managed to block a major effort to establish a biosafety protocol in Cartagena, 1999. The Convention Parties disagree on whether to aim for non-binding guidelines on biosafety or a legally binding protocol,[124] how to deal with the links to the WTO[125]

---

[123] Expert Panel IV concluded that there is a need for a protocol on Biosafety, against the sole vote of the USA. Expert Panel to follow-up on the Convention on Biological Diversity, Report of Panel IV (1993): *Consideration of the need for and modalities of a protocol setting out appropriate procedures including, in particular, advance informed agreement in the field of the safe transfer, handling and use of any living modified organism resulting from biotechnology that may have adverse effect on the conservation and sustainable use of biological diversity*, UNEP/Bio.Div./Panels/Inf.4, Nairobi, 28 April 1992.

[124] Second session of the Intergovernmental Committee for the Convention on Biological Diversity, Nairobi, 20 June – 1 July 1994.

and the scope of an eventual protocol concerning living modified organisms (LMOs). The need for a Biosafety Protocol stems from the realisation that the environmental consequences of releases of genetically modified organisms are non-reversible and unpredictable. Many devloping countries fear that the existence of strict regulations at "home" may lead transnational corporation to use developing countries, where legislation and administrative capacity is weaker, as testing grounds (Hindar et al., 1990).

Why was the United States so reluctant to sign the CBD? There may be reason to believe that the perceived threats have been exaggerated. At the May 1992 meeting in Nairobi where the CBD was adopted, the US delegation hoped and believed that the Convention would prove acceptable to the US government. One week later, however, the Bush Administration backed out in Rio. The remaining OECD countries did not follow this course, in spite of seemingly similar material interests in terms of access to genetic material in the South, and protection of own biotechnology sector. If we open the black box and look for domestic factors this may help explain the US behaviour.

First, Washington's behaviour may partly be explained by a *contextual interpretation*. The combined effect of the formulations in the CBD became too much for the Bush Administration, even though it might have been willing to swallow each article separately. The final package containing UNCLOS is an example of this same phenomenon. Similarly, having to deal with both climate change and biodiversity at the same time could also have worked as a turn-off. Contrary to the argumentation in section 5.5.3.1, the *timing* of signing the CBD at UNCED may have worked both ways, finally proving too much for the US central government. While the US delegations to climate change and biodiversity respectively were ready to accept the deals, the combined package may have led the central administration to sacrifice the CBD in the end.

Second, part of the explanation may be found in *political culture* factors at the domestic level (Porter, 1993). This concerns the US bureaucratic style in dealing with international agreements. Typically, the US government displays great caution about committing itself to international agreements. Once such a commitment has been made, however, the United States tends to score high on compliance. Moreover, the domestic decision-making process has been characterised by internal bargaining and considerable friction between different agencies. Illustrative of the internal strife within the Bush Administration was the US biotechnology industry and its main political supporters – Vice President Dan Quayle and his Council on Competitiveness – whose influence won through in the final round. In effect, the US government refused to sign the Biodiversity Convention, mainly because of its formulations on intellectual property rights (Rosendal, 1995c). These fears, however, appear to have been loosely founded.

There are several reasons why the CBD is unlikely to impose severe limitations on biotechnology companies. Patenting in biotechnology is becoming increasingly

[125] This concerns the issue of regulating trade in LMOs versus free trade.

widespread, and there are thus no indications that the Convention has had any prior diminishing effects on this practice. This is linked to the increasing privatisation in the agricultural sector (Barton et al., 1992), and may be observed in several DCs, even some of those most zealous in opposing the international patent quest in the GATT Uruguay Round. Today, most of the developing countries that can sustain a patent system, such as India, Brazil and Mexico, are well on their way to accepting parts of it (Van Wijk, 1991). As the patent-system gains world-wide acceptance, it will be increasingly hard to side with the US biotechnology industry in arguing that they alone will be harmed by the Convention's regulations on biotechnology.

When the United States did decide to sign the CBD, this was accompanied by a statement of interpretation, seeking to tone down articles that seemed to put restrictions on the biotechnology sector. Such re-interpretative statements are not uncommon in international treaties, but in this case it may have the serious effect of upsetting the delicate equilibrium of the CBD text. The re-interpretative statement may be seen as a way to circumvent the perceived threats in the CBD; i.e. refuting the costs which the USA fears might be put on its biotechnology industries, while retaining the benefits associated with the Convention. In light of the *power-based* perspective, the USA may be powerful enough to get away with this policy, as no other state can pressure Washington to ratify the Convention as it is.

This approach does not, however, explain why the USA should eventually care to ratify. If the power-based explanation were taken to its logical extreme, the USA might as well do without an international convention on biological diversity. What indeed are the benefits of the CBD to the United States, as well as to other powerful industrialised countries? The major benefits, as would follow from the Northern global environmental agenda pushed by the USA, should have been *conservation (or rather preservation) of biodiversity for future generations* and *free access to genetic resources*. As these objectives are in fact not reflected in the output, we had to turn to other explanatory perspectives.

When US President Bush refused to sign the Convention in Rio, a direct response came from Venezuela, who proclaimed that it would stop signing new agreements with US scientific institutions on collecting and screening biodiversity in the country (Reid, 1992). Reluctantly, the US did go as far as signing the CBD. This could indicate that the *interest-based* perspective might have explanatory power in the implementation phase. The gene-rich DCs have indeed increased their *basic game power* in this issue-area, to the extent that they to some extent are able to pressure a country such as the USA into the fold. Thus, the CBD may come to have some influence as an international instrument for affecting state behaviour. A similar inference can be found in the *ideational* perspective. In this view, a major benefit in ratifying the CBD would lie in the Contracting Parties being accepted as *legitimate partners* in international transactions with genetic resources. If the US government chooses to ratify, this implies that it does not appreciate the idea of being (solely) left out of the international agreement. It would hurt the US "green image" as well as leaving less scope for making an impact on further developments of the Convention.

In a short-term perspective, however, this is hardly an adequate picture of the power relations between gene-exporting and gene-importing countries. Users still have free access to genetic material in most of the world's gene banks, and most owner countries are still in a weak bargaining position in bioprospecting deals. Moreover, the experience from Cartagena and the Biosafety protocol is also evidence that the USA seems to be quite influential in spite of being a non-party to the CBD. This means that we must remain open to *power-based* explanations in the implementation phase.

6.2.2 Activities relating to IPR

Now I shall continue with a more general discussion of the relationship between the CBD and IPR systems.

- One line of investigation relates to how the CBD can function as a framework for bioprospecting deals. Here I shall cite examples of relevant bilateral arrangements and unilateral action in owner and user countries undertaken by both public and private actors.
- Secondly, I shall examine the multilateral arena – how the IPR question has been dealt with in the COPs subsequent to the Convention entering into force. We will also take a closer look at the relationship between the CBD and the WTO with regard to IPR systems.

In addition to presenting what *actual activities* are taking place, I shall refer to the *discussion of what may be done* to follow up the objectives of the CBD. In presenting the examples of implementation activities, there is a danger of giving a biased picture of the general situation concerning implementation level. While these are mostly cases where the CBD has made a positive impact on relevant policies, it should be noted that the examples of adverse or non-action still largely outnumber the examples of implementation activity. All the same, it is important to give a general overview of what seem to be activities spurred by the CBD. The extent to which the CBD has spurred these activities will be further debated towards the end of this section.

6.2.2.1 Unilateral activities

A highly relevant formulation in the CBD is linked to the principle of *prior informed consent*. This implies in legal terms that the country providing genetic resources (*owner country*) must provide national legislation regulating the appropriation of genetic material.[126]

---

[126] This article has been studied in detail by Frederic Hendrickx, Veit Koester & Christian Prip (1993). Veit Koester played a central role throughout the biodiversity negotiation process, acting both as Chairman and Vice Chairman during various parts of negotiations, leading Working Group II, among other things.

Specific implementation activities may involve the establishment of specially designated bodies to conduct the deals and information databases on the genetic material. Moreover, the providing country will have to set priorities with regard to compensation mechanisms for its genetic material in line with the objective of euqitable sharing. A weak point in this regulation is that in order to turn down a request for genetic material, the providing country, depending on its national legal system, may have to refer to such legal provisions. In the absence of such provisions, there is still a substantial risk that the gene-flow must continue free of charge *accross national borders* (Hendrickx, 1993). Obviously, this may represent an impediment to governments, especially in low income countries, which lack administrative capacity both to enact and enforce a legal framework vis-a-vis external actors.

With regard to *domestic sharing*, activities would for instance involve making clear the relationship to genetic resources held by local communities and gene banks. It should be noted that the CBD is primarily aimed at regulating equitable sharing *between countries*. It is only Article 8(j) that addresses this question at the *domestic* level.

Examples of *unilateral* implementation measures in *owner* countries include the Access to Genetic Resources of the Andean Pact (Decision 391, 1996) and the Thai draft bill on a *sui generis Plant Variety Protection Act* (1997). Both of these are seeking to protect the native plant varieties as well as the rights of farmers, local communities and indigenous peoples.[127] Thailand has not ratified the CBD partly because it does not have national legislation in place to govern access to genetic resources. GRAIN has carried out a survey and compiled a comprehensive report on domestic *sui generis* systems in the South (GRAIN, 1997). According to this report, countries currently developing access legislation and community/farmers' rights legislation are Brazil, Colombia, Ethiopia, India, the Philippines, South Africa, Tanzania and Thailand. In addition, several countries are developing access legislation or community rights only. These are Argentina, Bolivia, Costa Rica, Ecuador, Indonesia, Mexico, Papua New Guinea, Venezuela and Vietnam. The enhanced awareness regarding sovereignty over genetic resources is not confined to the South. After the signing of the CBD, the Australian authorities have banned any plant or micro-organism from being taken out of the country before they themselves have had it bioprospected.

Among *indigenous peoples' organisations* at the *local level* several declarations pertaining to the subject of IPR and biodiversity have been made (Posey & Dutfield, 1996). These include:

- *Mataatua Declaration on Cultural and Intellectual Property Rights of Indigenous Peoples* from the First International Conference on the Cultural and Intellectual Property Rights of Indigenous Peoples, 1993

---

[127] *Biotechnology and Development Monitor*, No. 32, September 1997, pp. 13–15, University of Amsterdam, ISSN 0924–9877.

- Statement and basic points of agreement from the COICA/UNDP Regional Meeting on Intellectual Property Rights and biodiversity, 1994
- Final statement from UNDP Consultation on the Protection and Conservation of Indigenous Knowledge, Sabah, Malaysia, 1995
- Final statement from UNDP Consultation on the Protection and Conservation of Indigenous Knowledge, Suva, Fiji, 1995; and
- *Declaration of the Indigenous Peoples of the Western Hemisphere Regarding the Human Genome Diversity Project* (HGDP).

Mention should also be made of the Crucible Group, largely consisting of "veterans" from the Keystone Dialogue on Plant Genetic Resources.[128] The composition of the group also resembles Keystone as it includes members from public and private research institutions, intellectual property rights specialists, agricultural research scientists, NGOs, governments, industry and academia from both South and North, each participating in an individual capacity. In order to establish effective *Community Intellectual Property Rights*, the Crucible Group (1994:67–68) suggests that: such rights would have to be embedded in national legislation, national IPR offices should investigate complaints by local communities and governments, and a tribunal should have the power to revoke IPR, fees derived from IPR should be used to fund national IPR-offices and to provide legal aid to local communities involved in disputes, plant varieties developed by communities should be deposited in germplasm banks along with registration data, community IPR applications would have to include information about registration data: date, place, and location of origin (including names and addresses of communities), an international database would have to be created for tracing germplasm – possibly through the CGIAR system, and a public defender would have to be appointed in the shape of an internationally recognised mediator or ombudsman. Community IPR are, however, subject to the same critique as for patents: IPR does not extend to wild or natural genetic material, due to the fact that this does not satisfy the criteria of novelty or inventive step. At the same time, it is hardly "fair" that indigenous knowledge is the only intellectual property not protected by such rights.

While problems such as these, relating to domestic sharing, largely remains outside the scope of the CBD, there are also considerable problems relating to equitable sharing at the international arena. In view of the problems facing DC

---

[128] Members of the Crucible Group are: Bo Bengtson, Sweden; Dr. Tewolde Egziabher, Ethiopia; Erskine Childers, USA; Carlos Correa, Argentina; Don Duvick, USA; Paul Egger, Switzerland; Chusa Gines, Canada; Sven Hammrell, Sweden; Jaap Hardon, the Netherlands; Geoff Hawtin, Italy; Bente Herstad, Norway; Henk Hobbelink (GRAIN), Spain; Amir Jamal, Tanzania; Klaus Lampe, Philippines; Francisco Martinez-Gomez, Mexico; Camila Montecinos, Chile; Pat Roy Mooney (RAFI), Canada; Katy Moran; USA; Andrew Mushita, Zimbabwe; Bob Phelps, Australia; Michael Pimbert, Switzerland; Sarojeni Rengam, Malaysia; Tim Roberts, UK; George Rothschild, Australia; Rene Salazar (SEARICE), Philippines; Carl-Gustav Thornstrom, Sweden; Hans Wessels, Netherlands; and Vo-Tong Xuan, Vietnam.

governments in connection with enforcing catch quotas for foreign fisheries under UNCLOS, the problems regarding regulation of genes are striking. In addition to the general administrative burdens, the non-exclusive character of genetic resources further complicates control of their movements. This is partly negated, however, by the need for a user to obtain information about the genetic material in question. Without this, it may be difficult to screen genetic material for potentially valuable and interesting traits in secrecy.

This brings to light another aspect in which gene exchange differs from deals on catch quotas: *the rights and knowledge of local communities*. The 200-mile exclusive zones, which include the fish stocks within them, are generally regarded as state property, so rights to control access and levels of exploitation of fish are usually vested exclusively in government. Genetic resources, on the other hand, may have been developed through the work of local communities of farmers; or their valuable medicinal traits may be known only to certain indigenous or local communities. Government authority over their utilisation may thus be more seriously questioned. Local people have often been victimised as global and national interest has been spurred in these resources. The Western ideology of traditional wildlife management views man as an alien element in preservation areas, and central governments might increase their control over natural resources and groups within the population by employing the ideology, legitimacy and technology of preservation.[129] The uncertainties and conflicts within the domestic sharing dimension are closely linked to activities and regulations at the international level and this may complicate the implementation process.

The CBD mentions explicitly that the Contracting Parties shall respect, preserve and maintain knowledge and practices of *indigenous and local communities*, and encourage the equitable sharing of the benefits arising from utilisation of such knowledge and practices (Art. 8.j). A probable interpretation of the CBD is that governments may regulate the activities of their citizens regarding export of genetic material. Enforcement is less clear, as this brings up the tricky question of interference in domestic affairs, as well as how to identify who should be rewarded. One approach to ensuring the interests of local and indigenous people may be to include and elaborate the FAO Undertaking principle of *farmers' rights* in a protocol under the CBD. This principle applies to collectives and not individuals, but it would have to be expanded outside the area of plant genetic resources and agriculture to include the forestry sector, among others. A more general approach

---

[129] A recent World Bank technical paper (Kiss, 1990) describes the issue as follows: "The establishment of national parks and reserves, which may attract tourists and foreign exchange for the government, exclude and have often directly displaced rural communities from land they have traditionally considered to be their own. Anti-poaching laws turn the centuries' old practice of subsistence hunting into a crime, and people are often even prevented from eliminating "problem" animals to protect their crops, their livestock and themselves. In the simplest terms, rural people bear the significant costs of living with wildlife but have progressively been excluded from obtaining any benefit from them".

could be to link the concept of compensation to capacity-building at the local level. In recognition of the value of the traditional knowledge of indigenous and local communities, COP4 decided to set up an ad hoc open-ended inter-sessional working group to address the implementation of Art. 8 (j).

6.2.2.2 Bilateral activities

Donor governments and private corporations tend to prefer *bilateral* arrangements – especially when these can be arranged unencumbered by international regulations. This position was also noticeable in the biodiversity negotiations. Several bilateral deals are already in effect in the biodiversity field. Frequently quoted is the 1991 contract between the US multinational corporation, Merck & Co. Pharmaceuticals, and the Costa Rican National Institute of Biodiversity (INBio). The latter provides plant and animal species for drug research, which Merck gets exclusive rights to develop. In return Merck pays US$ 1 million as well as royalties for any drug developed. The establishment of protected areas, training of local *parataxonomists* and institutional capacity-building are also part of the deal. In the environmental and developmental NGO community there is still dispute concerning the quality of the contract. All the same, this raises the question of whether the developing countries need a framework convention for concluding such deals, since the Costa Rican arrangement already existed prior to the CBD. On the other hand, there is the effect of anticipation. It is possible that the Merck/INBio deal would have been less environment- and development-friendly, had the CBD not been in the pipelines at the time of its conclusion.

The impression that the CBD is seen as an important instrument in this respect is further stressed by the reaction of Venezuela, when US President Bush refused to sign the Convention in Rio. As mentioned above, Venezuela proclaimed its intention to stop signing new agreements with any US institution wishing to conduct *biodiversity prospecting* in the country. The term *bioprospecting* signifies the activities of collecting and screening biodiversity for medicinal or agri-industrial use (see Box 6.1).

Box 6.1 Bioprospecting

What is it about? 'Bioprospecting' denotes the search for natural products of potential value to the pharmaceutical and agri-business industries. Such products may be in the form of genes that are useful to biotechnology, chemicals that are useful as for instance colors and enzymes, and structures and constructions from nature that can be used in industry.

The relationship between the CBD objective of equitable sharing and bioprospecting: Advocates of bioprospecting tend to stress the potential benefits to be derived from bioprosepcting. Some stress as preconditions the existence of appropriate policies and institutions, which may ensure that the commercial value from genetic resources is used to stimulate development and conservation. Others argue that such legal regimes to regulate access to genetic resources are based on unrealistic expectations of benefits and that they can hinder both bioprospecting and biodiversity conservation (Grajal, 1999). On the other hand, bioprospecting is sometimes referred to as 'biopiracy'

(box continued from previous page)
by those who fear that the owner countries, or rather local communities, will be unlikely to take part in the benefits derived thereof.[130] The basis for this argumentation is partly that poor countries are unlikely to possess legislation and institutions with the necessary strength to enter into equitable deals with external (transnational) actors. Frequently, however, biopiracy-arguments rest on problems relating to domestic sharing rather than international equity issues (see 6.2.2.1). Moreover, the biopiracy-worldview is more likely to be supported by examples from plant genetic resources in agriculture, than from the pharmacutical sector (see 6.2.2.3).

The CBD basically deals with international sharing and identifies owners by the terms "country of origin" and "country providing genetic resources". These terms are problematic because they assume that a particular species or variety can be traced back to a specific country. The difficulty of tracing owners may be especially difficult with regard to seeds in international gene banks (see 6.2.2.3). Problems of origin may also arise for medicinal plants, as species in general are as likely to cross national borders as not, and hence, they may be native to several countries. Even if this dilemma was to be resolved amicably, the potential level of benefits deriving from bioprospecting deals is questioned. The great expectations are partly negated by the extremely small odds against finding useful products from bioprospecting samples, and partly by the competition from synthetic alternatives such as rapid in vitro screening in combinatorial chemistry. The latter is not a widely held assumption as natural bioprospecting is generally believed to present the greater potential: As an illustration, fewer that 1% of flowering plants have been thoroughly investigated for their chemical composition (Sheldon & Balick, 1995:46).[131] Moreover, screening results from Shaman Pharmaceuticals have revealed that of the samples that displayed promising chemical activity, 74% directly correlated with the original ethnobotanical use (Sheldon & Balick, 1995:58-59).

Bioprospecting and conservation: One problem is that overharvesting may take place once a promising sample has been made. On the other hand, the conservation potential is likely to be small if the end products are derived from synthetic processes or from plantations. For instance, microorganisms can be cultivated through fermentation processes, so there is no need to return for additional collections. Some useful chemical compounds discovered within medicinal plants have not been reproduced synthetically (like digitoxin), others have been reproduced but are less efficient than the original material (vincristine), but in most cases synthetic substitutes do exist. Development of synthetic production/chemical substitutes or plantations will provide a production of the product without the incentive to conserve the original ecosystem where the species was found (Balick, 1998:77). The primary conservation incentives are hence likely to be linked to ecosystems where the large diversity of organisms and their potential benefits are still 'unknown'.

---

[130] For a notable analysis of the policy aspects of bioprospecting, see the forthcoming PhD of Hanne Svarstad (University of Oslo, Sosiology), *Exchanges in Bioprospecting: Shaman Pharmaceuticals in Tanzania*.

[131] Furthermore, the diversity of marine species is largely unknown; the deep sea may rival tropical forests in species diversity. Microbial diversity may rival that of all other diversity, and is still also largely unknown. Arthropod or insect diversity is another new and unexplored area in natural products chemistry (Aylward, 1995:105).

Bilateral activities among users

There are several examples that indicate how the concept of equitable sharing is gaining ground among collecting (bioprospecting) agencies in the private sector:

- The *UK Royal Botanic Gardens at Kew* now states that any net profits derived from collaboration will be shared equally between itself and the supplier. As Kew receives duplicates of collected plants from a large number of agencies around the world, it is uncertain how far this statement goes.
- *Shaman Pharmaceuticals* develops new pharmaceuticals from higher plants and is committed to returning a portion of the profits from its products to all communities and countries in which it works. Shaman seeks to provide fair compensation and also offers immediate benefits to communities, by asking what pressing needs communities have that could be met by the company. The company has nevertheless been criticised for not making firm commitments regarding the payment of royalties, nor sharing patent ownership with communities (Posey & Dutfield, 1996:38–39). At present, however, this is a hypothetical criticism as Shaman has yet to deliver any patented products.
- There is also the case of *Biotics*, a private British for-profit company that acts as a broker between companies and in-country collectors, granting the latter 50% of Biotics' royalties (Laird, 1993).
- Another example of the normative persuasion potential (with material implications) of the CBD can be found in the response from *Phytera*'s vice-president, John McBride, on charges of bio-piracy. After the Third World Network (a coalition of development and environmental groups from developing countries) had revealed that a certain contract signed between biotechnology company Phytera and Palm Garden in Frankfurt did not include the usual clause on giving "an appropriate share of the profits to the countries of origin", McBride's response was: "What can I say? I'm embarrassed".[132]
- International Cooperation Biodiversity Group (ICBG) is a network of bioprospecting projects sponsored by the US National Science Foundation and the National Institutes of Health (NIH). The 34 projects provide financial rewards to local people, investments in research of local and international concern and strengthened local institutions – out of a total of half a million US$ each. About half of the royalties would be invested in the host country.[133] The ICBG design includes: a) active participation of host country individuals and organisations, b) multi-disciplinary research on diseases of both local and international significance, c) local training and infrastructure development in drug discovery and biodiversity management, d) biodiversity inventory and monitoring, and e) equitable intellectual property and benefit sharing

[132] Rob Edwards (1996) "Biotech firm 'embarrassed' by leaked plant deal", *New Scientist*, 29 June, p. 7.

[133] Ehsan Masood (1998) "A formula for indigenous involvement", *Nature* 392, 539.

arrangements. Scheiber (1998) points out the irony in that while the U.S. Senate leadership refuses to ratify the CBD, the Clinton Adminstration seeks to implement selectively some of the main terms of the CBD by attaching conditions to government grants to the academic-government-industry consortium (like ICBG) that engage in bioprospecting in developing countries.

6.2.2.3 The multilateral arena: Opportunities and constraints

Opportunities: Effectuating prior informed consent in user countries

As an example, *user* countries may improve the effectiveness of the *prior informed consent* principle by enacting national legislation on the import side: Along the lines of the rules governing international trade in endangered species of flora and fauna (CITES), national legislation could be tailored to prohibit illegal import of genetic resources, i.e. collections conflicting with prior informed consent export rules in the providing country. Likewise, companies and other importers could be obliged to keep records of imported genetic material, in order to facilitate monitoring by government authorities. Another interesting proposal is to require patent applications to include information about how genetic material was obtained (Hendrickx et al., 1993). Efforts to fight the expanding scope of patent legislation have so far hardly been rewarded. Hence, a more promising approach may be to direct attention to what can be done with the patent legislation itself.

A striking example of implementation activities moving in the right direction can be found in the opposition to the EU directive on biotechnology patents. In 1997, nine years after its initiation by the EU Commission, the directive was subject to yet another amendment by the European Parliament. The Parliament voted to demand safeguards against "genetic piracy" in developing countries and wanted explicit commitments to honour the EU's obligations under the CBD. The amendments, largely supported by Denmark, the Netherlands and Germany, included demands that the EU's international biodiversity commitments under CBD be balanced against obligations to harmonise patent legislation in accordance with WTO and TRIPs. The amendments would require that when inventions involve biological material of animal or plant origin, the patent application would have to specify its geographical origin and evidence that the material was used in accordance with the legal access and export provisions in force in the place of origin.[134] In the end, the amendements did not make it further than the preamble when the directive was finally agreed to in 1998. In response, the Netherlands decided to challenge the directive, inter alia on grounds that it is in conflict with the CBD. This means that Member States are still obliged to convert the directive into national patent law within two years, but they may choose to await the Court's ruling (if it comes within two years).

---

[134] *International Environmental Reporter*, 23 July 1997, pp. 713–714.

Constraints: Overlapping international institutions

The CBD grants *owner governments* the sovereign right over their natural resources and establishes principles for a multilateral framework to regulate bioprospecting deals. Following the philosophy of the TRIPs, in contrast, would mean to counter this type of government control – leaving bioprospecting arrangements to individual owners and users and granting intellectual property rights to the *users* (see Box 6.1).

The Conferences of the Parties (COP) to the CBD have put much emphasis on examining the relationship between the CBD objectives and the strengthened IPR systems.[135] COP3 issued a declaration on intellectual property rights, encouraging governments and organisations to submit case studies on the impact of intellectual property rights in regard to the Convention's three main objectives (conservation, sustainable use, and fair and equitable sharing of benefits). As the Committee on Trade and the Environment of the WTO began discussion of the relationship between environmental protection and TRIPs in June 1995, it centred on the relationship between that and the CBD. The relationship between TRIPs and CBD will also be a central topic at the review of the TRIPs, as the Millennium Round of the WTO is starting in November 1999.

A crucial question is whether these are actually efforts to succeed in coming to grips with the diverging regulations of the two regimes. Mere co-ordination of the functional scopes of the CBD and the TRIPs is hardly sufficient to deal with the problems emerging from their opposing norms and regulations. There are some indications that the potential conflict is being taken seriously, at least by the Secretariat and the Parties to the CBD. A document from COP3 points out some regulatory options of possible complementarity:[136]

- First, *mutually agreed-upon terms* for access to genetic resources could allocate intellectual property rights as part of the benefits to be shared among parties to an agreement on genetic resources. Such IPR could be defined under TRIPs-compatible IPR-systems.
- Second, there is a proposal to require or encourage disclosure in patent applications of the country and community of origin for genetic resources and informal knowledge used to develop the invention.[137]

---

[135] In 1996, COP3 asked the Executive Secretary to 'liaise with the Secretariat of the World Trade Organization to inform it of the goals and the ongoing work of the CBD and to invite the Secretariat of the WTO to assist in the preparation of a paper for the COP that identifies the synergies and relationship between the objectives of the CBD and the TRIPs Agreement'. Decision II/2 on IPR, 1996 Buenos Aires.

[136] UNEP/CBD/COP/3/23.

[137] Countries implementing measures that implicate both agreements such as rules requiring patent applications to disclose the country of origin of biological material might report them to the TRIPs Council, while at the same time disclosing the same information to the clearing house mechanism for scientific and technical co-operation established under Article 18 (3) of the CBD.

However, there are also regulatory obstacles: COP3 points out that national measures to promote technology transfer under CBD Article 16 might raise WTO *most-favoured nation* issues if Convention Parties and non-Parties were treated differently. It might also raise TRIPs issues if owners of proprietary technology were compelled to license technologies on grounds other than those prescribed in the TRIPs Agreement.[138]

There are also potential *institutional obstacles*. Even though the conflicts are explicitly admitted and attended to, implementation of the CBD objectives may be hampered by the stronger compliance force of the TRIPs:

- First, patenting in the biotechnology sector is a contested, though relatively small part of a larger issue-area concerning international trade and patenting in all sectors. The driving forces in this much wider issue-area are powerful and gaining in strength as the opposition is declining with the economic developments in NIC countries.[139]
- Second, if WTO members refuse to sign up to TRIPs, they become liable to economic *sanctions*. This makes the WTO a more powerful instrument than the CBD, which carries no economic sanction mechanisms.[140]
- Third, the WTO is a stronger institution in terms of its compliance mechanism, incorporating different sets of *timetables* for countries to harmonise their patent legislation. In contrast, the CBD does not provide timetables for parties to comply with its objectives.

Constraints: IPR systems and germplasm in international gene banks

Finally, the parties to the CBD will have to consider the legal vacuum surrounding genetic resources in *international gene banks*, especially the IARCs. Biological materials collected *prior* to the Convention's entry into force are not covered by the Convention, as it could hardly be given a retroactive mandate. These materials are thus still subject to the principle of *free access* for all. These resources, obtained largely from farmers in developing countries, represent important breeding material for the improvement of world food production. A central question is how to avoid widespread patenting of this germplasm, which would to some extent undermine the

---

138 *Most-favoured nation* is a basic principle of the GATT/WTO regime, saying that any trade advantages conferred by one country to another must be given all GATT parties.

139 The NICs are increasingly accepting patenting as beneficial to their own economies, thus splitting up what used to be a strong and concerted opposition from the South. This development is not likely to be significantly restrained by the CBD objectives, although the CBD may have had a mitigating effect in providing a pressure for a period of grace for the DCs (that is, WTO members) in accepting patent systems.

140 The two regimes do not differ significantly in terms of membership, which is very broad for both. By 1998, the WTO had 132 members and 34 observers, of which all but four (the Vatican, Bhutan, Cape Verde and Ethiopia) have applied to join, and approximately 170 states have ratified the CBD.

DC achievements in the CBD: a scenario like that would bring the South back to square one.

In response, DC governments and environmental and developmental NGOs argue that this genetic material should be included in the FAO Global System on Plant Genetic Resources with the intention of developing a protocol to the CBD on *farmers' rights,* placing such material under democratic intergovernmental governance. A similar proposal was included in the Nairobi Final Act, adopted by the parties negotiating the Convention.[141] An agreement of 1994 between the FAO and the Consultative Group on International Agricultural Research (CGIAR) has been concluded on this issue, making the germplasm in international agricultural research centres part of the FAO international network of gene banks. Gene banks within this network are based on the principles of making genetic material available for breeding and research while respecting the rights of the providers of germplasm.[142] Prior to this decision, environmental and developmental NGOs had warned against the World Bank taking over leadership of international germplasm collections, fearing loss of intergovernmental control over this valuable material.[143]

An avenue suggested by the CGIAR itself is to advise that the IARCs employ *preventive patenting.* That way the germplasm in the international gene banks could be kept in the public domain (CGIAR, 1992). Another "patent-preventive" strategy may be to publish information on all germplasm in the IARCs. The rationale behind this strategy is that such publication could violate the patent criterion of novelty, thereby rendering the material non-patentable. Publishing could, however, be risky, as there might still be loopholes through which to claim a patent, and it would draw attention to valuable germplasm.

The FAO/CGIAR 1994 agreement, making the germplasm in international agricultural research centres part of the FAO international network of gene banks, is intended to operate on the principle of *free access* combined with the principle that users will *not proceed to patent* the material. In practice, several violations have occurred, with 40 infringements by Australian organisations and corporations alone. Two public institutions in Australia (Agriculture Western Australia and Grains Research and Development Corporation) have recently been patenting chickpea seeds collected from a gene bank in India (Aakvaag, 1998). These and other cases are currently under scrutiny by the FAO and CGIAR.

Summary and evaluation

National legislation is an important vehicle for enhancing implementation of the objectives of the CBD. Institution- and capacity-building in developing countries

---

141 Nairobi Final Act, Resolution 3, Nairobi, 22 May 1992.

142 FAO report to the Second Session of the Intergovernmental Committee on the Convention on Biological Diversity, Nairobi, 20 June–1 July 1994; "Progress Report on Resolution 3 of the Nairobi Final Act: Ex-situ Collections and Farmers' Rights".

143 Second Session of the Intergovernmental Committee of the Convention on Biological Diversity (Nairobi 20 June–1 July 1994).

would need to be strengthened in order to increase their ability to reap benefits from their own genetic resources. Ideally, according to the Convention, contracts on biodiversity prospecting would include components of technology transfers, institution-building and environmental concerns as compensation for access to genetic resources. The Merck/INBio deal in Costa Rica does integrate these components, although some still object that there should be more of each component. Payment mechanisms such as these will hardly suffice to secure full conservation and sustainable use of genetic resources (whatever this may mean). Bioprospecting deals highlight two important aspects related to the implementation of the three main objectives of the CBD:

- There is a renewed danger that the developing countries may become increasingly divided in the scramble for biodiversity prospecting deals.
- It would seem, however, that if the CBD succeeds in strengthening equity in international gene transactions, this would legitimise its other objectives with the developing world.

Regarding the relationship between the CBD and the WTO/TRIPs, the second is stronger both in terms of institutional mechanisms and in being controlled by the more powerful states. On the other hand, there is ample evidence that the issue of the two regimes' diverging objectives has been accepted as an important one in the WTO and the CBD alike. The need to deal with these overlapping institutions within the issue-area have been institutionalised by providing for continued discussions between the two, as well as by institutionalising representation in the respective fora. The final result of these deliberations is still far from certain, and the opposing norms and regulations between the two regimes are likely to remain contested international issues for a long time to come.

Within the framework of this study it is not possible to cite every action which may be related to implementation of the CBD objectives. I would maintain that the examples provided cover the main activities and that they are fairly representative of implementation activities. Another question is whether I can say with some confidence that the activities cited are causally connected to the CBD. Is it likely that they would not have taken place in the absence of the CBD? With some qualifications, I will answer that question in the affirmative:

- First, most of the activities make explicit reference to the CBD and its objectives.
- Second, subsequent to the conclusion of the CBD there has been a proliferation of such activities, while there are few similar examples of either public or private initiatives earlier. The one prior example is the Merck/INBio deal concluded in September 1991 – an arrangement generally believed to have been constructed in anticipation of, and much in the spirit of the Convention (WRI, 1993).

On the other hand, it may be argued that once the biodiversity issue had been brought to the attention of both owners and users of genetic resources by being put on the international environmental agenda, increased activities like those quoted would have been likely to appear whether or not a specific convention had actually been concluded by the negotiating parties. The way the issue was framed during the negotiations has undoubtedly made a significant impact on current decision-making in relevant and affected parties. This is most clearly illustrated by the increased consciousness among owner countries and local communities concerning their sovereignty and rights to genetic resources within their realm. This consciousness has been further legitimised by being set out in legal terms in an international agreement. Here, however, we should also note that the increased awareness about local and indigenous peoples' rights is part of a broader international issue advocated both within and outside the particular scope of the biodiversity issue-area.

Another criticism that can be raised against this section is its focus on the positive implementation activities. There are certainly numerous examples of non-action and contrary activities going on in the field relating to IPR systems and biodiversity. For instance, concerning the related international fora, mention has been made of the World Health Organization. The WHO *could* have made a positive contribution by establishing codes of conduct for medicinal research as this relates directly to bioprospecting for medicinal plants and IPRs.[144] Another example is the Multilateral Agreement on Investment (MAI) drawn up by the OECD countries, which might contradict several important principles of the CBD. If concluded in its present form, the MAI will give maximum protection to the property rights of foreign investors – including intellectual property rights – and prohibit governments from imposing requirements on foreign investors. States will be forbidden to impose conditions such as technology transfer and minimum levels of local share equity. A unique feature of the MAI is that an investor can sue a state, for instance if he feels that his intellectual property rights are not adequately protected. Although this treaty is being negotiated by the 29 OECD members, it is also aimed at the developing countries. It may become a part of the WTO Milliennium round (expected to run for three-four years after its onset in November 1999). The DCs have the resources and their governments welcome, but would also like to regulate, foreign investments.[145] Thus, the MAI reflects the core conflict of the CBD, and may well affect important aspects of future implementation efforts.

6.3 Prospects Concerning the Financial Mechanism

The possibilities for DC governments to achieve payment from external use of genetic resources may, at least for the time being, be limited. The major external

---

[144] This was pointed out by Pat R. Mooney from RAFI; at a seminar in Oslo, 22 April 1998.

[145] Martin Khor (1997) "What is the MAI?", *Third World Resurgence*, no. 90/91, pp. 5–8.

source of income for biodiversity conservation is still direct funding from bilateral and multilateral sources.

The North–South controversy over funding was barely resolved by the compromise that the Global Environment Facility (GEF) of the World Bank, UNDP and UNEP should be used for an interim period, and on condition that it become more democratic. The latter condition was a concession to the South, who would have preferred a new funding mechanism over which they could have greater control. This reflects the fundamental conflict regarding the South's concern about national sovereignty over its own natural resources, and the North's claim for some degree of conditionality tied to its donor activities in developing countries, including spending on global environmental projects. This conflict was hardly subdued by restructuring of the GEF (which entered into force on 1 July 1994). At the second COP meeting in Jakarta (1995), there was still no agreement between the parties to make the GEF the permanent financing mechanism. This is mainly due to the dominant role of the World Bank, and the weighted voting system in the Bank, which grants 62% to developed countries and 38% to the developing countries.[146]

One main reason why the GEF was accepted for an interim period despite certain shortcomings in its democratic structure was its assumed ability to attract major funding. The GEF has already been active in financing several global environmental projects; beside biodiversity, it covers climate change, international waterways and ozone depletion.

Table 6.1 GEF-financed Projects by Focal Area (as of June 1997)[1]

	Pilot	**Phase**	**GEF (FY Jun 97)**	**Feb 95 –**	**Total**	
	# Projects	$ m	# Projects	$ m	# Projects	$ m
Biodiversity	58	332	39	253	97	585
Climate Change	41	259	54	347	95	606
Int'l Waters	12	118	7	62	19	180
Ozone	2	4	9	109	11	113
Multi-focal	3	20	5	90	8	110
Subtotal	116	733	114	861	230	1594
Enabling Activ.[2]			107	22	107	22
Total			221	883	337	1616

[1] Excluding Project Preparation and Development Facility (PDFs)

[2] Expedited enabling activities in biodiversity and climate change approved by CEO

Source: GEF /C.10/9 October 3, 1997, Draft Annual Report of the GEF, FY97 (Fiscal Year 1997)

As Table 6.1 shows, projects on biodiversity and climate change account for the major shares of GEF funding, with 36% and 38% respectively of the US$ 1.6 billion

[146] World Bank, annual report 1986.

budget. As the recently introduced Enabling Activity projects concern either biodiversity or climate change, their combined share is even larger; and in terms of projects, their combined share is 89%. By 1994, nearly half of the GEF projects (44%) had been allocated to Asia, 25% went to Latin America and 18% to Africa (Fraz, 1996:373).

Will the GEF prove to be a suitable tool for the conservation of biodiversity? This question concerns both the quantity of funds allocated and the quality of how funding is being used:

- *Quantity* is likely to remain a formidable barrier to actual improvement of the situation. The UNCED Secretariat estimated the annual funding requirements for sustainable development activities in developing countries to be US$ 125 billion worth of international assistance (with US$ 600 billion needed for sustainable activities world-wide). In contrast, the annual GEF budget is only about US$ 0.5 billion. Although the "exact" figures are obviously uncertain, the magnitude of the financial needs is generally accepted, and far from being achieved.
- *Quality* concerns criteria for funding projects in the GEF portfolio and whether the GEF can be adjusted to reach, for instance, small-scale farmers – who are central in growing and maintaining a diversity of food crops in the fields.

As regards the *quantity*, a contentious issue concerns "mainstreaming", or integrating global environmental concerns into non-GEF operations of the three implementing agencies (the World Bank, UNDP and UNEP). So far, little progress has been made in this direction, as many World Bank projects may have significantly adverse effects on biodiversity and climate change.[147] Hence, the question of quantity rapidly involves quality criteria. The content of "mainstreaming" has yet to be defined; it may prove to be a two-edged sword, as neither donors nor recipients may in the end welcome a complete transformation of traditional official development aid (ODA) portfolios.[148] For one thing, World Bank loans for coal-based power plants and infrastructure projects cutting down forests would be terminated as a result of mainstreaming. On the other hand, "mainstreaming" could come to involve greater emphasis on country-driven efforts, based on national priorities and improved local participation.

As to the aspect of *quality*, the GEF has already been active in funding global environmental projects since 1991. Reporting on their review of the GEF biodiversity portfolio after four years of operation, a group of experts agreed there

---

[147] In a study conducted by an international expert panel, headed by Gareth Porter (1997) *Study of GEF Overall Performance*, the World Bank has been criticised for failing to work in tandem with GEF objectives.

[148] At the first GEF assembly (New Delhi, 1–3 April 1998), developing countries are reported to have been most apprehensive about "mainstreaming", with several European countries and NGOs pushing for it (Max Martin, "In Search of Vision 2020", *Down to Earth*, 15 May 1998, pp. 29–37).

were various shortcomings (GEF, 1992): *inter alia* that projects take insufficient consideration of the expertise and interests of local people, and that the GEF has been biased towards traditional biodiversity conservation through the establishment of protected areas. This bias is related to the traditional approach of the agencies who operate the GEF, which in approximately two-thirds of the cases is the World Bank[149]. The World Bank deals mainly with large-scale programmes, whereas what many believe is needed for biodiversity conservation are small-scale, community-based projects. Despite the "greening" of the World Bank, the standard operating procedures of the organisation are undeniably linked to economic development, rather than local sustainable development problems (Keohane & Levy, 1996). A crucial factor in the further development of the GEF mechanism would seem to be its *Small Grants Programme* for NGOs, which so far constitutes only 2%–3% of the total GEF portfolio.

A major concern at the first intergovernmental CBD follow-up meeting, held in October 1993, was to provide input to the GEF meeting the following December.[150] The parties did not succeed in this objective. There was some agreement that biodiversity conservation can scarcely be understood in terms of *global benefits* versus *national benefits*, but as yet no solution has been reached for dealing with the concept of *incremental costs* in the GEF. These incremental costs are intended as an incentive for developing countries to include in projects a global conservation benefit which may not be in their immediate national interest. This division seems to work fairly well in projects concerned with ozone depletion and to some extent for climate change, but in the biodiversity area, the separation of national and global benefits rarely makes sense.

Some examples may illustrate how it may seem that some benefits are "more global" than others:

- By establishing a wildlife reserve to conserve threatened or endemic species of trees or animals, a country may lose revenue from timber extraction, as well as contracting extra costs in terms of resettlement of the local populations originally living in the area. On the positive side, the country may gain revenue through increased tourism. The deficit in this budget (the difference between lost revenue from timber and gained revenue from tourism) constitutes the incremental costs, and will most likely be accepted as a global benefit component, for which compensation may be granted.
- Through continued use of a diversity of local varieties of rice plants, instead of the widespread introduction of uniform, high-yielding rice varieties, a country might experience an economic loss in terms of reduced yields. In this budget one must first subtract the lower costs of input factors in the first place: less costly seeds, less need for pesticides and fertilisers. The conservation

[149] GEF /C.10/9 October 3, 1997, Draft Annual Report of the GEF, FY97.

[150] Intergovernmental Committee on the Convention on Biological Diversity (ICCBD) in Geneva, October 1993.

component of the activity is the benefits derived from conserving a large variety of local cultivars in the field, securing breeding material for the future agricultural research and development: How much of this benefit can be counted as a national benefit, and what part of this is truly global?

As the debate proceeded, some of the G–77 governments began complaining that what were termed *global benefits* were in fact *Northern* benefits. Rather, they claim, global benefits in biodiversity projects should be defined as national benefits. This option would erase any semblance of conditionality – and is thus obviously out of the question for Northern governments. The central dilemma remains: How to define global benefits in biodiversity conservation in line with the precautionary principle, i.e. without risking a bias towards biological *hotspots*, at the expense of biodiversity of less immediate economic value? The CBD itself does not prescribe global benefits as a criterion; instead, funding is linked to fulfilment of the objectives of the Convention. What it does is to provide a basis for setting priorities in biodiversity conservation: including high diversity and high numbers of endemic and threatened species, as well as species or habitats of social, economic, cultural or scientific important. This comes closer than the notion of "global benefits" with regard to guaranteeing a diversity of projects in the GEF portfolio.

Currently, the bottom line in the GEF is to maintain the division between development aid and biodiversity conservation. This issue was much debated during an April 1997 intergovernmental meeting on scientific aspects of the CBD.[151] India, Brazil and Malaysia argued against scientific input providing a basis for setting priorities about conservation projects (primarily in terms of GEF financing). As the most resource-rich among the developing countries, they insist on having political control over GEF money in terms of development, rather than accepting what they claim to be Western scientific priorities. Furthermore, Malaysia regarded the rules concerning indigenous people as untoward meddling in domestic affairs, and denied the presence of such communities within its own borders. In the end, the parties to the Convention are expected to have the final say regarding programme priorities and eligibility criteria for GEF funding. This question is further debated at the annual Conferences of the Parties (COP).[152]

---

[151] The open-ended Intergovernmental meeting of Scientific Experts on Biological Diversity (ISE) in Mexico in April 1997.

[152] The third COP gave several recommendations concerning guidance to the criteria of the Enabling Activities, including:

(a) capacity-building to assist Parties in the implementation of Art. 6 (national plans) and 8 (benefit sharing with indigenous and local communities);

(b) supporting efforts for the conservation and sustainable use of biological diversity important to agriculture;

(c) capacity-building to implement measures on access to genetic resources; and

(d) an examination of support for capacity-building projects for indigenous and local communities embodying traditional lifestyles.

Source: GEF /C.10/7 October 3, 1997, Relations with Conventions.

6.4 Concluding Remarks

Here, I make a rather extensive summary in order to provide links to the next main part of the analysis. I will focus on the following two questions:

- What is the outlook for the implementation of the CBD – the principles, norms and regulations that were found largely to represent a breakthrough for the South?
- Are institutional factors likely to have an impact in the implementation phase?

6.4.1. Moral and material breakthrough?

The first question concerns the translation of norms and principles into practical politics. The answer to this question seems to be of a dual nature. On the one hand, DC representatives were reported as having summed up the Rio Conference as follows: "Climate Change was theirs – Biodiversity was ours". On the other hand, the CBD has been interpreted as a rather hollow victory for the DCs: "We won the moral debate, but we never got around to operationalising it".[153] This pessimistic outlook is primarily tied to the situation concerning plant genetic resources and the agricultural sector. It shows the frustrations experienced within the environment and development movement, in the face of the many difficulties connected to the concept of *equitable sharing* and the establishment of a workable Protocol for access to and control over plant genetic resources. Especially in the seeds issue, where exchange has had such a long and widespread history, the operationalisation of *country of origin*, *country providing* and *mutually agreed terms* are still a long way from finding a feasible solution in practical politics.

On the other hand, the principles of *equitable sharing of benefits* and *country of origin*, and the financial and technical mechanisms already set out for their implementation, seem to have a broader scope for application within the wildlife

---

COP4 provided additional guidance to GEF; deciding that it should support:

(a) "assessment of current legislative, administrative and policy measures on access to genetic resources and benefit-sharing.."

(b) "formulation of access and benefit-sharing mechanisms at the national, subregional and regional levels.."

(c) " capacity-building on measures on access to genetic resources and sharing of benefits, including capacity-building on economic valuation of genetic resources."

(d) ".. specific benefit-sharing initiatives such as support for entrepreneurial developments by local and indigenous communities,..".

Source: texts of the decisions adopted by the Fourth Meeting of the Conference of the Parties. http://www.biodiv.org./cop4/FinalRep-/13.htm.

[153] Personal communication from Rene Salazar, leader of SEARICE (an NGO working with in-situ conservation and use of plant genetic resources, in co-operation with small-scale farmers in Southeast Asia; also in collaboration with the Norwegian NGO: U-fondet), the Philippines, 3 April 1997.

sector.[154] This can be observed primarily in connection with biodiversity prospecting and screening for the utility of genes in new medicinals. In this sector, the CBD goes further than a moral victory, as formal regulations are being put into place among relevant actors in the gene-exchange business. The new policies on equitable sharing of the Royal Botanic Gardens at Kew and of Biotics are indications that these ideas are catching on. Perhaps paradoxically, it is the *private sector* which has taken the largest steps in this direction, while the public sectors, even in "like-minded countries", lag behind.

A crucial question is whether Northern *governments* will comply with the new regime regulating access to and exchange of genetic resources and technology. As far as mechanisms to achieve *equitable sharing of benefits* in relation to access to genetic resources are concerned, it is now up to national governments to enact appropriate legislation. Whether *equitable sharing of benefits* accruing from use of genetic resources will become a viable concept depends on whether the new dual property rights regime of the CBD will take hold. The amendments introduced by the European Parliament in the EU patent directive could have been a move in the same direction as that of Kew and Biotics. A central question is whether OECD countries will actually establish corresponding legislation. As yet, even the *bridge-builder* Norway has not made *equitable sharing* part of its national biodiversity plan, in terms of specific legal measures. The real test of Norway's pusher and bridge-builder role will come when the country must decide on whether or not to accept the EU patent directive on biotechnological products.[155]

As long as the developed countries have free access to germplasm through international gene banks, they have an incentive to remain *free-riders* in reaping the benefits from genetic resources. Tracing germplasm back to its origin is scarcely feasible in all cases, so what is needed is a general system of recording how the material was obtained. This could be developed in a protocol to the CBD. An important incentive for Northern governments to enact legislation in line with the principle of prior informed consent, however, would be for them to be considered *legitimate* purchasers of genetic resources. This is basically a question of the strength of the new Bioregime in establishing *common norms* which the actors in gene transactions can find acceptable and fair. In terms of the larger, related picture regarding prospects for a new economic world order, this idea might seem far-fetched, but it could have a viable chance within a limited issue-area. This aspect

---

[154] Basically, the new international regime for access to and exchange of genetic resources does not distinguish between wild and domesticated material, but we should not disregard the possibility that the two areas may still be influenced by different factors, or be influenced differently by the same factors. We return to this aspect in the case-study of Ethiopia (Chapters 7 to 9).

[155] Norway has supported the Netherlands in its action for the annulment of the patent directive (Directive 98/44/EU) in a Statement in Intervention Submitted by the Government of Norway in Case C-377/98; the Kingdom of the Netherlands against European Parliament and Coucil of the European Union.

represents a link to the study of national biodiversity policies and the analysis of CBD implementation in Ethiopia.

As demonstrated in Chapters 5 and 6, the normative input was not, however, restricted to the South and the environmental / development NGO community. The principles of *local capacity-building* and *indigenous people participation* became catchwords on the global agenda with the UNCED process, and also constitute integral parts of the CBD. These concepts may be more closely linked to the democratisation agenda of the West, and are likely to reappear in the guise of conditionality in e.g. GEF financed projects. As noted above, Malaysia was not enthusiastic about the reference to *indigenous people*. On the other hand, this is an unrefined judgement; I return to the application and use of these concepts in the next part of the analysis – the case-study of Ethiopia.

Biotechnology regulations, equitable sharing and criteria for financial transfers for conservation projects remain controversial issues. With regard to biotechnology regulations, a central question is whether the US government will keep blocking the development of a Biosafety Protocol under the CBD. If so, this will indicate that the USA is still able to influence the CBD negotiations, even though it is not a Party to the Convention. If so, this would represent an important support for H0.

Finally, the relationship between TRIPs and the CBD objectives is still a major issue on the COP agenda, indicating that the CBD Secretariat and the Parties do not intend to let this issue slip, nor leave it to a "business as usual" development. There are as yet no indications that TRIPs regulations will (be allowed to) take precedence over the objectives of the CBD. On the other hand, the TRIPs regulations are supported by a stronger institution compared to the biodiversity objectives. The WTO Millennium round that started in November 1999 is expected to provide answers to how to deal with divergencies between environmental and trade regimes.

6.4.2 Scope for institutional factors to strengthen implementation?

The second question remains the most elusive at present. Several factors associated with the *interest-* and *ideational-based* perspective were assumed to strengthen implementation. It is largely premature to judge their actual effects, but some comments can be offered.

Implementation seems to be spurred by increasing legitimacy of common norms.

In either case, moral or material victory, the objectives of the CBD represent more of a breakthrough than merely coinciding with DC interests. The second part of the analysis will focus on the implications of this breakthrough in terms of national implementation. Here let me offer some points on the possible effect of such breakthrough. As already mentioned, a breakthrough for DC national interests and ideas does not necessarily imply that an IEA will have an impact on behaviour among the successful negotiators. Looking for instance at the LRTAP (Convention on Long-Range Transboundary Air Pollution) regime, it is obvious that the pusher countries (among them Norway and Denmark) won through with their interests in the agreement. The regime did not, however, make a significant impact on air pollution control in these countries, *as they had already fulfilled the agreed standards unilaterally*. Similarly, it could be surmised that if the CBD merely codifies the kind of behaviour that DCs would have practised anyhow, there would be little background for assuming any major effect of the CBD on national biodiversity policy in the South.

On the other hand, the biodiversity issue, as seen from a developing country perspective, *does not provide a wide scope for unilateral action*. For those countries seeking co-operative solutions in this issue-area, unilateral action is not useful. For DCs in the biodiversity negotiations, it was clearly more important to promote behaviour change in the developed countries. It may also be conjectured that the DCs needed the CBD as a framework on which to attach the kind of policies which they deem necessary, such as stronger administrative control over access to and use of their own genetic resources. This kind of activity would clearly be useless without a *compatible legislation* in both owner and user countries. I shall return to this issue in the Ethiopian case-study.

The *ideational* perspective may shed light on the relationship between owners and users of genetic resources (the IPR/CBD relationship). The examples of the EU patent directive, Venezuela, Kew Gardens and biotechnology company Phytera all indicate that the CBD increasingly provides a "new normative framework" guiding exchange and transactions with genetic resources. Only a few years ago, the name of the game was still *common heritage of mankind* – or rather, free access to the genetic resources of the South. Today, a contract that does not include clauses about shared benefits with the country of origin is likely to be "exposed" – and the "culprit" is likely to admit embarrassment. This may be interpreted as showing a high degree of *legitimacy* concerning the objectives of the new bioregime – at least among actors in the private sector.

These examples also point to the importance of the national and international NGO community as watchdogs for upholding IEA objectives. The example of Phytera directs attention to the possibility that *informal monitoring* mechanisms may be as important as the formal ones, in a very similar manner: First, in terms of increased transparency and hence reduced distrust among the contracting parties, and second, by exposing non-compliance. As far as DC implementation is concerned, *developed country compliance* was presumed to be an important condition, and the effect of informal monitoring could thus prove to be of great importance.

Implementation seems to be spurred by monitoring and incentive mechanisms.

The effect of *formal monitoring* remains uncertain at present. The majority of Contracting Parties submitted their national reports at the 4th Meeting of the COP in April 1998, but it is still early to conclude whether this will work as a compliance mechanism or not. In preparation for this, the 175 Contracting Parties (countries that have ratified) are now in the process of developing national strategies, plans and programmes for conservation and sustainable use of biodiversity.

The picture is confused with regard to the development of the *financial (incentive) mechanism*, the GEF. The post-CBD situation is similar to the pre-CBD situation with regard to the conflict level between developed and developing countries. The GEF provides funding for development of such national plans in developing countries. Is it reasonable to interpret this (in realist terms) as "proof" that the GEF is being applied to promote the environmental agenda of the North? We return to this aspect and the role of the GEF in our discussion of biodiversity policies in Ethiopia.

The *funding* for conservation and compensation for use of genetic resources may lead to discord among developing countries, as they differ in terms of gene and species richness and also with regard to technological capacity. In the aftermath of UNCED many fear that the least developed, gene-poor countries will be left increasingly at the losing end – in two respects: limited funding for biodiversity conservation, and reduced access to improved breeding material. As the GEF evaluation has demonstrated, there is a tendency to disregard the potential importance of genetic resources in arid and semi-arid areas, and focus on tropical rainforests instead. The explanation is most likely found in the Western paradigm of wildlife conservation, which still links *global benefits* to biodiversity *hotspots*.

This situation underscores the elusive character of *negotiation power* compared to *basic game power*. For biodiversity conservation in a long-term perspective, this does not augur well for the implementation of the CBD objectives. Unlike the FAO debate, there are few rallying points available to the G–77 within this issue-area, in the form of NGOs. As NGO representation gained legitimacy and increased during and subsequent to UNCED, so has the plurality of their voices, and thereby their difficulties in rallying round common goals. The schism between wildlife conservation and agricultural biodiversity (such as on-farm conservation and use of traditional food plants) still constitutes a hurdle here. If environmental and developmental NGOs are to have a positive impact on biodiversity conservation in a long-term perspective, the old dispute between wildlife conservation and agricultural biodiversity will have to be laid to rest. In Chapters 8 and 9 we return to the role that these actors may have had in creating links to the national level.

Discontent with the *financial* aspects of the CBD is not restricted to gene-poor countries. Biologically rich countries such as Brazil and Malaysia are expected to become the greatest beneficiaries from the North's focus on global environmental issues, but they have no intention of being told what to do with this money. The sovereignty issue looms large in this part of the follow-up of the CBD, as the North is unlikely to accept financial obligations without conditionality. Donor governments cannot be expected to abandon their demand for scientific criteria as a basis for setting priorities in biodiversity conservation. A pertinent lesson from the forestry sector seems to be that the temperate countries of the North must commit themselves to the same standards of conservation and sustainable utilisation as they request from the tropical countries, in order to gain legitimacy and credibility for their argumentation. In this way the conditionality/sovereignty trap may be avoided.

Applying a long-term perspective to biodiversity conservation, a well functioning system for *equitable sharing and payment mechanisms* for use of genetic resources is probably more important than direct funding of piecemeal wildlife conservation measures. On a similar note, it may be more important to integrate biodiversity perspectives into the large social sectors of agriculture, forestry and fisheries, than to establish small islands of preservation areas. Biodiversity conservation is essentially about the ability to engage in a wide range of activities at the same time. Considerable scientific uncertainty remains regarding the relationship between species and ecosystems, and about what happens when small or larger parts of diversity disappear. Some priorities will certainly have to be made; but with the present scientific uncertainty, it may prove just as dangerous to neglect, say, plant genetic resources in Mali, as animal species of the Amazon jungle. Additional resources resulting from UNCED as well as conventional development aid need to be strengthened to integrate sustainable use of genetic resources. This may include the use of traditional plants in agriculture, as well as the development of incentives for integrated methods for agriforestry and aquaculture.

In this perspective, *GEF funding* is indeed necessary, but far from sufficient. Even though resources for overall conservation are strained, the GEF should still be made available also for projects outside the traditional preservation portfolio. The GEF principles of incremental costs and cost efficiency must be treated with utmost care in relation to biodiversity projects. Moreover, the very nature of the biodiversity complex makes it natural to concentrate on small-scale projects rather than big, top-down ones. Likewise, the GEF should fulfil its obligation to fund projects for capacity-building. The overall benefits from helping developing countries to secure revenue from utilisation of genetic resources may prove to be the best investment for ensuring implementation of the CBD objectives in the South.

Finally, it is important to recognise the two-level nature of the distributive question: *international* distribution of benefits from genetic resources, and *domestic* distribution. The CBD encourages action on both levels, but the latter is essentially a matter of national politics and legislation. Within the international institution there are no enforcement mechanisms to reach this objective. Furthermore, it is doubtful whether national governments will perceive such legislation to be in their direct interest.[156] If national legislation does make provisions for local communities of farmers and indigenous people, this could hence indicate acceptance of the norms developed within the new biodiversity regime.

---

[156] For instance, throughout the international negotiation process, Malaysia maintained that there were no indigenous communities in the country. Moreover, the developing country parties to COP4 are reported to have made little effort to respond to the major theme of the Conference: a discussion of the significance of traditional knowledge systems and their role in biodiversity conservation and use (Sumita Dasgupta, "Knowledge Seeks Power", *Down to Earth*, 15 May 1998, pp. 22–23).

Turning to the Ethiopian case-study

In the case of a developing country like *Ethiopia,* the interest-based model is partly concerned with the national will to implement the objectives of the Convention, and partly with whether incentives emanating from the Bioregime may increase Ethiopia's ability to do so. We may approach the first part by asking: How is the CBD perceived from an "Ethiopian" point of view – do they perceive it as a victory for their main interests in the negotiation phase, and do they view the CBD as a legitimate instrument for improving conditions in the issue-area? If Ethiopia succeeded to a large extent in getting its interests through in the international agreement, then it will presumably be more inclined to implement. The second part concerns how Ethiopia perceives the trends in the international community with regard to complying with the new bioregime regulating access to and exchange of genetic resources and technology. If other parties, notably the North, fail to comply with these obligations under the agreement, it would hardly seem rational for Ethiopia to do so.

Both of these dimensions necessitate an outline of what may be anticipated to be the national interest of Ethiopia in the issue-area. The background for posing this question is the belief that an enhanced conservation and utilisation of biological diversity in a developing country like Ethiopia, will very much depend on whether these activities can be supported by external factors, e.g. by improved compensation for external use of Ethiopia's genetic resources and by financial mechanisms. Two factors are assumed to be particularly important in this respect: internal acceptance of output, and external compliance with output. In light of the limited capacity associated with DCs, incentives are assumed to be more important than sanctions in influencing actor behaviour in the implementation phase.

This brings us to the next main section of the thesis, Chapters 7 to 9. In Chapter 7 an analytical framework for studying environmental policies in DCs is developed; in Chapter 8 I present the empirical data underlying the analysis. In Chapter 9, the national and international links are examined.

PART THREE:
IMPLEMENTATION OF THE CBD IN ETHIOPIA

Lines by a Humanitarian

Be lenient with lobsters, and ever kind to crabs,

And be not disrespectful to cuttle-fish or dabs;

Chase not the Cochin-China, chaff not the ox obese,

And babble not of feather-beds in company with geese.

Be tender with the tadpole, and let the limpet thrive,

Be merciful to mussels, don't skin your eels alive;

When talking to a turtle don't mention calipee –

Be always kind to animals wherever you may be

ANON.

CHAPTER 7

A DOMESTIC IMPLEMENTATION MODEL

7.1 Introduction to Analytical Framework

The broad topic in this part of the analysis concerns the second part of the question to be dealt with in this thesis:

> *How have international institutional factors had an impact on domestic biodiversity policies in a developing country? Are the biodiversity policies of Ethiopia* consistent *with the CBD objectives, and is this consistency a result of deliberate efforts to* implement *the CBD?*

In the first main part of the analysis (Chapter 3) I formulated several propositions concerning how a breakthrough for DC interests would be possible. Evidence was found that strengthened the proposition that institutional factors have facilitated the development of agreement between developed and developing country interests, and also that learning from repeated interaction may have enhanced the scope for including the norms and ideas of developing countries in the CBD.

My main hypothesis in this second part of the analysis is as follows:

> *H1: International institutional factors have facilitated and enhanced implementation of the CBD objectives in Ethiopia.*

If this is shown to be so, I also want to study how – through what causal mechanisms – such influence has come about. Recalling the discussion on implementation in sections 1.2.2 and 1.2.3, I shall primarily focus on implementation output: domestic legislation and policy formulations directed at following commitments under the CBD. The potential for analysing implementation outcome (enforcement of domestic output, leading to behaviour change in target groups), was assumed to be small so shortly after the establishment of the CBD, although some indications may be found. According to H0, biodiversity policies in Ethiopia would *not* have been different in the absence of the international environmental agreement.

Two operations are essential in order to examine whether H1 can be strengthened and H0 refuted:

1. First, domestic factors must be controlled for. In section 1.2. I argued that domestic *compliance* to international environmental agreements may come about "by default". This could for instance be due to economic recession or growth, or

technological developments, leading to consistent domestic policies even in the absence of deliberate *implementation* activities. If, when domestic factors are controlled for, it emerges that relevant and significant changes in some of these factors coincided with compliance, this would seem to strengthen H0, and vice versa. Such a finding would not exclude H1 altogether, as I do not subscribe to one-factor explanations; but it would make it harder to establish a credible link. It must further be noted that international regimes may have a direct effect on *compliance*, through establishing incentive mechanisms for economic and technological transfers to DCs. In that case, we could still speak of regime effects, although these would have less direct links to domestic factors through *implementation* activities. This leads to the second operation:

2. Second, to establish a causal relationship between international institutional factors and domestic implementation (regime effects), it must be possible to identify *causal mechanisms* linking factors at the international and national levels.

Hence, in order to control for and identify links to domestic factors, I need to clarify *which* are these domestic factors (relevance) and *how* they can be expected to work in relation to compliance or implementation (significance). This is the subject of this chapter.

In Chapter 3, on the international factors, I started out with an overview of the broad theoretical approaches to international relations, and then ordered these into three main perspectives with different propositions as to whether and how institutions matter. In this chapter on domestic factors I intend to proceed in a similar manner:

First, I briefly discuss the rationale for applying a domestic approach to implementation of IEAs, specifically in developing countries. This leads to a discussion of the applicability of current implementation research, which has been alleged to be primarily Western in its approach, in a developing country context. As an extension of this discussion, I shall also present the more general development theories in brief – partly in order to relate my study to this theoretical domain, and partly to consider the utility of these theories with regard to my own project.

Second follows an overview of the implementation literature.[157] On the basis of this exploration, I go on to present three dimensions of factors believed to be central in boosting environmental policies: state capacity, non-governmental actors, and state-society relationship.

Finally I discuss the causal mechanisms linking factors at the national and international levels, seeking to draw up some general propositions.

[157] It should be noted that *implementation* here is more generally defined, compared to my specific use of the concept when comparing national policy to international commitments. See section 1.2.2 and 1.2.3.

7.2 Applicability of a Domestic Implementation Model in DCs

Do we need a domestic model for analysing implementation in less developed countries? To answer this question we must first ask:

1. What is the general rationale for and what are the implications of introducing a domestic model for analysing implementation of international environmental agreements?
2. How may the rationale and the implications express themselves in the case of DCs? What advice can development theory provide?
3. Is it fruitful to apply the same model in North and South alike?

7.2.1. Rationale for and implications of introducing a domestic model

Basically, there might be two types of *rationales* for utilising a domestic model in analysing implementation: to achieve improved explanations and conclusions, or to acquire improved data leading to a better understanding of important mechanisms. For the practical application of lessons learned from implementation studies, a better understanding of causal mechanisms may be equally important as a model's explanatory force.

The *first implication* of introducing a domestic model is departure from the idea of states as unitary, rational actors (URA model). The model implies that we look at governments as complex organisations that do not have full control over "their" societies. The *second implication* is that we move from structural-level explanations (associated with the international distribution of power and national interests) to actor-level explanations.

These implications are commonly put forth as an argument for the parsimony of URA models in the study of international relations. Specification of policies and actors at the sub-systemic level requires a great amount of information, as compared to explanations based on the constraints imposed on states by the international system. Secondly, it is argued that sub-systemic factors pose greater difficulties in terms of generalisation, as these factors are more likely to be idiosyncratic (Keohane, 1984; Zürn, 1993:286). The first argument has its main relevance in the regime formation process, while it is regarded as less problematic to introduce domestic factors in the implementation phase (Zürn, 1993). This builds on the assumption that policies tend to become more specified as they move from problem identification to the stage of implementation. Sub-systemic actors are more likely to become affected, and mobilised, in the implementation phase (Underdal, 1998a). Still, it may be prudent to examine how to approach domestic factors in the regime formation phase, as this may provide advice for the implementation phase as well.

The idea that domestic factors are absent from the foreign policy phase (of regime formation) and only enter the domestic policy phase (of implementation) is, moreover, largely an artificial distinction. Michael Zürn's (1993) advice for including domestic factors also in the study of the regime-formation process, is to limit the focus to states which have professed either regime-conductive or regime-preventive activities (rather than endeavouring to examine the domestic factors of all participant states). Likewise, it may be more pertinent to focus on the activities of *pusher* and *laggard* countries in the implementation phase as well.

The idea of *pushers* (regime-conductive actors) and *laggards* (regime-preventing actors) brings us to the *third implication* of, and a central rationale for, introducing a domestic model in the study of implementation. A rationale for including domestic factors is that the main focus is moved from studying the state's *willingness* to implement, to its actual *ability* to implement (Underdal, 1998). The question of ability is assumed to be especially important for implementation in developing countries in general. This would seem to be the case not least with regard to the implementation of international environmental agreements.

7.2.2. Rationale and implications in the case of DCs

A crucial issue is whether DC ability is determined mainly by external factors, rendering a domestic model redundant. As pointed out by Adil Najam (1994), DCs are seldom among the pushers in international environmental negotiations, and the treaty texts frequently "excuse" them from the more onerous burdens of implementation. This would seem to indicate a general lack of external pressure towards DC implementation, as other parties would not be sensitive to their activities. Biodiversity is, however, a deviant case in this respect, as it is precisely in the tropical countries that the most critical conservation challenges are found. This would seem to bolster the importance of *implementation ability* in DCs, although there are clearly great differences among developing countries with regard to the urgency accorded to particular conservation measures by the international society – consider, for example, the contrast between the rainforests of Brazil and the arid lands of Mali.

At the same time, recent global environmental treaties (ozone, climate change, biodiversity) tend to include *incentives* in terms of financial or technological transfers linked to DC implementation. This could lead to the assumption that the domestic *ability* to implement IEAs in DCs largely depends on other parties' fulfilling their commitments (such as meeting their financial obligations) under the treaty. Rather than constituting a reason for not looking at domestic factors, I will argue that this assumption opens for an interesting line of investigation in the study of domestic policies in DCs: Is the assumption valid?

Moreover, if we admit the possibility of variance in the implementation score both between DCs concerning the same IEA and within DCs regarding different IEAs, and if we want to study effects of international institutional factors in the domestic implementation process, this means that we need to check for effects from domestic factors as well. Finally, as section 1.2.3 suggested (indicating also variance between the different sectors of biodiversity conservation), the need for more complex models for analysing the dependent variable becomes apparent.

Advice from Development Theory

In relating my investigation to the more general *theories on development,* I will argue that these broad frameworks are generally concerned with projects and questions which differ substantially from my line of study. Secondly, to the extent that the frameworks do give rise to relevant implications, these are not substantially different from those provided by more general theories of international relations.

We must first ask: within the large body of development theory, what contributions may shed light on the implementation of international environmental agreements (IEA) in DCs? An obvious starting point is the traditional or *classical modernisation theories*, which dominated the field of development sociology until the 1950s. These schools have their roots in, for example, Tönnies' dichotomy of *Gemeinschaft and Gesellschaft*, and Durkheim's dichotomy of *organic and mechanical solidarity*. Modernisation theory would use such ideal models, e.g. specifying a low and a high degree of division of labour in society, on the general assumption that development proceeds from low to high. Fred Riggs, in his *Theory of Prismatic Society*, located developing countries somewhere midway between the two extremes, embodying both *modern* and *traditional* characteristics (Riggs, 1964).

Partly as a result of decolonisation, modernisation theories have been criticised for being deterministic, simplistic, ethno-centristic, and for giving rise to functionalistic tautologies (Østerud, 1978). Development was explicitly or implicitly defined as a change towards closer and closer correspondence with the Western industrialised state, with economic growth as the ultimate aim. One reason why I have chosen not to pursue the wide range of development theories further, is thus the general difference between their "projects" and mine. The aim in this study is to examine the implementation of a specific IEA – not to explain or analyse a modernisation process, whether in terms of political or economic development.[158]

Inherent political and cultural aspects within the developing countries themselves were rarely deemed important in early modernisation theories. Thus, hardly anyone questioned drawing heavily on Western expertise (for example World Bank personnel) in designing development strategies (Martinussen, 1994). To some extent we may thus identify a parallel between the thinking associated with modernisation theories and that of the liberal idealist approach to international relations: the world-wide adoption of Western institutions and ideals would ultimately lead to progress and order. The practical ramifications of these and later neo-liberal theories in terms of multilateral development assistance have been to downplay the role of the state and rather focus on deregulation and on private initiative – as articulated, for instance, in the 1980s in the Structural Adjustment Programmes of the World Bank and IMF (Leys, 1996). Again, we may discern a parallel in international relations literature and in studies focused on the developed world. This era was also a period of looking at deregulation and the role of non-state, transnational actors. With a

[158] It could, of course, be argued that my thesis question is somehow embedded in the larger questions regarding political and economic developments. I shall still argue that that line of investigation would be confusing rather than revealing, as well as largely outdated.

view to my research question, a relevant implication from this theoretical tradition would seem to be:

- "As policies in DCs to a greater extent are formulated in a 'vacuum', separately from societal impact, there is less need to introduce a general domestic model to study policy formulation (implementation output) here than in developed countries."

Alternatively, an opposing view with a similar conclusion could be anticipated:

- "As DC governments are assumed to have less control over their societies than governments in developed countries, there is a greater need to develop idiosyncratic domestic models to study implementation (outcome) here."

Correspondingly, many policy studies in developing countries have been cast in terms of idiosyncratic case-studies.

Another possible source of theory input could be gathered from the *dependencia school*, represented largely by Samir Amin (1976) and André Gunder Frank (1967), and expressed for example in the concept of *unequal exchange*. Where modernisation theories generally predict progress and development, *dependencia* theories give rise to largely deterministic assumptions of backwardness and underdevelopment. Modernisation theories have generally downplayed the role of conflict for understanding social and political change in the Third World, whereas their main opponents, Marxist-based theories, generally miss conflicts other than the class-based ones.[159] These classes of theories are consistent with *structural theories* in international relations in their emphasis on the state as the main actor. Their impact in practical terms should not be disregarded, as much of the impetus behind G–77 demands for a New International Economic Order may be traced back to this line of thought.

Like other structural theories, the *dependencia* school has been criticised for largely ignoring domestic factors (Østerud, 1978). It places DCs within the international structural power relations, predicting outcomes largely in the same direction as that of Realist perspectives. Several scholars still argue that complex domestic models are less useful in analysing the foreign policy of developing countries.[160] It may be argued that domestic policies in DCs are dominated by external actors; hence, domestic factors are of scant importance in the implementation phase as well. The implication of this line of thinking is that the environmental agenda in developing countries is seen as externally driven – a product of international structural power relations. If such is the case, the *realist regime model* will suffice to account for compliance with IEAs in DCs. A *dependencia*-school proposition would thus seem to be:

---

[159] For a discussion of this, see Joel Migdal (1988: 31).

[160] See Arild Underdal (1984).

- "As DCs are assumed to be highly subject to influence by external factors, the focus on structural, international level factors is sufficient to explain both domestic output and outcome."

This approach makes no distinction, however, between regime factors and other external factors. Hence, despite the possible links to my project, it does not seem to add any new perspective outside what may already be found in general international relations theory.

Summing up, there are two main reasons for my choice of not pursuing these development theories further within the scope of this thesis:

- Partly, their "projects" differ from mine in focusing on different research questions. Basically, these approaches do not seem useful in trying to single out and assess the effects of international institutions.
- To the extent that they do predict implications of relevance to my research question, these implications do not differ significantly from those already predicted by other, more general, theories in international relations.[161]

All the same, the early efforts to construct "Grand Theories" make an important contribution also concerning my line of investigation. Do we need a common framework for analysing public policy "throughout the world"? Indeed, it is possible to construct such a framework? Does it make sense to apply the same concepts and measure effects by the same standards? These issues are the subject of the next section.

7.2.3. Common framework for implementation studies in North and South

What are the main obstacles and the main opposing arguments to applying a common framework for studying policy formation and implementation in the North and the South alike? Robert Fried (1990) points to three practical obstacles to establishing such a "Grand Theory". The first two are aimed against comparative implementation studies in general, and the last is against comparison between North and South specifically:

- First, there is the nature of public administration which Fried characterises as an *evasive prey*. There is usually a high, but still differentiated, reluctance to transparency, making access to comparative data difficult.

[161] Again, I must stress that this is obviously not to say that I deem either of the development theories as inherently fruitless or inapplicable for the research focus in this study. Rather I am arguing that as long as I can draw relevant implications from more universal theories, which are not refuted or greatly added to by development theory, I shall refrain from confusing the larger model by a multitude of partly overlapping hypothesis. Parsimonity advises us to choose the more general approach.

- Second, Fried notes the need for multi-disciplinarity in the study of public administration, which also causes obstacles to comparative and cumulative research.
- Third, Fried claims that public research funding institutions (national research councils) have tended to prefer studies of administration in developing countries, rather than focusing on their own administrative institutions. On this point the literature disagrees, as other authors point to the prevalence of Northern implementation studies.

The general reluctance towards applying a common framework in North and South may rest on the dogma that "our bureaucracies and institutions are better able (i.e. more rational/ competent/ efficient/ effective/ legalistic/ honest/ less corrupt) than theirs (which are thereby viewed as irrational/ less skilled/ inefficient/ ineffective/ clientelistic/ corrupt)". In Fried's view the development administration approach to comparative public administration thus tends to "exaggerate the irrelevance of Third World studies to the understanding of the advanced world, and downplays the potentialities for irrational-bureaucratic performance in the advanced world" (Fried, 1990:328–29). I subscribe to this view, and will argue that while there are obviously substantial differences in capacity both within and between bureaucracies in developed and developing countries, as well as between different sectors of the administration, this does not necessarily mean that the similarities are too few to permit fruitful use of the same concepts. The argument concerning ethnocentricity may hence also be less relevant, as to a great extent, "the damage is already done"; DCs have largely adopted Western bureaucratic ideals by now.

7.3 Theoretical Sources – the Growth of Implementation Studies

> "Instead of a trap door, what about a trap window? The guy looks out it, and if he leans too far, he falls out. Wait. I guess that's like a regular window."
>
> – more deep thoughts, ANON

The question of particular interest in the following two sections is:

How can we make relevant choices within the many-faceted implementation literature for the study of IEA implementation in DCs?

Although implementation research may trace its sources back to the classical works of Weber and Taylor, the main growth of implementation theory did not commence until the early 1970s. The first stage (approx. 1970–75) was characterised by primarily a-theoretical single case-studies,[162] followed by the debate (about 1975–85) between the two major schools generally referred to as *top–down* and *bottom–up*. Top–down scholars (e.g., van Meter & van Horn, 1975; Mazmanian & Sabatier, 1981; 1983) focused on the central decision, using goal achievement as the main evaluation criteria for implementation success. The top–down models identified from four (Edwards, 1980) to seventeen (Sabatier & Mazmanian, 1980) independent factors held to affect implementation. This focus on the central decision-makers brought heavy criticism from the bottom–up school of thought, who pointed to the important role of local and *street-level bureaucrats* in forming policy outcomes (Elmore, 1978; 1979; Hjern & Porter, 1981). Along with this criticism of the top–down approach came a rejection of the *stages heuristic*, the conceptualisation of the processes of policy formation and implementation in terms of ideal-type phases (Sabatier & Jenkins-Smith, 1993).

The bottom–uppers on the other hand, have been criticised for mixing empirical descriptions with normative statements. This is an argument which in my view could be turned the other way around as well. The notion that street-level bureaucrats have an impact on goal definition may thus either be conceived of as an implementation problem (top–down: "distorting the deliberate efforts by central authorities to follow up their international commitments") or as a virtue in itself (bottom–up: "as this may result in improved problem-solving"). Related to this is the question of identifying the most "relevant actors". "Relevance" may involve the aspect of power and influence over the implementation process, as well as the aspect of being affected by the process. These aspects are not necessarily overlapping, and the actors involved may change during the process (output/outcome/impact).

Basically, this brings us back to the discussion of whether to focus on implementation output and outcome or problem-solving (section 1.2). Clearly, as discussed in section 2.2.3 (ideational perspective) one may question the theory on which a central decision is build in order to solve the problem at hand: There is a possibility of a mismatch between what the elite (learn to) perceive as environmental problems and what local people perceive as environmental problems. If this is the case, it is possible to have perfect compliance and implementation of the international environmental obligations and still be far removed from actual problem-solving.

Incidentally, this also points to the usefulness of a *stages heuristic* conceptualisation, as dividing the implementation process into separate phases allows for an analytical separation between output and outcome on the one hand and impact (problem-solving) on the other. There has been a long discussion among regime scholars concerning the possibility of measuring effectiveness of IEAs in terms of real problem-solving. Various proposals have been set forth – including relying on expert advice – but a frequent conclusion is that impact assessment of environmental problem-solving remains outside the field of social scientists alone.

---

[162] See for instance Pressman & Wildavsky (1973).

Also, it necessitates a much wider time-frame. Hence, the stages heuristic can help us to demarcate our field of investigation.

Moreover, the stages heuristic helps us clarify the distinction between those implementation activities relating to legislation and policy formulation designed to conform with international commitments (output) on the one side, and on the other, those relating to efforts to bring about behaviour change in the target groups for this policy (outcome). This is in my view the main purpose of making these distinctions; contrary to the critique raised against it, the stages heuristic does not imply one-way causality. Furthermore, I would argue that the choice of studying the impact of international institutions on domestic policy-making, implies the choice of a predominantly top–down approach. This does not mean that studying interaction among local participants in reaction to a central directive could not have been interesting, but it would have been the answer to a different research question. On the other hand, the focus on such a relatively recent topic as DC implementation requires a cautious and flexible approach, in which lessons from the bottom–up tradition may prove useful.

In what may be termed a third generation of implementation literature, there is now movement towards a synthesis between the top–down and bottom–up perspectives (see Elmore, 1985; Sabatier, 1986). So far, however, this does not seem to have led to a greater consensus on which are the crucial factors in determining implementation success or failure.

A related but not overlapping dichotomy observable within implementation literature is the divide between primarily *state-centred* and *society-centred* approaches. The former view factors related to the state apparatus as the major dimension by which to analyse implementation success or failure, whereas society-centred studies regard sub-domestic groups as the most important obstacles or pushers for policy formulation and enforcement. The difference in focus may often be ascribed to methodological choices, as a strategy of elite interviews will probably result in a state-centred approach, while extended field studies may give rise to a society-centred approach. In constructing an analytical framework, I shall endeavour to include both dimensions. For methodological reasons, such as access to documentation, it may be easier to obtain state-centred information, as I rely heavily on elite interviews. On the other hand, I also draw on several field studies from the Ethiopian society, and this should mitigate the state-centred bias. Finally, the focus on implementation output (policy formulations) will also tend to bias the analytical framework towards a state-centred approach.

Apart from the often "strawman-like" debate between top–down and bottom–up advocates, implementation research is generally held to be bogged down by conceptual confusion, hampering the accumulation of research findings. Using a rather far-fetched analogy to the field of biology, it nevertheless seems striking that biologists estimate the existence of between 5 to 100 million species, and have described 1.9 million of these scientifically – and are *still* to able to make scientific progress in their field; while students of public policy invest a major share of their academic effort in debating over different concepts that often describe the same phenomenon. The fragmented field of implementation literature thus abounds with different ways of ordering and labelling factors which may be relevant in explaining implementation.[163]

From this prolific field we now need to make pragmatic, rational choices for analysing the case in point. I believe it is possible to pin down a limited number of essential dimensions and factors, even though they may be presented by a variety of concepts. A "trap window" may seem like a good concept for describing and analysing a specific situation or phenomenon, but closer scrutiny may sometimes be warranted.

7.4 Implementation and Domestic Factors

On this basis, I chose to draw on the work of three scholars:

- First, I chose the much-quoted work of Merilee Grindle (1980), as her focus is on DC implementation, and as she draws on top–down literature.
- Second, there is the recent work of Adil Najam (1995), with its focus on DC implementation of international environmental agreements (IEAs). He presents an overview of the implementation literature and proceeds to combine lessons from top–down and bottom–up contributions.

The combined contributions of these two scholars should increase confidence that major factors have not been left out here. Moreover, their contributions show that when we synthesise contributions from more or less idiosyncratic case-studies, both from developing and from developed countries, we end up with a rather limited and similar set of factors.

- Third, I have found valuable insights in the work of Martin Jänicke (1995) who offers an analytical framework for the concept of *capacity for environmental policy and management*, assumed to be important in developed and developing countries alike. Interestingly, Jänicke starts out from a developing country perspective, drawing lessons for studying environmental policies in developed countries. This approach should help to counter the risk of methodological problems related to transferring analytical concepts from North to South.

These combined contributions thus seem to offer useful introductions and insights for the purposes of this study. I will need to focus mainly on those factors which can be assumed to have an effect on policy formulation (*domestic implementation output*), and tone down those factors primarily connected with the enforcement phase (*domestic implementation outcome*). On the basis of their contributions I shall identify three dimensions of variables generally agreed to be important for the formulation (and enforcement) of environmental policy:

---

[163] For an overview; see Najam (1995).

- State capacity
- Non-governmental environmental proponents and opponents: within this dimension I shall also discuss policy content.
- State–society relationship.

7.4.1 A Top–down model for DC implementation: Grindle's model of content and context

Building on implementation models of van Meter and van Horn (1975) and Pressman and Wildavsky (1973), among others, Grindle draws a distinction between factors related to *policy content* and *policy context.*

Policy content includes:

- the type of benefits associated with the policy: collective benefits are more easily implemented, while distributive costs and benefits may lead to competition among the affected parties;
- the type of change expected: long-range objectives are harder to implement, as benefits are less apparent to target groups;
- the site of implementation: a limited number of decision units will facilitate implementation, while widely dispersed units make it harder;
- clarity of policy goals is presumably better than ambiguous goal statements;
- the capacity and expertise of the programme implementers are important;
- the more resources committed to follow up the policy, the higher the likelihood of successful implementation.

To my mind, the main drift in Grindle's four first assumptions is captured in Underdal's (1998a) concept of *vertical disintegration*, indicating how target groups are affected by policy content, once these policies become specified or "translated" into policy programmes, and the scope for domestic resistance when policies are found to carry explicit costs to specific sub-national target groups.[164] These factors are hence mostly concerned with the implementation *outcome* phase. The final two factors seem to belong within the state capacity dimension, and may thus be more relevant with regard to the policy formulation (*output*) phase. As the last assumption in addition may be related to the crucial issue of external impact on DC domestic policies, I shall return to this in a later section.

---

[164] In his article, "Explaining Compliance and Defection: Three models", Underdal's Domestic Model B is mainly concerned with *ability* to implement, while his URA Model A is more concerned with *willingness*: cost/benefit calculations (Underdal, 1998a).

Grindle goes on to discuss the policy context*, which is defined by:*

- The *capacity* of the political regime and the administrative organisation seeking to enforce their policies.
- The calculation of interests and power capabilities between the actors involved. This seems to relate to questions regarding society's demand for environmental policies and government supply of same; and also the role of *public pressure groups*.
- The balance between compliance and enforcement versus responsiveness and flexibility. This would seem to relate to the *state-society relationship*.

These factors are clearly important in the policy formulation phase (*output*).

7.4.2 Combining developing and developed country implementation: Jänicke's "Political System Capacity for Environmental Policy"

Martin Jänicke (1995) identifies two broad categories of actors which may be regarded as proponents of environmental policies:

- government institutions for environmental protection,
- non-governmental proponents (environmental organisations, the media and ecologically modernising firms).

A potential weakness in Jänicke's model is the focus on environmental proponents, which may tend to downplay the effect of opponents and conflict in the formulation and implementation of environmental policies. In constructing an analytical framework, I shall stick to the notion of societal groups, which may either assist or resist the formulation of particular policies.

Jänicke goes on to describe three main *capacity* dimensions which are relevant conditions for the state's capacity for environmental action:

- *Cognitive-informational framework*: environmental knowledge and public awareness
- *Political-institutional framework*: capacity for participative, integrative and strategic action
- *Economic-technological framework*: per capita GDP and R&D expenditures for environmental purposes.

Using a number of indicators for each of these dimension, Jänicke provides a further operationalisation of *capacity*, which is clearly a central concept for analysing environmental policy and management.[165] To his model Jänicke also adds

---

[165] Jänicke also mentions external shocks and the character of the problem as important factors of environmental change. As these factors are treated elsewhere I shall not repeat them here.

an element of content, which he terms *character of problem*, as well as the possible impact of external factors such as sudden crisis or shocks (*situative context*). The use of these concepts (as well as his references to the work of Haas, Jacobsen/Brown Weiss, and Böhmer-Christiansen) indicates how Jänicke draws on international environmental literature in constructing his model.

7.4.3 Combining top–down and bottom–up lessons: Najam's 5 Cs

Adil Najam (1995) finds a basis for a synthesis of factors from top–down and bottom–up perspectives in his five Cs:

- *Content:* of policy itself, what it sets out to do (goals), how it problematises the issue, how it aims to solve perceived problems (methods).
- *Context*: the institutional context (often structured as operating procedures) through which policy must travel.
- *Commitment:* of those entrusted with carrying out the implementation at various levels to the goals and methods of the policy.
- *Capacity:* administrative capacity of implementers to carry out the changes desired.
- *Clients and Coalitions:* support of those whose interests are enhanced or threatened by the policy, and the strategies they employ in strengthening or deflecting its implementation.

Najam's Capacity and Context may, following Jänicke's model, be subsumed under state capacity; while Commitment seem closely related to both state capacity and what I chose to call the *state-society relationship*. Clients and Coalitions clearly equals non-governmental environmental proponents and opponents. These factors may all be important for policy formulation (*output*) and enforcement, while the impact of Content is mainly related to the *outcome* phase.

On this basis, I shall examine more closely the three dimensions which seem to account for the major independent factors and which seem relevant in terms of IEA implementation in developing countries. While this is not a theme with the above contributors, I shall also initiate the discussion of the relationship between factors at the international and national level:

7.4.4 A synthesis in three main dimensions

1. State Capacity

The *political and administrative capacity* of a government to formulate and enforce policies will probably depend on institutional resources, financial (monetary) resources and human resources.

* *Institutional resources* can be identified as the general efficiency of the bureaucracy, – as expressed in Graham Allison's (1971) concept of *standard operating procedures (SOP)*. The presence or absence of SOPs in a policy area will indicate whether the country has traditionally performed activities in the issue-area or not. We may expect the administrative capacity to formulate and implement government policies to be higher in countries or sectors in which there are established SOPs. SOPs allow for co-ordination among individuals in large bureaucracies by providing rules according to which things are done. SOPs are assumed to influence how new (international) obligations are received into the domestic sphere. Western bureaucracies are believed to have an added efficiency because they have such standard procedures of handling new policy measures, procedures that channel them into a well functioning system. DCs may be assumed to have less stringent SOPs, making implementation less predictable, and probably less efficient.

On the other hand, the lack of rigorous SOPs may enable greater flexibility in the implementation output phase, since one drawback of SOPs is that they may be ill suited to handle *non*-standard situations. Lack of SOPs may furthermore give room for increased influence by individual efforts to promote certain policies.[166] Hence, there may be a wider scope for coalitions between international epistemic communities and domestic interest groups when SOPs are weak or lacking in an issue-area. At least in the national output phase of formulating legislation, this may be assumed to boost implementation activities. In the outcome phase, however, it may mean that the central authority may lack the necessary means to enforce its regulations.

* *Human resources* can be expressed in terms of environmental *expertise* and *commitment*. Expertise indicates the cognitive capacity to deal with environmental problems and commitment concerns the willingness to pursue such policies. This factor may be affected by an international institution through learning processes, for instance by participation in transnational epistemic communities developing within the regime.[167]

* The *monetary state capacity* in a DC may most directly be influenced by external factors: Awareness raised by international environmental agreements may help to increase this aspect of capacity by providing technological and financial transfers. A crude assumption would be that the more help given by the institution to improve domestic capacity, the higher the likelihood of achieving a domestic output in compliance with the international agreement. The flip-side of this coin is related

---

166 This view seems to be supported by Putnam and Henning (1989:104–14) who argue that dispersed national decision-making is important in enhancing international co-operation, as it allows for transgovernmental coalitions.

167 When we define human resources as part of state capacity, this indicates that capacity concerns both ability and will. This definition may be ambiguous in light of the misgivings from the South about complying to "the Northern environmental agenda", which touches on the problematic relationship between power and knowledge.

to conditionality and sovereignty, and the possible mismatch between external and domestic problem perception. We must distinguish between material mechanisms working as an incentive, trying to boost the will to implement, and mechanisms working as an enabling factor, when ability is lacking. The latter can be expected to have more effect than the former, provided there is a high degree of willingness.

Indicators:

Within the dimension of state capacity, we will examine the seven indications set out below, which are based largely on Jänicke's work. A high score on these indications is believed to enhance the capacity to formulate and enforce environmental policies:

Administrative capacity (conditions for SOPs):

- a central ministry/agency for environment
- environmental institutions on all levels (regional, local)
- environment divisions within other ministries
- a national environmental plan

Environmental knowledge and awareness:

- existence and production of relevant data
- independent research institutes, role of universities
- environmental funding (green GDP)

2. Non-Governmental Actors

As the *state capacity* factor is obviously state-centred, we will need to examine the possible role of other actors in society. In what may be termed society-centred approaches, *society pressure groups* are assumed to play a central role both in the policy formation and in the implementation outcome process.[168] These groups may include domestic NGOs as well as NGOs with external links, international donor agencies, the media, scientific communities and industries in the private sector.

Public pressure may be assumed to work in two directions: either facilitating or hampering the state's ability to comply with and enforce the international regulations in question:

* *Environmental proponents*: In countries with a large public environmental movement, government policies that are in line with IEAs may find domestic allies and thus be encouraged. Recent studies show that *public environmental concern* is not limited to the affluent part of the world, either as a phenomenon or as a possible channel for political pressure. [169] In several developing countries, local

[168] See Andresen et al. (1995).

[169] "*The Health of the Planet Survey. A Preliminary Report on Attitudes on the Environment and Economic Growth Measured by Surveys of Citizens in 22 Nations to Date*", the George H. Gallup International Institute, (Princeton, NJ: Princeton University Press), 1992.

environmental problems have been found to have much greater salience among citizens than in the North, as well as serving as a basis for electoral activism. Such groups may also be directly strengthened through transnational coalitions, e.g. with epistemic communities adhering to the international regime.

* *Environmental opponents*: Alternatively, enforcement of government programmes and policies may be hampered by *vertical disintegration* – indicating the scope for domestic resistance when policies are found to carry explicit costs to specific sub-national target groups. A central assumption in Underdal's notion of vertical disintegration is that, as environmental ideas are developed and specified, it will become clear that they will have significant consequences for particular sectors of the economy or segments of society. Regulatory policy tends to impose costs upon those whose behaviour is to be changed; since they are the ones who in effect control the activities that are to be modified, governments are dependent upon their co-operation to ensure compliance (Underdal, 1998a:15).

The risk of *vertical disintegration* depends on such factors as the character of the political system and the relationship between state and society: Firstly, vertical disintegration tends to be greater in open, pluralistic political systems; secondly, vertical disintegration tends to be greater when states have weaker control over "their" societies. At this point Grindle's characteristics of developing countries seem useful for discussing vertical disintegration tendencies (Grindle, 1980). She points out how the policy-making process seems remote to most individuals, and that conflicting interests tend to appear at the output stage rather than input stage of decision-making: – interest aggregating structure is often weak in developing countries, as these are seldom open, pluralistic societies. Thus, it is in the *implementation outcome* phase that public demands are made – consequently, it is in this phase that vertical disintegration may be expected in DCs. On the other hand, according to Grindle, the public sector is often a strong and extensive institution in developing countries, and this factor serves to increase the control of the central state over "its" society. This latter characteristic suggests that there may be less scope for vertical disintegration to influence *implementation output* in developing countries than in most developed ones.

Resistance may come about not only as a reaction to added costs, but because there is a gap between the perceived environmental priorities of the state elite and local communities of people. This is where we may discern a difference in thinking about implementing environmental policies in North and South:

- Vertical disintegration is identified as an obstacle in the North, and we may look for ways in which central authorities may deal with domestic resistance from target groups who become affected by new environmental policies.
- In the South, the widely spreading concepts of *indigenous and local community participation* and *local capacity building* draw attention to how these domestic groups may become empowered vis-à-vis central authorities.

Thus in the North implementation is primarily (and, it seems, legitimately) seen from a top–down perspective, while in the South legitimacy is often linked to a bottom–up perspective – without necessarily questioning the implications for environmental impact.

Indicators:

Jänicke suggests that the main indicators along this dimension relate to the domestic NGO community. A high score on autonomy vis-à-vis government and high score on professionalism are assumed to have a positive effect on environmental policies. Other non-governmental actors which may play important roles are the media, the "green" business sector and international NGOs. At this point I deviate from Jänicke's list of indicators as I assume that, in DCs, INGOs will play a relatively larger role than the media and a "green" business sector, as compared to OECD countries. The main issue is whether the central government sees and approaches these INGOs as environmental allies and proponents, or as competitors.

The risk of *vertical disintegration* is also closely linked to *policy content,* and this needs further clarification:

Policy Content and the Issue-Area

The specific nature of the biodiversity issue may itself foster and aggravate a situation of *vertical disintegration*: This is based on the need to develop policy programmes that put restraints on human utilisation of natural resources in order to conserve habitats and threatened species. As conservation principles are turned into practical policies, these regulations may carry *specific costs to specific domestic actors*, for instance restrictions on land use for the local people living close to wildlife reserves. While the establishment of wildlife reserves may benefit central governments in terms of increased revenue from tourism, there may be resistance from sub-national groups who see their livelihood affected. When costs are thus concentrated to specific sub-groups, domestic implementation of national legislation may be harder to achieve, and it will thus be harder for members of international regimes to meet their obligations. This line of thought is central in Wilson's (1973) conceptualisation of policy areas – which is the basis for Table 7.1.[170]

Table 7.1: Distribution of costs and benefits for target groups in policy areas

		benefits: concentrated	distributed
costs:	concentrated	1	2
	distributed	3	4

[170] This may lead to what Arild Underdal terms "vertical disintegration": a state of affairs where the aggregate thrust of micro-decisions deviate more or less substantially from what higher-order policy goals and "doctrines" would seem to require. See Arild Underdal & Kenneth Hanf, (1998).

While policies in category 3 (concentrated benefits, distributed costs) are assumed to be the easiest to implement, environmental policies typically fall in category 2 (distributed benefits, concentrated costs), which is assumed to be most difficult.

Nevertheless, the relationship between policy content and context may vary across the two main different sectors within biodiversity policies: *Wildlife policies and domesticated genetic resources (seeds) policies.*

First, an aggravating factor in the former is that domestic target groups are likely to be highly *fragmented*, as policies directed at enhanced *wildlife management* must address local administrative bodies, and then again local communities of people throughout the country. In contrast, through their national gene bank, the government (in our case: that of Ethiopia) may be expected to control a significant share of the country's plant *seeds*; more directly than they may be expected control rural wildlife habitats. This will add to the government's political and administrative *capacity* to manage the *seeds sector*.

Secondly, there may also be a higher level of personnel *expertise* associated with the *seeds sector*, as Ethiopia has long conducted an active policy in this area, being conscious about controlling the country's plant genetic resources.

Third, the need for *financial resources* may be more pronounced in the *wildlife policy area*: As noted above, the establishment of wildlife reserves may benefit central governments in terms of increased revenue from tourism, but resistance may come from sub-national groups who see their livelihood affected. Moreover, revenue from tourism is generally highly uncertain, and the question of who should carry costs of conservation activities is a contentious issue in DCs: External funding often has strings attached, and developing countries are becoming increasingly adamant in demanding national sovereignty over natural resources. As mentioned in section 1.2.3, there may be more immediate benefits associated with *plant genetic resources*. There is, on the other hand, also a scope for conflict in the seeds area, as considerable uncertainty attends the domestic *distribution of potential benefits* from external marketing and transactions with genetic resources. This basically concerns whether revenues should go to local communities of farmers or to the central authorities. Moreover, the initial *costs* of establishing a control system for access to genetic resources may also be a cause for apprehension in DCs.

> *On the whole, we may assume both a higher level of will and ability in Ethiopia with regard to domestic policy formulation and enforcement within the seeds sector, compared to the wildlife sector.*

The nature of the issue-area is an important control factor, both in analysing international and domestic politics. First, it is possible that the environmental situation has taken a turn for the worse, and this could affect policies in the area. Second, increased or new scientific and technological knowledge may go some way in accounting for domestic output. Moreover, the way actor interests link up with aspects of the issue-area may change, for instance through changing property rights to resources.

One element not covered in Wilson's table is the situation where the central authorities are in dispute with domestic environmental grassroots movements – even to the extent of regarding them as opponents. This may partly reflect the "Western bias" in Wilson's table, as it does not address situations in authoritarian states. Even in fully fledged democracies, however, there is certainly considerable scope for environmentalists to demand more radical environmental policies that what the state is willing to supply. The opposite criticism was raised against Jänicke's approach, as he is concerned with pointing out factors that strengthen state capacity for environmental policies, rather than looking at obstacles. It is important to check for those domestic actors which are resisting environmental policies, as well as those who want stronger actions in the field.

3. State–Society Relationship

When the *state–society relationship* is singled out as a different dimension from the two others (which are also clearly linked to matters of state–society relations), this is done in order to capture yet another aspect of domestic policy formulation and enforcement: *the institutional relationship and link between the first two dimensions.* State–society relations will have implications for the balance between an effective central bureaucracy and local participation in policy formulation (Evans, 1989). This also relates to the balance between enforcement and responsiveness (flexibility) (Grindle, 1980). It is assumed that in order to have a positive effect on compliance, this factor must be in some kind of equilibrium. Some degree of central control is needed for co-ordination and efficiency in central policy-making, while too strict central control may reduce the legitimacy of policy initiatives with the regional administration and with local people. Here it is also important to look at the types of capabilities the various agencies have at their disposal. Even though central authorities may have strong political control by way of commanding police and military forces, they may have weak enforcement control if the environmental sector agencies are weak.

Indicators:

This dimension may represent the greatest challenge in terms of finding valid empirical indications of the theoretical concepts. Again I shall rely on the indicators in Jänicke's list, as they seem to capture central elements in evaluating the institutional relationship and links between state and society. High scores on these indicators are again assumed to strengthen the capacity to formulate and enforce environmental policies:

Participative capacities:

- participation in general elections
- participation in formulation of national environmental policy
- participation in ecologically relevant long-term planning

Integrative capacity:

- intrapolicy co-ordination of environmental policy at different levels
- interpolicy co-ordination across sectors of conflicting policies
- external integration of environmental institutions and NGOs

The Three Dimensions

The three dimensions cannot be seen as independent of one another. Rising public concern may, for instance, have an impact on the state–society relationship, as pressure is increasing for delegation of authority to lower levels of administration. Public environmental concern may also affect capacity, to the extent that non-governmental actors influence policy decisions on allocating scarce resources. Moreover, the state–society relationship may have implications for the scope and development of the administrative capacity of the central government. This also points to the problems relating to making sharp analytical distinctions – for example, an increase in state capacity in terms of financial resources may eventually strengthen NGOs, which in turn may try to promote environmental awareness among the public and the government.

A central question is still whether the proposition that "regimes matter" is equally relevant in developing countries. Do regimes matter differently in developing countries, compared to in developed countries? Are there differences in the way domestic factors link up with factors at the international level? Even though the three dimensions – state capacity, non-governmental actors, and state–society relationship – are likely to be important both in developing and developed countries, they can, as we have seen, be expected to be important in different ways. The most important differences seem related to such features as *vertical disintegration* and *standard operating procedures*. Regimes may affect implementation in developing countries indirectly: first, by helping DCs get a breakthrough for their interests in the IEA; second, by increasing other parties' compliance with the IEA, and in effect, increasing the likelihood of DC implementation. Against this background I shall proceed to make assumptions about the links between factors at national and international level.

7.5 Implementation and Factors at Different Levels: Main Propositions

Developing the analytical frameworks has so far been a matter of sorting out the most relevant among a multitude of domestic factors and seeking to link them up to coherent theoretical perspectives. I shall now develop propositions regarding how causal mechanisms may link institutional factors at the international level to domestic implementation. This last step has been relatively less explored, and I will have to rely more on using the case-study in an explorative manner.

A widespread underlying assumption concerning environmental policies in DCs is that they, unlike industrialised countries, lack the necessary capacity to establish and enforce effective environmental policies. A basic assumption of this study is that this situation may be mitigated by international environmental regimes:

Regimes may be assumed to affect both the willingness and ability of governments to follow up their international environmental obligations.

I shall here identify four **mechanisms** through which international institutional factors may have a role in enhancing domestic implementation:

1. Moral mechanism:

The scope for implementation willingness may primarily be determined in the regime formation phase, as indicated for instance in the discussion of breakthrough for DC interests in the IEA. The moral aspect is meant to indicate the normative or symbolic level as expressed in the objectives of the IEA. The regime may enhance implementation through normative language for two reasons: either because the content of the IEA is perceived as legitimate, or because an agreed international norm is perceived as an obligation *(pacta sunt servanta).*

2. Material mechanism:

The regime may provide material mechanisms in the form of financial and technological transfers. This mechanism may function in two different manners, either as incentive or as enabling: an *incentive* may spur implementation activities despite the absence of will, while *enabling* indicates that the will is there, but so far the ability has been lacking. In the latter case, the effect would seem to have a greater potential, and both functions are assumed to be of greater importance in DCs than in OECD countries.

3. Mutual reassurance mechanism:

Institutional factors are assumed to reduce uncertainty about free-riders among the parties through information, monitoring and sanctions. These mechanisms may have an impact in stabilising mutual expectations among the parties, by reducing transaction costs and by providing a frame of reference for future interactions.

4. Empowerment mechanism:

Environmentally oriented coalitions between domestic and international actors associated with a regime may be developing within *epistemic communities*; strengthening development of consensus within an issue-area.[171] The regime may

[171] An international regime may strengthen non-governmental domestic actors through other channels than epistemic communities. I focus om epistemic communities, partly because the other channels are treated more generally under the moral and material mechanisms, and partly because the empirical identification of such effects would demand a different methodology (extended field studies) than what the time frame for this investigation has allowed.

policies. It should be noted that the mechanisms for acting out the empowerment role may also here consist of moral and material argumentation, but the addressees will be individuals or non-governmental organisations rather than governmental agencies.

The relationship between the mechanisms

The relationship between the four mechanisms may be pictured along a continuum, where the existence / absence of one mechanism may strengthen / weaken the effect of the next. While moral mechanisms may represent the weaker form of institutional effect, they may still enhance the effect of material mechanisms. Here we should recall the distinction between *coinciding* interests and a *breakthrough* for domestic interests in an IEA:

- Moral and material mechanisms may be mutually reinforcing if the norms employed in the IEA also happen to *coincide* with domestic interests.
- The moral persuasion may increase in the event of a *breakthrough* for domestic interests, as the party may then feel increasingly obliged to adhere to IEA objectives and regulations.

While both moral and material mechanisms may be regarded as necessary but not sufficient conditions for implementation, they may still be important conditions for the activation of the third mechanism – mutual reassurance. The mutual reassurance mechanism is probably the most important one, not least in this issue-area, where unilateral action has been found to be of little use. The empowerment mechanism may well be the condition least necessary for implementation, but, under certain conditions, it may nevertheless have a strong effect in terms of enhancing the scope for implementation. The relationship between these mechanisms and the conditions for their coming into play will be analysed in Chapter 9.

7.5.1 Preliminary empirical foundations of the main propositions

The propositions linked to the four mechanisms link up with the main preliminary conclusions and questions raised in the analysis in part one (Chapters 3 to 6). In the following section I shall point out these links as well as make further assumptions regarding how these institutional mechanisms may affect the various domestic factors presented in this chapter. So far, the propositions have been mainly related to the *positive* effects, in terms of how institutional mechanisms may *enhance* domestic implementation. In the following, I shall also point to the potential *negative* effects. Negative effects will refer to two aspects:

- *Scope*: reduced scope for implementation,
- *Content*: generally implying that implementation activities will be concerned with the first objective of the CBD (conservation), and hardly encompass the two latter objectives (sustainable use and equitable sharing). This standard links up to the discussion of whether the breakthrough for DC interests in the

international output will remain largely of a symbolic nature (see also Table 3.2 on scope and content of implementation).

1. Moral mechanism (norms and symbolic language)

In Chapter 6, I offered several preliminary conclusions regarding the effect and viability of what may be termed the moral mechanism. The analysis of the CBD left little doubt that:

> *DCs have had a* breakthrough *for their norms and interests in the international output.*

The main question to be dealt with in this part of the Ethiopian case-study is how the CBD is perceived from an "Ethiopian" point of view: Is it seen as a victory for their main interests in the negotiation phase, do they view the CBD as a legitimate instrument for improving conditions in the issue-area, and/or do they perceive implementation as obligatory? The general assumptions tied to the moral mechanisms give rise to three case-specific propositions:

- If the symbolic language in the CBD text is perceived as catering to DCs norms and interests, this could enhance the *commitment* of central actors in the bureaucracy towards the obligations in the international regime. Without, or in addition to, symbolic language catering to DC interests, this effect could come about in terms of a national perception of common, legitimate norms and *pacta sunt servanta*.
- Likewise, this could strengthen the *commitment* of non-governmental environmental proponents to advocate environmental policies.
- If the symbolic language in the CBD text is perceived as catering to DCs norms and interests, this could also go some way towards *mitigating the effect of vertical disintegration*. Governments would presumably find it easier to convince their constituencies of the need to accept stricter environmental policies, if they could show a high level of national success in the international environmental agreement.

On the flip-side of the coin, there are two potential dangers associated with relying on the moral mechanism to enhance domestic implementation.

> *One obvious stumbling block regarding symbolic language is that it may end up as no more than that: a play to the grandstand, without material content.*

There is indeed a lack of concrete measures regarding the practical implementation of CBD concepts like *equitable sharing*, *country of origin*, *prior informed consent*, and *mutually agreed terms*. It is left to the nation-states to establish relevant legislation. For coastal states of the South, the gap between symbolic and material level has been particularly apparent in connection with the UN Law of the Sea (UNCLOS). While it was relatively easy to make oral concessions to grant coastal states of the South their "rights" in the fisheries regime, this has scarcely been followed by practical policies in the North to enforce these rights in material terms. Compared to the problems of controlling catch quotas and enforcing fisheries agreements with long-distance fishers, the problems regarding control of access to genetic resources are even greater. Still, the user will need information about the properties of the genetic resources. Without such information, the selection process becomes highly time-consuming, and this could broaden the scope for control.

The same elusiveness is true for the concepts of *local capacity building* and *local and indigenous community participation*, which became catch-words on the global agenda with the UNCED process, and also constitute integral parts of the CBD. These concepts may, however, be more closely linked to the democratisation agenda of the West, and they are likely to reappear in the guise of conditionality, for instance in GEF-financed projects. These examples highlight another potential flip-side of the coin of the moral mechanism, which was discussed in the ideational perspective:

> *Norms as authoritative persuasion, as part of a hegemonic agenda (*content: *focus on conservation only).*

The latter stumbling block is hardly a significant scenario in this case, as the analysis revealed that the CBD text was scarcely induced by hegemonic activity. Whether the coin shows up as head or tail, the result may nevertheless be a mere play for the grandstand.

2. Material mechanism (financial and technological transfers)

As shown in Chapter 6, the picture is confused with regard to the development of the financial mechanism, the Global Environment Facility (GEF). The North–South conflict on funding of biodiversity projects basically concerns the perennial question of sovereignty contra conditionality. The GEF provides funding for development of national reports and environmental plans in the South – which could be interpreted as an indication that GEF is used to further the environmental agenda of the North. As far as compensation mechanisms for access to genetic resources are concerned (equitable sharing), it is now up to national governments to enact appropriate legislation regarding *mutually agreed terms*. Such a system will depend on the capacity of DC governments to enact and enforce appropriate legislation. While the GEF has a stated mandate referring to the objective of local capacity building, it has not been particularly successful in meeting this goal.

A central question concerning the domestic scene is whether the quantity of international funding is likely to match conservation needs. I shall study the extent to which conservation and utilisation of biological diversity in Ethiopia is supported by external factors, e.g. by improved compensation for external use of Ethiopia's genetic resources and by financial transfer mechanisms. What would the presumed effects of material mechanisms be?

- First, this would strengthen *state capacity* to formulate and enforce environmental policies through added *financial resources*.
- Second, increased financial resources could go some way in *mitigating the effect of vertical disintegration*, as there would be a larger scope for compensating target groups affected by environmental policies.
- Depending on the type of material mechanisms, there could also be positive gains in terms of enhanced *capacity building* in general.

The flip-side of this coin is that financial and technological transfers may be part of a hegemonic scheme, in line with the power-based proposition (P1b):

- Dominant regime members tend to induce compliance, in accord with their own interests, in weaker states. This is achieved through structural leadership, as they have the capacity to use sticks and carrots to achieve their goals.
- In light of the scarce resources of international environmental funding, this may also be seen as an instrument with which to divide and rule. Hence, compliance with an IEA in DCs may be expected to be induced by dominant Northern states (*content*: focus on conservation only).

3. Mutual reassurance mechanism (conjoint implementation)

Mutual reassurance concerns the interactive and interdependent aspects of implementation. A breakthrough for the national interests and ideas of the South does not necessarily imply that an international environmental agreement will have an impact on behaviour in these countries. Rather, this is basically a question of the strength of the new bioregime in establishing common norms which the actors in gene transactions can find acceptable and fair:

- The biodiversity issue, as seen from developing countries' perspective does *not provide a wide scope for unilateral action*. In order to meet the obligation of *equitable sharing* relating to gene transactions, compatible legislation is necessary in industrialised countries.
- On the same note, it may be conjectured that the South needed the CBD as a framework onto which to pin the kind of policies which they deem necessary, such as stronger administrative control over access to and use of their own genetic resources. Such legislation would clearly be useless without compliance from all parties, including user countries.

The main question here concerns whether other parties, notably the industrialised countries, will comply with their obligations under the agreement, especially the third main objective of the CBD (equitable sharing). Chapter 6 gave a number of examples which could be interpreted to mean that international actors, through learning processes and participation in international fora, choose to comply to international norms because they perceive the norms as legitimate. More specifically for the domestic level, this question will concern how Ethiopian actors perceive the trends in the international community with regard to complying with the new bioregime.

- If other parties honour their obligations, this would presumably enhance Ethiopia's *commitment* to the CBD, and
- strengthen the development of *standard operating procedures* in the area.

If, on the other hand, other parties, notably the OECD countries, fail to fulfil their obligations under the agreement, it would hardly seem rational for Ethiopia to do so. This option represents the flip-side of the mutual reassurance mechanism, as it would show that the symbolic language agreed to was no more than a grandstand play.

This is by far the most subtle mechanism in terms of measurement. Its effect will mainly be sought established through interviews with central actors and by looking into project documents describing the behaviour of external actors involved in biodiversity projects in Ethiopia.

4. Empowerment mechanism (strengthening domestic actors)

The mechanism in focus here is an alliance forged by epistemic communities linked to the regime. Chapter 5 discussed the epistemic community role of the Keystone Dialogues in developing consensus among widely different and opposing actors in the issue-area. The potential role of such actors in creating links to the national level will be central to the analysis in Chapters 8 and 9. Chapter 6 concluded that if environmental and developmental NGOs are to have a positive impact on biodiversity conservation in a long-term perspective, the old dispute between wildlife conservation and agricultural biodiversity will have to be laid to rest. This dispute constituted a major barrier to agreement on the CBD in the initial phases, as conservationists feared politicisation of the issue-area.

Also, it is important to recognise the two-level nature of the distributive question: international distribution of benefits from genetic resources and domestic distribution of benefits. The CBD encourages action on both levels, but the latter is essentially a matter of national politics and legislation, and there are no enforcement mechanisms within the international institution to reach this objective. Furthermore, it is doubtful whether national governments will perceive such legislation as being in their direct interest. Moreover, there are no feasible solutions concerning systems for protecting the rights of e.g. local communities of farmers. Hence, if national legislation makes provisions for local communities, this could indicate acceptance of the norms developed within the new biodiversity regime. A central question here is whether we can identify links between an international epistemic community and domestic groups, with a bearing on biodiversity policies in Ethiopia. This would make possible an investigation of the central proposition linked to this mechanism:

- The international regime may foster alliances between international and domestic actors, and thus

- strengthen domestic actors who advocate environmental policies, which in turn may increase their *commitment* and *capacity* to call for more far-reaching implementation of the CBD objectives.

In order to identify such links, two conditions should be met: First, it must be possible to distinguish a relatively strong and autonomous domestic NGO community. Second, it must be possible to distinguish the activities of an issue-specific epistemic community from those of other transnational organisations active at the domestic level. If neither of these can be made out, it will be much harder to identify causal mechanisms between an international regime and empowerment of domestic actors, even though such an impact may not be precluded. This is shown in Table 7.2.

Table 7.2: Scope for identifying effects of empowerment

	IEA epistemic community	**other / none transnat. org.**
strong domestic NGOs (or autonomous elite)	largest scope for identifying empowerment alliances	
weak domestic NGOs		least scope for identifying empowerment alliances

The empowerment mechanism provides an important point concerning the link to domestic factors: If a breakthrough of interests (or at least coinciding interests) is found to have taken place, it was assumed that the government would meet with less resistance in enforcing environmental policies at the national arena. This could happen either:

- because of a large degree of consensus at the domestic level about the objectives to be implemented, or
- as a consequence of the government using the international agreement as a lever for stronger environmental regulations “against” domestic opponents.

We could also envisage a third situation where there has been little breakthrough for the “national interest” in the international output. Contrary to what has so far been assumed, this does not fully preclude the possibility for international institutional factors to affect domestic policy formulation in the area:

- The international regime may foster alliances between international and domestic actors, demanding more far-reaching implementation of the IEA objectives than what the government had expected to supply. In other words, the IEA goes a step further than what the government would have wished, and the interplay between actors at the international arena and domestic pressure groups *enhances the scope* for successful implementation.

The flip-side of this mechanism concerns the nature of such alliances: As discussed in the ideational perspective in the first part of the analysis (section 3.2.3), elite talent from peripheral countries might become co-opted into international institutions, internalising the ideas of the prevailing hegemony (*content*: focus on conservation only).[172]

The main links which may be envisaged between domestic factors along the three dimensions and the mechanisms of the international regime can be summed up as shown in Table 7.3.

Table 7.3:
Links between domestic factors and the mechanisms of the international regime

	Moral	Material	Mutuality	Empowerment
1.1 Institutional resources			+ SOPs	
1.2 Human resources	+ commitment		+ commit-ment	+ commitment
1.3 Financial resources		+ capacity		
2.1 Environmental proponents	+ commitment			+ strengthened alliances and capacity
2.2 Environmental opponents	- commitment (to own cause)	+ compensa tion		
3.1 Participation capacity		+ capacity building		
3.2 Integrative capacity		+ capacity building		

Compared to the fields of international institutions (regimes) and implementation studies in general, little work has been done on the more narrow and specific field of studying links between factors at international and national level in the implementation of IEAs in DCs. However, this topic has been addressed by a few other scholars. In order not to create another "trap window", I shall contrast the four mechanisms to what seems to constitute the major contribution in this specific field: *The three Cs*, developed in the work of Haas, Keohane, and Levy (1993), and Keohane and Levy (1996):

In *Institutions for the Earth* (Haas, Keohane & Levy, 1993), the authors point to *concern*, *contractual environment* and *capacity* as the three main pathways (three Cs) through which international institutions may affect domestic environmental policies. In *Institutions for Environmental Aid* (Keohane & Levy, 1996), the three Cs are developed and studied further. The specific focus here is the field of financial

[172] See also Robert W. Cox (1983).

and technological transfers for enhancing environmental protection in developing countries and in Europe. In short, *concern* relates primarily to the national preference or will to fulfil international environmental obligations, *capacity* relates to the ability to do same, and *contractual* refers to the trust among the parties to an IEA that others are able and willing to honour their commitments.

The **moral mechanism** most closely corresponds to *concern*, and hence to some extent to how IEAs may affect actor preferences or willingness. The main point concerning my choice of concept is to emphasise the often symbolic nature of this stage in the policy process: There is a widespread use of symbolic or normative language in international agreements.

Second, the **material mechanism** obviously closely resembles *capacity*, and this is hence concerned with the potential effect on ability. I have chosen a different concept, partly due to semantic concerns: Unlike the more general approach of Keohane and Levy (1996), I have developed a domestic implementation model for the analysis in my case-study. As this model operates with *state capacity* as one of the major dimensions, the use of *capacity* as a causal mechanism linking domestic and international level could be confusing. Moreover, **material** constitutes a natural counterpart to **moral**, focusing attention on whether policy formulations are indeed followed by actual allocations directed at practical action.

Third, the concept of **mutual reassurance** is closely related to *contractual environment*, but in my opinion it is semantically better suited for capturing the post-IEA output phase – that of implementation. This is because "contractual" directs attention to the regime formation phase (reaching agreement on a contract), while "mutual reassurance" may be applied both to the regime formation phase and the regime implementation phase without causing confusion.

Finally, in the three Cs scenario, possible regime effects through non-governmental actors are subsumed under *concern*. I have chosen to single out this aspect and introduce yet another mechanism through which international institutions may influence domestic environmental policies. This is done in order to emphasise the specific potential of non-governmental actors: The **empowerment mechanism** directs attention to how international institutions may affect domestic policies even though the IEA text does not directly reflect the "national interest". Again, this partly reflects the analytical approach embedded in my domestic implementation model. The model shows how it makes sense to treat these groups of actors independently, and a central purpose is to point out that the "national interest" and "will" may not be an issue which is easily determined.

On the whole, the cases reviewed in *Institutions for Environmental Aid* are likely to provide interesting and relevant conclusions for which to compare and contrast my own findings in the field, not least because there are no African case-studies in the former. This may strengthen the scope for generalisations; I return to that issue towards the end of the analysis in Chapter 9.

CHAPTER 8

BIODIVERSITY POLICIES IN ETHIOPIA

8.1 Analytical Approach and Introduction to the Domestic Scene

In Chapter 3 I offered several propositions concerning how a breakthrough for DC interests in the CBD can be possible. Evidence was found that strengthened the propositions that institutional factors have facilitated the emergence of agreement between the interests of developed and developing countries. Thus, the CBD output coincides with DC interests to a greater extent than the "terms of trade" prior to its conclusion, and DCs have had a real breakthrough for their interests in the international output. Against this background I made the proposition that DCs are more likely to implement their obligations under the regime. The H0 is that domestic (biodiversity) policies in DCs would not have been different in the absence of the international environmental agreement.

In this chapter we shall see whether this means that developing countries are more likely to implement their obligations under the CBD, as well as looking into the effects of and links to domestic factors:

- A thorough analysis of and control for domestic factors is necessary, in order to try to falsify H0.
- Another reason for analysing domestic factors is the need to examine the capacity of the state for formulating and carrying out biodiversity policies in Ethiopia.
- Furthermore, we need to understand the role and relative importance of the various domestic factors, in order to study their linkages with the international institutional mechanisms.

Hence, there are four central steps in this analysis:

1. Making inferences about implementation based on the theoretical perspectives.
2. Examining whether Ethiopian biodiversity policies are in compliance with the objectives of the CBD.
3. Examining whether compliance in national biodiversity policy might have come about primarily as a consequence of coinciding and significant change in domestic factors (e.g. technological or economic change).

4. Examining the links between domestic and international factors: through what mechanisms may international institutional factors influence domestic implementation activities (as identified in step 1)?

The examination of domestic factors will follow the analytical framework presented in Chpater 7, but I shall start out with a general historical introduction to Ethiopia. First, I shall present the inferences about possible links between domestic policies and international institutional mechanisms. These propositions will not be dealt with directly until Chapter 9, but they should be briefly introduced here, to prepare the ground for the subsequent analysis:

Box 8.1 Step 1. Inferences about implementation:

1. Moral mechanism: The regime may enhance implementation through normative language for two reasons: i) the content is perceived as legitimate, or ii) an agreed international norm is perceived as an obligation *(pacta sunt servanta)*.
Propositions: This mechanism may increase state capacity by enhancing the *commitment* (with governmental and with non-governmental actors alike) to international obligations. The government would presumably find it easier to convince their constituencies of the need to accept stricter environmental policies, if they could show a high level of national success in the international environmental agreement.
Question: How is the CBD perceived from an "Ethiopian" point of view? Is it seen as a victory for Ethiopia's main interests in the negotiation phase? is the CBD viewed as a legitimate instrument for improving conditions in the issue-area, and/or is implementation perceived as obligatory?
2. Material mechanism: The regime may provide incentive and enabling mechanisms (financial and technological transfers), and this was assumed to be of greater importance in enhancing implementation in developing countries.
Propositions: This mechanism may increase state capacity through added *financial resources*. Second, increased financial resources could go some way toward *mitigating the effect of vertical disintegration*, as there would be more scope for compensating target groups affected by environmental policies. There could also be positive gains in terms of enhanced *capacity building* in general.
Question: Is conservation and utilisation of biological diversity in Ethiopia being increasingly supported by external funding, and may this be linked to the CBD? (This might be by improved compensation for external use of Ethiopia's genetic resources, or by use of financial and technological transfer mechanisms of the CBD.)
3. Mutual reassurance mechanism: Institutional factors are assumed to reduce uncertainty about free-riders among the parties through information, monitoring and sanctions or incentives. These mechanisms may have an impact in stabilising mutual expectations among the parties, by reducing transaction costs and by providing a frame of reference for future interactions; and thus enhance implementation.
Propositions: If other parties honour their obligations, this can be assumed to enhance Ethiopian *commitment* to the CBD, and strengthen development of *standard operating procedures* in the issue-area.
Question: To what extent has the institutional framework of the CBD contributed to reduce uncertainty about other parties complying with the new bioregime?
4. Empowerment mechanism: The regime may facilitate coalitions that may strengthen domestic actors who advocate environmental policies, and thus enhance implementation.

Propositions: The international regime may foster *alliances* between international and domestic actors, increasing their *commitment* and *capacity* to call for more far-reaching implementation of the objectives of the CBD. This could strengthen state capacity to formulate appropriate legislation, or the demand could even go further than what the government had expected to supply.
Question: Can we identify links between an international epistemic community and domestic groups or individuals, with a bearing on biodiversity policies in Ethiopia?

8.1.1 Political history and ethnic factors

When the United Nations was created in 1945, Africa was represented by four states only: Egypt, Ethiopia, Liberia and South Africa. Only Ethiopia and Liberia among the African states have not been former colonies. Liberia, however, was established by the colonial powers, whereas the Ethiopians may trace their history some 3000 years back. The first imperial ruler accepted as a historical fact in the Ethiopian Constitution of 1955 is Menelik I – son of the Queen of Sheba and King Solomon, who ruled in the 10th century BC.[173]

During the 4th century AD, (Coptic) Christianity was introduced, due not least to the country's contacts with Byzantium and its Roman emperor. External contacts included moreover Egyptian and Hellenistic influences, as well as Semitic and among these Judaic immigrants from southern Arabia. These contacts were largely severed by the spread of Islam in the 7th century, which established itself to the north of the Tigray highland as well as in eastern Shoa. Islamised populations contended with the Amhara-Tigre peoples in the north.[174] Dating from the 1st to the 7th century AD, the Axumite empire at its peak stretched from the northern half of Ethiopia through to Yemen, parts of Sudan, Somalia and Saudi Arabia. The state of Axum issued its own coins, it was a literate society, and it was the second country in the world (after Armenia) to declare Christianity its official religion (in AD 330, 50 years before the Roman Empire).

In 1896, an Ethiopian army led by Emperor Menelik II inflicted a crushing defeat on the Italians at Adwa, and Ethiopia emerged as the one African state to have defeated a colonial power. This constituted a great boost for nation-building in Ethiopia and basically coincided with the time when the present borders of Ethiopia were established.

Marking the beginning of an effective central government and modern administration, Ras Tafari was crowned in 1930 as Emperor Haile Selassie I. He was to remain in power for more than 40 years, broken only by the short-lived Italian occupation from May 1936 to May 1941. The Selassie reign was marked by a centralisation of state functions, the development of national institutions and the

---

173 *Area Handbook for Ethiopia*, Foreign Area Studies, The American University, Washington DC, 1971.

174 Religious discord is still a source of domestic tension. It also played a role in the Eritrean struggle to separate from Ethiopia (which was achieved in 1991). By a UN resolution Eritrea, a former Italian colony, was federated to Ethiopia in 1952.

introduction of Amharic as the official language. State power was absolute, based on feudalism and ethnicity, with the northern part practically "colonising" the rest of the country (Human Rights Report, 1995).

This Amharicised nation-building and the tight control by the central government did not change appreciably during the Mengistu regime, from 1974 to 1991. Through most of this period, Ethiopia was afflicted with civil war between the Derg military junta and the TPLF/EPRDF in Tigray and Eritrea. It was the Mengistu Derg regime that established the Peasant Associations (PA).[175] These were originally meant for representatives of the peasants, but were turned into tools of government control as politically trained cadres entered the leadership of the PAs (Human Rights Report, 1995).

Ethiopia is home to a diverse mosaic of ethnic groups, of which Amhara, Tigre and Oromo constitute the largest. The Amhara, with about 25% of the population, have traditionally been politically and socially dominant, followed by the closely related Tigre group (with 8% of the population). Amhara-Tigre belong to the Semitic language group, and adhere primarily to Ethiopian Orthodox (Coptic) Christianity, although a number of Northern Tigre speakers have converted to Islam. From the restored Solomonid dynasty in AD 1270 and until 1991, all emperors have been Amharic.

Ethiopia has been led by the (Transitional) Government of Meles Zenawi, leader of the Ethiopian People's Revolutionary Democratic Front (EPRDF), since July 1991, when the country cast off the despotic regime of President Mengistu Haile Mariam. The EPRDF is a coalition of liberation groups from the civil war, which is still believed to be dominated by the Tigray People Liberation Front, the TPLF. With the new government, political power in Ethiopia is mainly vested in the Tigray ethnic group. In May 1994, Ethiopia voted on its new Constituent Assembly. More than half of the eligible voters had been registered, and the government had put quite some effort into educating the voters.[176] Still, the general verdict by international observers was that there was hardly free competition among political alternatives for the votes, as the major opposition groups boycotted the election (Human Rights Report, 1994). The first general (parliamentary) elections were held on 7 May 1995. The President has symbolic status, while the Prime Minister enjoys extended authority. Again the elections were boycotted by all major opposition groups, leaving no realistic alternative to the government coalition – the EPRDF (Human Rights Report, 1995). The EPRDF currently holds more than 90% of the seats in the 556-member parliament. This "landslide" on the part of the government coalition is due partly to the majoritarian "winner-take-all" election system. This system, also known as "first past the post", favours few parties and enhances governability, whereas a proportional election system often tends towards multiple parties and

---

[175] The *Derg* is the Amharic word for the military committee which succeeded in dethroning Haile Selassie.

[176] *The Economist* (1994) "Voting, of a sort", 4 June, p. 50.

greater representativity (Lijphart, 1971).[177] Multi-ethnicity has made the majority system a favourite one with many African governments seeking to manage heterogeneous societies, but the price is often political discontent among the public at large.

The outlook for the political situation in Ethiopia may at best be characterised as unstable. The main, and seemingly only, political cleavage lines are constituted by ethnic differences. There are no important cross-cutting, and thus no conciliating, cleavage lines to unite and strengthen the political web of the country. Religion might have become such a line, as Ethiopia is split in fairly equal proportions between Islam and Christianity, but these persuasions seem more likely to strengthen than transgress the ethnic segmentation. After a political history dominated by the Amhara ethnic group, often in collaboration with the Tigre, it is now the Tigre group that has risen to political power. A number of high standing political and administrative positions are claimed to be changed from Amhara to Tigre. A schism thus seems to be developing between the Tigre and the Amhara groups. Shortly preceding the 1995 elections, the government sacked 40 university teachers, many of them Amhara. The dominance of the Amhara and the Tigre is becoming increasingly illegitimate in the eyes of the other main ethnic group, the Oromo.

The Oromo, although constituting 40% of the population, have never had the necessary internal coherence to attain political power. In view of growing Oromo separatist tendencies, as well as the independence of Eritrea on 24 May 1993, the central government may have reason to fear a further splitting up of the country. The ongoing decentralisation process is basically an attempt to preserve unity, while at the same time granting some degree of autonomy to the regions. The drive for decentralisation may basically come as a reaction to the centralised Derg regime – but in a worse-case scenario it may constitute a reaction against the Amhara/Tigre dominated nation-building project of Ethiopia as such. As set out by the Constitution, Ethiopia is now emerging as a federation, originally divided into 14 regions. Each region has an elected regional council, with the five largest corresponding to the Tigrayan, Afar, Amhara, Oromo and Somali ethnic groups. Five regions subsequently have joined together to compose the Southern Region, reducing the number of regions to ten.[178] Each region is to have the right to its own language and culture.[179] The situation is currently undecided, as uncertainties remain regarding the exact division of powers between central and regional levels of government.

---

[177] See also A. Lijphart (1969) "Typologies of Democratic Systems", in A. Lijphart (ed) *Politics in Europe* (Englewood Cliffs, New Jersey: Prentice Hall).

[178] Region 1 - Tigray, Region 2 - Afar, Region 3 - Amhara, Region 4 - Oromia, Region 5 - Somali, Region 6 - Benshangul, Region 7, 8, 9, 10 & 11 - Southern Ethiopian Region, Region 12 - Gambella, Region 13 - Harar, Region 14 - Addis Ababa.

[179] *The Economist* (1994) "Voting, of a sort", 4 June, p. 50.

8.2 Ethiopian Biodiversity Policies prior to the CBD

In this section we shall look at the factors associated with what constitutes the administrative and political capacity of the government to formulate and enforce environmental policies. This capacity was assumed to depend on i) institutional resources, operationalised by SOPs, which in turn can be captured by the presence of environmental agencies; ii) human resources, expressed by environmental expertise and commitment (knowledge and awareness); and iii) financial resources available for and allocated to biodiversity policies.

Here we will examine the status of, and look for changes in, environmental agencies. This involves the organisational set-up, as well as the capacity of such environmental agencies to control and enforce policies within their respective sectors. I shall concentrate on two sectors and the two main agencies traditionally responsible for their management:

- *Wildlife management (national parks and wildlife reserves)*: these come under the auspices of the Ethiopian Wildlife Conservation Organization (EWCO).
- *Plant genetic resources management (the seeds sector)*: this is primarily run by the Ethiopian Genebank.

I will examine their legal mandate for managing their respective sectors, their resources in terms of staff and expertise, as well as their reputed track record. If these agencies are found to have been strengthened along the way, then we can assume that this has served to enhance capacity for enforcing environmental policies.

The second empirical component in this section concerns the situation regarding biodiversity in Ethiopia:

- *Biodiversity*: First, we look at the ecological situation concerning the two main components of biodiversity: wildlife and domesticated plants (seeds sector).
- *Biodiversity loss*: Second, I discuss the main driving forces behind biodiversity loss in Ethiopia, looking into land-cover change and the legitimacy of conservation policies. In this section, the factors within the state–society dimension will be important. We will need to see how the responsible agencies have dealt with these sectors, and how the issues have developed over time. This understanding is necessary for the subsequent evaluation of whether there are policy changes brought about by, or in anticipation of, the CBD. If, for instance, Ethiopia has experienced a sudden environmental crisis, this may go some way towards accounting for changes in biodiversity policies.

Finally, the relative amount of financial resources allocated to environmental policies will be examined. To put the latter factors in perspective, I shall start out with a general introduction to the economic situation in Ethiopia.

On the basis of studying these factors, I will then offer a preliminary conclusion regarding the will and ability of Ethiopian authorities to pursue policies in the biodiversity sector. If both will and ability are found to be virtually non-existent, it could immediately be concluded that there is scant scope for implementation activities relating to the CBD. I shall also return to these factors in the analysis of the post-CBD phase, when describing the legislation process and output in section 8.3.

8.2.1 The state of the economy

Together with human and institutional resources, *financial resources* make up the three main factors affecting state capacity for formulating (output) and enforcing (outcome) environmental policies. The impact of financial resources will come into play mostly in the outcome phase: There may be a larger scope for formulating symbolic policies, compared to the funds available to carry out the plans and programmes envisaged. Here we must distinguish between the availability of *financial resources* – which is a precondition for conducting environmental projects and programmes, on the one hand, and the possible harmful environmental externalities of economic activities on the other.

If the economical situation in Ethiopia has changed significantly during the time of my investigation, this could in part account for compliance with the CBD. It should be noted that issues relating to pollution control may be "positively" affected by *economic recession.* This is because less production may lead to reduced emissions, and hence a higher score on compliance with IEAs. With regard to biodiversity conservation, it is also possible to conceive of a "positive" relationship: Economic recession may reduce human activities such as road-building and land-cover change, which may then lead to reduced loss of biodiversity. At the same time, however, the relationship between economic recession and biodiversity conservation may be more likely to go in the other direction: Dwindling financial resources may reduce the ability to establish and control natural parks and sanctuaries, while also leading to increased pressure on marginal land areas, hence threatening vulnerable habitats. Hence, I shall expect *economic growth* to have a positive impact on compliance with the CBD in Ethiopia.

The most marked change in this time period is obviously the transition from the Mengistu Regime's 17 years of planned economy to a policy aimed at creating a marked-oriented mixed economy. Economic performance has varied greatly in the 1990s, from a high growth rate of 12.3% in 1992/1993, to a low growth in GDP in 1993/1994 of 1.3% (Hansson, 1995:2). The fall in growth rate was predominantly due to drought, while there were still high growth rates in manufacture, building and construction, and in services (ibid.). Both exports and imports have increased in 1993/94, with total export value increasing most, primarily due to increases in coffee prices (1995:3). During these years inflation has been between 12% and 15%, which is not alarming, but still creates problems for civil servants who are not fully compensated in their wages (1995:5). In the UN–NADAF Mid-term Review, GNP per capita in Ethiopia declined from 116 in 1991 to 101 in 1995. Among African countries only Mozambique, Somalia and Tanzania show lower levels in GNP per capita.[180] Moreover, estimates from the social sector in 1993/94 show that only 4.4% of GDP was allocated to social services (IMF, 1994).

Ethiopia is not a member of the successor to GATT, the World Trade Organisation (WTO), and has been reluctant to join. In a statement by the Ethiopian Ministry of Trade to the 1994 Tunis Conference on the Uruguay Round Agreement (URA), the government presented several critical views of the URA and WTO, including: i) not compatible to the subsistence structure of the the Ethiopian economy, ii) immense losses due to reduced role of GSP and Lomé, iii) higher international agricultural prices will hit Ethiopia hard, as it is a net importer of food (food production per capita in Ethiopia is the lowest in Africa), iv) lack of institutional and negotiation capacity, v) the URA rules on TRIPs will mean a drastic reduction in the access to foreign technologies, and vi) unfair distribution of gains from the new international trade system. As noted in section 6.2, the WTO now has 132 members and 34 observers. All observer countries have applied to join, except the Vatican, Cape Verde and Ethiopia.

Since the fall of the Derg regime in 1991, and the introduction of economic and political reform programmes, foreign aid to Ethiopia has increased. It has also changed from the "traditional" emergency assistance to programme aid and balance-of-payments support. Official grants increased from around 3% of GDP (1989/90) to approximately 7% in 1993/94 (IMF, 1994). Such grants are commonly linked to the demand that the funds shall be openly included in the total financial budget and used to promote social and economic development, primarily *education* and *health* (Hansson, 1995:22).

- While there have been recent changes in the political and economical system, there is *little reason to expect any major increase in revenue for allocation to the environmental* sector in Ethiopia.
- Thus (as little increase can be expected), *change in economic factors is unlikely to account for consistency* (if such is found) between the CBD and biodiversity policies in Ethiopia.

I now proceed with a closer look into the two main biodiversity sectors of wildlife conservation and domesticated genetic resources.

8.2.2 Cultivation and land characteristics

Ethiopia, with an estimated area of about 1.15 million square kilometres, has a population estimated to be between 55 and 75 million, 88% of which is rural. Agriculture is the dominant sector of the economy, accounting for about 50% of GDP and 85% of export revenue, and provides livelihood for 85% of the

[180] FAO; ECA Secretariat 1993.

population.[181] Land-use can be divided into three categories: intensive, mainly cultivation in the highlands; extensive, mainly pastoralism in the lowlands; areas that are little used, due to extremes of aridity or altitude (Hillman, 1993).

Before the revolution of 1974, the system of land-ownership was dominated by oppressive feudal lords. Emperor Haile Selassie did little to mitigate human suffering during the famine of 1973–74, and this contributed to sparking the revolution that put an end to feudalism as well as to the Selassie regime. In his biography on Haile Selassie, Ryszard Kapuscinski may have captured some of the sentiments from this period in Ethiopian history:

> The event that hastened the confrontation between the army and the Palace was the starvation in the northern provinces. Usually it is said that periodic droughts cause bad crops and therefore starvation. But it is the elites of starving countries that propagate this idea. It is a false idea. The unjust or mistaken allocation of funds and national property is the most frequent source of hunger. There was a lot of grain in Ethiopia, but it had first been hidden by the rich and then thrown on the market at a double price, inaccessible to peasants and the poor. Figures about the hundreds of thousands who starved next to abundantly stocked granaries were published. On the orders of local dignitaries, the police finished off whole clans of still-living human skeletons. This situation of intense evil, of horror, of desperate absurdity, became the signal for the conspiring officers to go to work. After a short period of bewilderment, shock, and hesitation, H.S. began to realize that he was losing his most important instrument (the army). Ryszard Kapuscinski (1984); *The Emperor*

The Mengistu post-revolution era was marked by radical redistribution of land property rights, with the aim of allocating one to five ha. to each farmer. The distribution is vested in the *Public Ownership of Rural Lands Proclamation* of 29 April 1975. Today land-ownership is vested in the state, but control lies in the hands of Ethiopia's local Peasant Associations to which most farmers belong. In the north, where feudalism had been less prevalent, there are still some farmers opposed to the collective ownership of land and the Peasant Associations; it is not so in the middle and in southern parts of the country (Harrison, 1987).

Currently, political action and decision-making are constrained by the unstable political situation, which again is aggravated by the economic and ecological difficulties threatening the country. In 1984, the rains failed again, causing a food crisis that affected eight million people. The recurring droughts are partly explained by the pronounced environmental degradation. Some 60% of the Ethiopians live in mountainous highlands that comprise 40% of the total land surface of Ethiopia. Due to the high population density, more than half the highland areas show signs of

[181] Ethiopian National Report to UNCED, 1992; *National Report on Environment and Development*. A Report prepared for the United Nations Conference on Environment and Development, Rio de Janeiro, Brazil. Addis Ababa, Ethiopia.

erosion, and crop yields are steadily decreasing. As a result, cultivation is taking place on steeper and steeper slopes. While as much as 40% (there is scientific dispute concerning this estimate) of Ethiopia may have been forested in historical times, only 4% remains today. The widespread deforestation has also led to a shortage of fuelwood, and the population is now felling trees in the dry savannah of the Rift Valley (Harrison, 1987). Some of the responsibility for the disastrous agricultural situation has also been put down to the failure of the state farm system introduced by the Mengistu regime. The situation is ascribed primarily to lack of credit to buy seeds, lack of marketing, transport, price incentives and storing facilities (Cheru, 1992). According to the World Bank, most of the needed resources were going to the state farms, although they constitute only 3.5% of the cultivated land and in spite of their alleged inefficient and bureaucratic profile (World Bank, 1990).

The land tenure system, including the state farms, is currently in a transitional phase. The land issue is contested, and a main political divide between the government and the opposition has been the latter's demand for private ownership of land and natural resources. The EPRDF's land policy is partly that land belongs to the state and to the people, and partly that farming rights will be distributed to the peasants by the state. This means that land tenure follows the rules of the Mengistu regime and farmers are guaranteed user rights to land, but not exchange rights (Constitution 1994, Article 40: 4, 5, 6). There is a widespread belief that the Peasant Associations will carry out this distribution at the local level, making the political control over the PAs an uncomfortable issue (Human Rights Report, 1995).

8.2.3 The Wildlife Sector

As the Ethiopian and Eritrean highlands constitute about 80% of all areas above 3,000 m. altitude in Africa, there is a high number of endemic plant and animal species in Ethiopia.

Accordingly, Ethiopia is recognised as an important biodiversity centre, but there is still a limited involvement from international science and conservation communities.[182]

Ethiopian wildlife policies date back to 1909, when Emperor Menelik II passed the first legislation to regulate hunting, especially of elephants. Then came the establishment of a hunting licence office in 1944. In response to increasing pressure on land and wildlife, the *Ethiopian Wildlife Conservation Organization* (EWCO) was established in 1964, under the Ministry of Agriculture. EWCO is responsible for the effectuation of all aspects of wildlife policy. In the late 1980s, the work on a National Conservation Strategy and a Forestry Action Plan was initiated. This came about at the request of the World Conservation Union (IUCN), and the Mengistu regime did not refuse. Part of the explanation for Mengistu's goodwill is reputed to

[182] World Conservation Monitoring Centre (1991) "*Biodiversity Guide to Ethiopia*", Cambridge.

be fear of a similar fate as Haile Selassie, whose downfall has been ascribed partly to inaction regarding environmental degradation. This move was clearly not enough to prevent Mengistu's fall, however. In the next chapter I present a detailed description of the NCS legislation process.

In 1992, the (then Transitional) Government set up a Ministry of Natural Resources Development and Environmental Protection (MNRDEP)[183], which formerly constituted a department within the Ministry of Agriculture. EWCO was located under MNRDEP. Prior to the establishment of this environmental ministry, most conservation policies were handled by the Ministry of Agriculture, where EWCO was attributed a limited status and where conservation was hampered by production-driven vested interests (Hillman, 1993). In 1995, EWCO and MNRDEP were returned to the Ministry of Agriculture. Little is known about the background for this move, as the decision-making process in Ethiopia is generally closed to public debate. A new minister, from the Oromo, has been installed as the chief leader of the department for Natural Resources and Environmental Protection.

Ethiopia has very little revenue from tourism, and money for implementing wildlife policy has come partly from external donors, such as Wildlife Conservation International (WCI) and the World Wide Fund for Nature (WWF). In the 1960s external non-governmental organisations took an interest in Ethiopian natural conservation, exemplified by missions from UNESCO and the New York Zoological Society, as well as Swiss and Japanese scientists. Recommendations from such missions paved the way for a system of National Parks and Controlled Wildlife Areas. Still, the capacity for wildlife conservation is much constrained by inadequate funding and staffing, and a lack of functional equipment and infrastructure – much of which was moreover destroyed during the political upheaval of 1991 (Hillman, 1993). Hillman points to the lack of a colonial past as an explanation for the relatively short lifetime of Ethiopian wildlife policies. While EWCO dates back to 1964, Kenya may trace its conservation policies another fifty years back in time. Moreover, Kenya earns a significant proportion of its foreign exchange from tourism, at 37%. While Kenya's estimated gross revenue from tourism for 1989 was US$ 420,000,000, Tanzania earned $40,000,000, and Ethiopia only $3,338,296 (Hillman, 1993:54).[184]

Hillman also estimates the expenditures on and earnings from wildlife per average year in the 1980s: The figures[185] are 3,826,200 Birr from the Ethiopian government and 1,150,607 Birr from assistance aid, total support for wildlife being 4,976,807 Birr; and income from wildlife at 5,541,569 Birr (Hillman, 1993:245). Income earnings come from gate fees, hunting fees, wildlife utilisation, crocodile products, Ethiopian Airlines, Ethiopian Hotels Corporation, wildlife viewing and

---

[183] *Feasibility study for a wildlife conservation project in Southern Ethiopia*, figure 1, p. 10, Final Report, Agroconsulting, Italy, an EU assistance project (UNDP, 1993).

[184] Reference in Hillman (1993): PC Globe, 1992; Woodford, J.D. 1990, "The Status of Wildlife in Ethiopia, its Conservation and Utilisation", Addis Ababa, 77 pp.

[185] In Ethiopian Birr, approx. the same as NOK.

professional hunters. As such, the figure should probably be halved, as the estimated income from crocodile production alone accounted for 2.5 million, and this project has yet to produce any revenue.[186] These estimates are obviously highly uncertain, as tourism in 1989 could hardly have accounted for from three to six times as much as wildlife earnings, including tourism, for the average 1980s. The estimates give no indication as to whether any of these resources are re-invested in wildlife management.

According to Ethiopia's National Conservation Strategy, protected areas in Ethiopia consist of protected forests, nine National Parks, 11 Wildlife Reserves and Sanctuaries and 17 Controlled Hunting Areas (NCS, Vol. I, 1994). In his comprehensive report on wildlife management from 1993, Hillman lists nine National Parks, 11 Wildlife Reserves, four Sanctuaries and 18 Controlled Hunting Areas. Together Parks and Sanctuaries encompass 3% of the surface area of the country. Out of the nine National Parks that are given total protection, only Awash and Simien mountains are legally gazetted (1969). In the Parks and Sanctuaries there may be no settlement, no exploitation of natural resources, no grazing of livestock, and no mining. In practice, however, even the legally identified Simien National Park has long been beyond EWCO's control, and in the Awash National Park the livestock of pastoralists causes considerable land degradation (Hillman, 1993). The situation for most of the proposed National Parks and Sanctuaries is characterised by conflicts over land use between local pastoralists and farmers and the central wildlife authorities. The remaining areas are given partial protection: Wildlife Reserves allow some human activities but no hunting, and Controlled Hunting Areas allow human activity and some hunting. In reality the proposed Wildlife Reserves exist on paper only, with no development and no wildlife conservation activities occurring in them.

The sharing of responsibility for the protected areas has long been an uncertain item. In 1993 the Ethiopian government proclaimed that the administration of national parks would continue to be the responsibility of the central government, leaving *game reserves* administration to the regions.[187] According to one interpretation (that of the local authorities in the Oromiya region), responsibility for all non-gazetted areas was to be conferred to the regions. This would mean that only Awash and Simien would remain under central government "control". One possible outcome of this situation could be that the regional authorities might decide not to recognise protected (but not gazetted) areas lying within their jurisdiction. The verdict now seems to be that the regions will be given responsibility for all protected areas, including Awash and Simien. As it is, less than half of the protected areas are manned with field staff and all suffer from poor infrastructure, again reflecting EWCO's serious lack of resources. There are some 300 staff allocated to EWCO, of

[186] Per Ole Syvertsen, Biology Institute, University of Oslo, personal communication, March 1996.

[187] Transitional Government of Ethiopia, Proclamation No. 41/1993, Negarit Gazeta No. 26, Addis Ababa.

which about 25% work at the Headquarters in Addis Ababa. The rest are spread through the National Parks and Sanctuaries, rarely meeting senior staff at the HQ, and receiving little training (Hillman, 1993). Further aggravating this situation, there is very limited scope for scientific monitoring of trends and movement of wildlife populations. On the other hand, there do exist rather extensive data concerning wildlife inventories and protected areas (Hillman, 1993).

Ethiopian wildlife policies have been dominated by traditional Western preservation ideals, not least as a result of the 1933 London Convention for the African States. Herein lies part of the explanation for the recurrent conflicts with local people over land use. An equally important factor in relation to the way in which local people see the concept of natural resources management is that the Ethiopia's national parks and protected areas were closely linked to the Derg regime of Mengistu. When the Mengistu regime fell, protected area installations were destroyed and many animals were killed by people living in and around these areas, as the conservation policies in these areas had enjoyed no legitimacy with the local communities.

8.2.4 The Seeds Sector

Ethiopia is regarded as an important primary and secondary *centre of diversity* for several domesticated plants. This makes the country a useful source of germplasm (in the form of seeds) for important crop traits, such as resistance to diseases and pests, tolerance of different climatic conditions and a variety of tastes. The international "discovery" of such centres of diversity was made by the Russian botanist N.I. Vavilov. In the 1920s, Vavilov identified 13 *centres of origin* for most of the world's economically important crops, one of them in Ethiopia.

The great diversity in Ethiopian crops is believed to result from the highly dissected highlands and ravines, which act as natural barriers. Here crop plant species have evolved in isolation, with heterogeneous environmental conditions varying over small distances, and a very diverse cultural history of the people has provided thousands of years of artificial selection. Among the most important crop species displaying great diversity are barley, sorghum, wheat, finger millet, sesame, chickpea and lentil. Even more important is the high variability of coffee (*coffee arabica*), which originated in Ethiopia (Smartt & Simmonds, 1995:439). Until 1989, coffee accounted for some 58% of foreign exchange revenues. The economy is still highly vulnerable to the rapid fluctuations of coffee prices. In addition, there are approximately 800 medicinal plants in use, many of which are threatened with extinction.

The main threats to plant biodiversity in Ethiopia are, according to a government White Paper: "population pressure and over-exploitation, introduction of plant varieties with narrow, uniform traits displacing indigenous cultivars, economic systems discouraging conservation, lack of proper resource conservation and lack of proper national development policy; drought and other natural calamities and war".[188] When Ethiopian authorities can take such a critical look at the national conservation policies, this is clearly made easier by the recent shift in the political regime.

The government is also acutely aware of the need to increase agricultural production, so as to meet the needs of the population. In 1996, this led the government and Meles Zenavi to approve of the large-scale "Sasakawa Global 2000"–programmes, supported by the Sasakawa Foundation and the Jimmy Carter Center. This is a dominantly traditional Green Revolution programme, including large-scale introduction of higher yielding varieties (HYVs) of plants as well as massive and expensive input factors in terms of fertilisers. The Ethiopian Genebank has a smaller role within this sector, in trying to rescue traditional land races before these are ousted by the HYVs. The HYV introductions have been limited to certain climatic zones with sufficient precipitation. In the dry areas, it is primarily only the traditional varieties that can survive.[189]

Ethiopians have long traditions in on-farm conservation and development of food crops, as well as long traditions of community gene banks. These gene banks offer a way of providing communities with the genetic resources needed for building on traditional practices, and they play a significant role in conservation and use of crop biodiversity. They also provide security for locally adapted materials, and provide a source of such material for on-farm breeding and crop improvement. Community grain pits are made in which different crop seeds are stored (Heywood, 1995:947). Traditional agricultural methods include crop rotation and selecting for drought resistance in crops. Even though this knowledge has not prevented the several droughts (five severe ones in the last few decades alone) in Ethiopia, such knowledge nevertheless form part of the public awareness and consciousness in the country. Efforts to reduce soil erosion and mitigate drought have been part of the education system, and information has also been spread through the PAs.

Ethiopia has several institutions for research and training in agriculture and plant breeding. The Addis Ababa University (AAU), the oldest and largest of the institutions of higher education, has the best developed research capacities, although these were retarded by the civil war during most of 1974–91. The Ethiopian National Herbarium, under the Department of Biology of AAU, is responsible for various smaller projects in ecology and ecophysiology, such as producing the Flora of Ethiopia. The Alemaya University of Agriculture, a constituent college of AAU, is regarded as the best developed higher educational institution dealing with agriculture. There are also close ties with external educational institutions, including the doctoral programme in collaboration with NORAGRIC at the Norwegian

---

[188] The Transnational Government of Ethiopia, *National Policy and Strategy for Plant Genetic Resources Conservation and Development*, September, 1993, Addis Ababa.

[189] Nevertheless, several farmers doubt whether the new, improved varieties will adapt to local environments. Setbacks have already occurred due to el Niño, with great losses for farmers who can ill afford the increased price of fertilisers as it is. Interview with n3 (see Appendix 1).

Agricultural University. Like AAU, NORAGRIC also has close working relations with the Ethiopian Genebank (PGRC/E), which stands out as a very important agency in this sector, and to which we now turn.

8.2.4.1 The Ethiopian Genebank: Plant Genetic Resources Centre (PGRC/E)

One of the most important gene banks in Africa is located in Addis Ababa. An evaluation of gene banks in developing and developed countries by the International Plant Genetic Resources Institute (IPGRI) in 1987 gave the Ethiopian Genebank its best recommendations and top score.[190] Moreover, in the FAO Report on the State of the World's plant genetic resources (1996), the PGRC/E obtains top scores on characterisation, evaluation and regeneration of the seeds collection. After ratifying the CBD, the government of Ethiopia allocated to PGRC/E the task of co-ordinating the African follow-up of the work on biological diversity. In the course of the development of Ethiopia's National Conservation Strategy, the PGRC/E has been renamed the Biological Diversity Institute (BDI), in order better to indicate its central role in this sector.

The Ethiopian Genebank was established in 1976 as a collaborative project between the then Federal Republic of Germany (FRG) and Ethiopia. In return for economic support for the establishment and operation of the gene bank, it was agreed that the FRG authorities should be given seed duplicates, in particular of barley, from Ethiopia. The Ethiopians, however, being highly aware and vigilant when it comes to national sovereignty over gene resources, stopped the transfer of gene material to the Germans. This story contributed to Ethiopia's fame or notoriety respectively in various parts of the international seeds sector. According to Dr. Melaku Worede, director at the Genebank at the time and later a prominent member of the Keystone Dialogues, the deal was originally forfeited by the Germans: They had been granted the right to collect duplicate seeds for storage, but not for use in their own breeding programmes. When Dr. Melaku Worede realised that the latter activity was taking place, he put an end to further access to Ethiopian seeds. A large picture of N.I. Vavilov, centrally placed in a corridor in the Genebank, is a clear indication of the Ethiopian appreciation of their genetic heritage (see Box 5.6 on Vavilov).

The Ethiopian Genebank currently contains a total of 52,000 deposits from 101 different seed and plant varieties, including various corn, oil, leguminous, spice, edible roots and medicinal plants. Like Kenya, Ethiopia practises in-situ conservation of coffee plants, due to the impossibility of storing this type of plant seed in ordinary gene banks. Apart from the usual cultivated plants, about 15% of

190 IPGRI was previously known as the IBPGR: International Board for Plant Genetic Resources, located with FAO, Rome. In the IBPGR Report from the 14th Meeting of the Board of Trustees, Rome, 25–27 February 1987, the Ethiopian Genebank is given the following evaluation: "The PGRC meets nearly all the IBPGR standards. Minor recommendations have already been suggested to this gene-bank and most will have already been implemented by the PGRC".

the deposits in the Ethiopian Genebank comprise less common, but potentially significant plants. Special programmes have been drawn up for preserving traditional land races in-situ in everyday farming practice. There are some 30–40 people in the Genebank involved in plant selection and breeding, 20 more at secondary level, and 30 engineers. In all there are 105 employees; of these, 45% are technical personnel: 4 Ph.D., 10 MS., 17 Diploma and 18 Certificate holders. Areas of specialisation of the senior technical staff include plant and seed physiology, biochemistry, biosystematics, genetics, molecular biology and genetic resources conservation.[191]

The major objectives of the Genebank are to:

- promote collection, evaluation, documentation and scientific study of crop germplasm in Ethiopia, East Africa and adjacent regions;
- preserve/conserve germplasm using various conservation strategies;
- provide germplasm and information pertinent to it for national breeding programmes;
- introduce new and appropriate germplasm to the country;
- survey and document Ethiopia's genetic diversity, and the indigenous knowledge associated with it.[192]

8.2.5 National report of Ethiopia to UNCED, 1992

There is little documentation on policies and institutional frameworks in Ethiopia. To get an idea of the situation immediately prior to the establishment of the CBD, we may use as an indication the National Report on Environment and Development of the (transitional) Government to UNCED.[193] Clearly, at this point in time, the general orientation of the objectives of the CBD was known. Hence, Ethiopian policies may already have been adjusted to anticipate these goals, either symbolically or in defiance.

With regard to Ethiopia's expectations from UNCED, these include:

- Recognition of the specific environmental and developmental problems of developing countries
- Additional financial resources to developing countries
- Technology and knowledge transfer to developing countries
- International fund for genetic resources

---

[191] UNDP, Project Document, ETH/93/G31.

[192] Report to FAO, ICPPGR - International Conference and Programme for Plant Genetic Resources, "Country Report for Ethiopia".

[193] Ethiopia, *National Report on Environment and Development*, A Report prepared for the UNCED, Rio de Janeiro; Addis Ababa, Ethiopia, January, 1992.

- Special world attention to tropical mountainous areas and to halting their soil and genetic erosion
- Special world attention to arid and semi-arid areas and desertification[194]

Expanding on the need for an international fund for genetic resources, the National Report states:

> The world very much depends on the plant and animal genetic wealth of developing countries for its agriculture and industry. The widespread use of biotechnology has emphasised rather than obviated this dependency. And yet the poor peasant farmers of the third world get no share from the economic benefits accruing from the genetic wealth that is rightly theirs. They even face the grim prospect of having to buy back from the industrialised countries this wealth of theirs recombined in specific ways and patented as specific crop, domestic animal and industrial micro-organism varieties. If the world is to house societies interacting on just and mutually beneficial bases, the industrialised countries making use of genetic resources from developing countries should establish a fund and, together with the developing countries, work out mechanisms of using the fund to compensate the peasants of developing countries.[195]

8.2.6 Summary

The environmental crisis in Ethiopia, notably expressed by the recurrent droughts, can best be described as chronic, with cyclical peaks. In other words, the problems relating to the issue-area are not characterised by acute shocks. The various components of biodiversity in Ethiopia are threatened from a wide range of sources, including:

> *population pressure, deforestation, land-cover change and overgrazing, factors that also lead to soil degradation, which in turn is closely associated with biodiversity loss in Ethiopia.*

While population pressure is often portrayed as a main source of the problem, the inadequacies of the political system to respond may represent an underlying cause. Both during the reign of Haile Selassie and the Derg regime, the Ethiopian political system has been characterised by strong central control with decision-making very much isolated from the people at large. As there is no tradition of open, pluralistic interest aggregation, this indicates a smaller scope for vertical disintegration, at least in the decision-making process (implementation output).

Common to both wildlife and the seeds sector is also the challenge involved in the uncertain position of Ethiopia's environmental ministry (MNRDEP). An

---

[194] op.cit., p. 6.

[195] op.cit., p. 104.

important aspect of the capacity of the state to initiate and enforce successful environmental policies is the existence of independent environmental agencies. That MNRDEP was returned to the Ministry of Agriculture can only be regarded as a setback. As will become clear in the presentation of the National Conservation Strategy, however, this was followed by the establishment of an independent Environmental Protection Agency.

As assumed in Chapter 7, there are many differences between the two sectors. The agencies involved in *wildlife management* seem weak in terms of all factors within the state capacity dimension. This is indicated by EWCO's rather unsuccessful administration of national parks and sanctuaries, which is explained partly by lack of human and financial resources, and partly by limited institutional resources, due to the minor status of EWCO within the Ministry of Agriculture. There is also a lack of standard operating procedures (SOPs) to enforce and operate management policies.

The oppressive regimes must have had some control over peoples' behaviour, as indicated by the public outburst of anger against the parks and wildlife that accompanied the fall of the Mengistu regime. Wildlife policies have traditionally enjoyed very little legitimacy among local communities. Various factors account for this situation:

- The centralised decision-making system, which has traditionally excluded local communities from influencing policies with a direct bearing on their situation.
- Little or no revenue from tourism, further reducing the scope for compensation to local communities affected by conservation policies. This factor is both suited to enhance vertical disintegration while also weakening the political and administrative capacity of the government agencies involved in wildlife management.
- The content of the environmental agenda in Ethiopia: One the one hand, the general lack of environmental SOPs in the central bureaucracy. On the other hand, (Western) preservation ideals have dominated the environmental agenda in the sense that wildlife preservation has had precedence over the needs and interests of the local people.

State capacity in the *plant genetic resources sector* seems much more efficient:

- There is a long tradition of political awareness regarding these biological resources.[196]
- This sector is well supplied in terms of personnel, expertise and equipment.

---

[196] This awareness about the value of plant genetic resources is apparent in the Ethiopian National Report to UNCED, where the South's position on the biodiversity issue is clearly spelled out.This is not surprising, as Dr. Tewolde Egziabher is one of the most central entrepreneurs, both in the international negotiations and in policy-making at the national level in Ethiopia. I shall return to Dr. Tewolde's role in subsequent sections.

- The ideals of on-farm, in-situ conservation have long traditions, which should make it easier to integrate people at different levels in the policy process.[197]

In the following sections, I will go through the legislation process associated with Ethiopia's National Conservation Strategy (NCS), and evaluate the contents of the NCS in light of the CBD.

8.3 Legislation Process and Output: National Conservation Strategy

The *National Conservation Strategy* (NCS) constitutes Ethiopia's overall environmental plan. It was initiated in 1989, three years prior to the establishment of the CBD, and was set out in five volumes of documents in 1994. Two circumstances should be made clear at the outset:

- During this time period, a large number of DCs were in the process of formulating similar environmental action plans, and most of these plans were being constructed by external international organisations, such as the World Bank, the UNDP or the IUCN.
- The scope of Ethiopia's NCS goes beyond that of biodiversity, as it addresses the full range of environmental issues including pollution, water resources and energy. Although I shall concentrate on those aspects relating to biodiversity, it will be useful to compare the attention and resources devoted to this issue with other environmental issues, as well.

In this section I focus on the *factors associated with state capacity*. As guiding principles for the evaluation of the legislation process relating to the National Conservation Strategy, I shall use Jänicke's state-capacity indicators: These included the presence or absence of a central ministry or agency for environment, environmental institutions on all levels (regional, local), and environment divisions within other ministries. Furthermore, capacity is seen as dependent on the quality of and scope for environmental knowledge and awareness, indicated by the existence and production of relevant data, the use of independent research institutes and the role of universities, as well as the scope for environmental funding (green GDP).

In addition to state capacity, I will briefly touch upon factors within the two other dimensions of state–society relations and non-governmental actors, as they relate to *participation* in the legislation process itself: This means that I restrict myself to asking to what extent *governmental agencies at various levels and non-governmental actors* have played a role in the formulation of the NCS. In two subsequent sections I shall look further into the factors associated with these dimensions.

---

[197] A factor which may, however, have a detrimental effect on policy enforcement within this field is the uncertainty surrounding the land tenure issue.

8.3.1 Legislation process: National Conservation Strategy (NCS) [198]

What has been termed the *first phase* of the NCS process starts with its initiation in 1989 and ends with the downfall of the Mengistu regime in 1991.

The recurrent droughts played a decisive role in putting environmental problems on Ethiopia's political agenda in the late 1980s. The World Conservation Union (IUCN) approached the Mengistu government, which was receptive to the idea of developing conservation policies due to the environmental crisis. One of the factors held to have hastened the downfall of Haile Selassie was his inaction in the face of droughts and the widespread famines accompanying them, and this is believed to have made the Derg regime more susceptible to public demands for environmental action.

The National Conservation Strategy process was initiated in 1989, and an agreement was set up between the IUCN and the Mengistu government on 15 March 1990 for assistance in drafting the NCS. In addition, the Derg regime agreed to initiate an Ethiopian Forestry Action Programme and a National Disaster Prevention and Preparedness Programme. These and other programmes are currently included in the NCS. Some of them have been gazetted, while others remain proclamations.

The first phase of the NCS was also marked by a National Conference in May 1990. The National Conference gathered close to 200 participants from various sector departments, the scientific community and the regions, as well as delegates from other African countries with experience in NCS-related work. Twenty-five papers addressing scientific and socio-economic topics related to the environment and development were presented. These were later collected to form the first volume of the NCS. This work was carried out by a Secretariat placed within the Ministry of Planning and Economic Development (ONCCP). Then the process came to an abrupt stop with the civil war that was to end the Mengistu regime.

- At that point in time, environmental policies were dominated by government programmes, partly financed by external donors. A national civil society in environmental terms hardly existed at this stage, much less was it included in biodiversity policy-making.

The initiation of the NCS legislation process also marks the initiation of broader interactions between the international and national level. Due to the seclusion of the past political regimes, Ethiopia has had a relatively limited association with international organisations and agreements. Ethiopia is a member of the Environmental Liaison Centre International (ELCI), International Council of Voluntary Agencies (ICVA), International Planned Parenthood Federation (IPPF) and the IUCN, all long-standing organisations with broad membership. Ethiopia joined the Convention for the Control of the International Trade in Endangered

---

[198] The presentation in this section is based partly on my own reading of the five NCS documents and partly on interviews with central actors in the NCS legislation process. As some of the interviewees expressed the wish not to be quoted directly, I refer to the codes in the confidential list of interviewees (see Appendix 1).

Species of Wildlife (CITES) in 1989, and has since ratified the Climate Change Convention, the Montreal Protocol and the Ozone Layer Convention, as well as the Convention on Desertification. It is also a member of the World Heritage Convention, the FAO Code of Conduct on the Distribution and Use of Pesticides, and the FAO Undertaking on plant genetic resources. Ethiopia ratified the Convention on Biological Diversity in April 1994. (Bergesen et al., 1997.)

The *second phase* of the NCS process spans the years from 1991 to 1995, and it is hence moving into the post-CBD phase.

The NCS process was resumed in November 1991 by the Transitional Government of Meles Zenavi. A new direction was brought about by the *decentralisation process*, with the need to formulate Regional Conservation Strategies. Forty inter-ministerial Zonal Task Forces were established in the regions, and reports were compiled on environmental and developmental needs and challenges at the local, zonal and regional levels. Conservation Strategy Steering Committees were established in eight of the regions for co-ordination and guidance at the local, zonal and regional levels (delays in staffing the Harar and Somali Regional Bureaux had so far prevented their creation). Regional Conferences were *planned* to give priority to development and management issues and to reach consensus on strategies to be adopted.

A National Environmental Co-ordinating Committee, consisting of representatives from relevant Ministries and chaired by the Prime Minister's Office, is also involved in co-ordinating this process. Similar bodies have been established at the regional and local levels (zonal and woreda). Five of the regions have submitted their strategies for resource management, based on available information. The regional conferences, however, have failed to take place. The Ministry of Planning is responsible for approving all environmental projects.

- A concept found throughout the NCS documents is *sustainable development*. A common theme is the inclusion of *indigenous knowledge* as a basis for developing the NCS. Another central idea in the NCS process is the *participation of people at all levels* in policy-making. As the decentralisation process has not been concluded, it is still difficult to judge the impact of these changes on policy-making.

The NCS process has been directed from *shifting institutional set-ups*, but the central persons involved have been more stable (more below):

Box 8.2 Shifting Institutional set-up

- The scope and integration of environmental management in Ethiopia were until 1995 addressed by the new, and short-lived, Ministry of Natural Resources Development and Environmental Protection (MNRDEP), and the National Conservation Strategy Office.

(*box continued on next page*)

- Also important for conservation policies in the central government is the Ethiopian Forestry Action Plan, located with the Department of Planning in the MNRDEP (Hillman, 1993).
- In February 1993, the MNRDEP established a National Environment Protection Authority (NEPA), to implement the UNCED Agenda 21 among other tasks. NEPA later changed its name to EPA in response to Ethiopia emerging as a Federation composed of ten nation-states (14 regions).
- When MNRDEP was relocated within the Ministry of Agriculture, EPA initially went the same way. During this short period, EPA staff found their work to be constrained.[199] As it is food security that attracts the main attention within the Ministry of Agriculture, developmental goals have priority over environmental goals within the Ministry.

This second phase of the NCS process in Ethiopia coincided with a world-wide effort by the World Bank *to hasten the pace of conducting environmental action plans in DCs*. The World Bank and IMF had set June 1995 as deadline for DCs to complete their environmental action plans, this being a condition for future grants and loans. Financial resources and external experts were offered to help DCs meet this deadline. In Ethiopia, however, this offer was turned down by national experts, notably Dr. Tewolde, who were of the opinion that, in order to be sustainable, the NCS must be developed by the country's own human resources. Following a study of the Ethiopian NCS by a World Bank Expert Committee, the Ethiopian version was "accepted" by the Bank without further complications. This account is corroborated by Dorm-Adzobu and Hoben (1995:2), who conclude that, "Unlike many such projects in Africa which are primarily donor-driven, the NCS process was a sincere effort on the part of the Ethiopian government which realized the need for action because of recent droughts and associated natural disasters."[200]

This phase, according to resource persons central in the legislation process, was also characterised by a considerable increase in *environmental awareness* – an awareness that is, however, still rather *limited to the academic and political elite.*[201] This is not to say that environmental concerns were unknown in Ethiopia prior to the initiation of the NCS process. Within the plant genetic resources (domesticated genetic resources) sector, there are long traditions of awareness about the need for conservation and sustainable use of local cultivars of plants.

199 Interviews with several high-ranking EPA staff (C7, C8), February 1996.

200 Several interviewees also pointed to the tendencies of overlap in the multilateral donor community: This includes the NCS processes as advocated by the IUCN, and the Agenda 21 and NEAP (national environmental action policy) as pushed by the UNDP and the World Bank respectively. A general sentiment was that these are but different labels for the same policies and it diverts a lot of attention from the content of the legislation process.

201 Interviews with N4 and C4, February 1996 (App. 1).

The *third phase* of the NCS process, commencing in 1995 and still in progress, is devoted largely to a further examination of national strategies and priorities in the regions, as well as environmental capacity building. With *external assistance* from (possibly) the Norwegian Directorate for Development Aid (NORAD), the Swedish International Development Agency (SIDA) and the IUCN, the regions will develop their own action plans, and also strengthen capacity building in EPA for environmental management.

The institutional set-up is being currently taking shape, indicating a potential strengthening of the *institutional resources* factor:

Box 8.3 Institutional set-up

- In 1995, EPA was located directly under the Prime Minister's Office, rather than being linked to one specific ministry, and this move seems to indicate a wider scope of leverage for EPA. EPA is now accountable to the Ethiopian Council of Ministers.
- There are 15 ministries and agencies involved in co-ordinated implementation of the Federal Policy on the Environment. Those with direct involvement include the Ministry of Agriculture, the Ministry of Mines and Energy, the Ministry of Public Works and Urban Development, the Ministry of Trade and Industry, the Ministry of Water Resources, and the Ministry of Transport and Communication.
- Overall co-ordination of all federal programmes is undertaken by the Ministry of Economic Development and Co-operation.

Moreover, the Conservation Strategy of Ethiopia (CSE) is now in place, based at the Environmental Protection Authority (EPA). The NCS is expected to be phased out in 1998, while the CSE has given birth to the *Environmental Policy of Ethiopia*, which was approved in April 1997. EPA is responsible for the follow-up of the policy aspects of the Strategy. EPA also has the mandate to prepare environmental protection policy and laws, follow up their enforcement, handle the overall co-ordination of environmental matters in the country and also follow up the implementation of international treaties on environmental protection to which the country is a part. EPA is assigned to develop the regulatory framework of the NCS. In this work, EPA gets input from several actors and agencies:

- Co-operation with government institutions from various sectors: among them EWCO, the Forest & Wildlife Conservation & Development Department and the Biological Diversity Institute.
- Input from the regions, and assistance from experts and scientists from the Ethiopian Wildlife and Natural History Society (EWNHS) and the University in Addis Ababa.

EPA thus co-operates both with the public and the private sector. Concurrently, EPA continues the process of working out the details of a legal and institutional framework with regard to the central institutions. This involves such matters as clarifying the mandates of the Biological Diversity Institute (Ethiopian Genebank) and EWCO.

With regard to *human resources*, EPA consists of about 60 approved positions, 40 of which are professional[202] and 20 are support staff. These positions are filled by staff who were transferred from various government institutions, but many have limited qualifications in terms of what their present assignments require. The situation regarding equipment and supplies is critical, as everything from paper to computers is lacking. EPA is led by Dr. Tewolde B. G. Egziabher. Dr. Tewolde is the director of the National Herbarium, and he was employed by the IUCN to direct the first phase of the NCS process. Dr. Tewolde has also headed several of the Ethiopian delegations to the international biodiversity negotiations since 1992. The NCS Secretariat, also led by Dr. Tewolde, consists of six persons responsible for furthering the conservation process.

The Norwegian (NORAD) input has been envisaged as a co-operative arrangement between EPA and the Norwegian Directorate for Nature Management (DN).[203] The project may involve the establishment of an environmental laboratory and environmental information system, among other things. The background for this support is primarily the critical need for equipment and financial resources within the environmental agencies in Ethiopia. For example, the lack of paper makes it difficult to make copies of the NCS documents for distribution to the various ministries and regions. The main goal will be to strengthen EPA, relying as little as possible on external experts, but rather building on *domestic human resources* in Ethiopia. A central concern is also the serious differences in economic and political capacity between the regions.[204] So far, however, the project application, which was sent in early 1996, has not been approved.[205] The background for this delay is, according to NORAD, problems with Ethiopian authorities, who resist following up alterations in the project proposals.[206] According to other Norwegian experts involved, this delay is due to the wish of the Ethiopian authorities to rely as much as possible on strengthening own human resources and capacity building.[207]

---

[202] That is, holding M.Sc., B.Sc. or Ph.D. degrees.

[203] The Federal Democratic Republic of Ethiopia, Environmental Protection Authority (EPA). Draft: *Project Proposal submitted to NORAD on capacity building and implementation of selected projects*. Addis Ababa, May, 1997.

[204] Interview: n4, 13 February 1996 (see Appendix 1).

[205] According to NORAD, project approval is now in the pipeline. The project will eventually include NOK 18 million over three years, and focus on capacity building in EPA. (personal communication from n1, 18 May 1998 (see Appendix 1).

[206] Interview: 18 May 1998. (Appendix 1: n 1).

[207] Interview: 18 May 1998. (Appendix 1: n 2).

The Peasant Associations have had some part in the NCS process. Increasingly, however, these are becoming politicised and part of the government bureaucracy, rather than representing the population at large.

Summary

Until we have evaluated the NCS legislation with a view to its compatibility to the objectives of the CBD (next section), there are little grounds for drawing conclusions. Here I shall merely point out the main features characterising the legislation process, which may later be linked up with the evaluation of the legislation output. Four elements stand out as significant, novel or both, in characterising the NCS legislation process:

- First, the initial phases of the NCS process have coincided with a significant change in the political system. Ethiopia is currently embarking on a democratisation process, albeit hampered by several complications and subject to critical reviews.
- Second, there has been a frequently changing institutional set-up to deal with the NCS process. This is most significantly shown by the fate of the Ministry of the Environment, even though the central actors (elite) mostly remain the same. This elite is characterised by a high level of expertise, most particularly in the seeds sector. Although the elite has thus kept a firm grip on decision-making, it would seem that the institutional resources available to policy-makers in the environmental sector are weak. This might well be deliberate, but it sheds light on the rather limited scope for state capacity within environmental policy. On the other hand, it should be noted this the lack of SOPs may go both ways, as it may also enable a more flexible approach to a relatively novel policy area.
- Third, there has been a great emphasis on utilising domestic human resources in the legislation process. The ideal of building on the country's own human resources can also be seen as an integral aspect of the decentralisation process, although the actual participation at lower levels of government may still be limited. The focus on human resources was shown by the decision to turn down the World Bank's offer to construct the environmental action plan for Ethiopia. It is also illustrated by the problems involved in getting the NORAD project off the ground. These episodes may be interpreted as a symbolic gesture to demonstrate Ethiopia's independence vis-à-vis the external donor community, but the whole idea of independence clearly has deep roots in Ethiopia.
- Fourth, there has been a coinciding increase in environmental awareness in Ethiopia, but this is also primarily limited to the central elite.

An important question concerns the role of the Ministry of Agriculture, which has alternately been given top responsibility for environmental affairs:

- On the one hand, agriculture represents the main economic sector in Ethiopia (involving some 85% of the population), a fact that would seem to provide a rationale for treating environmental questions within this ministry.
- On the other hand, there are obvious conflicts between the need to increase agricultural production, and conservation concerns. The friction is illustrated by the fate of the short-lived Ministry of the Environment, but that potential conflict seems to have been resolved by the establishment of the largely autonomous EPA. This means that control over environmental affairs has been removed once more from the Ministry of Agriculture.

The relatively new and unstable political situation may work both ways: On the one hand making it easier to criticise former policy inadequacies, but on the other making it harder actually to execute new policies.

8.3.2 General contents of the NCS

Volume I of the NCS focuses on environmental and developmental challenges in Ethiopia. The concept of participation and the relationship between environment and development needs are dealt with extensively in Chapter 1 of the NCS.[208] An example of what is termed *vertical disintegration* in this thesis is elaborated here:

> When we conserve something we must use it or keep it in such a way that it is obvious to everybody that it is being kept because it will be of better use in the future. Even then, we can sell the idea of future use only if the people affected by the conservation measures are well-off to afford to look to the future. For example, protecting the elephant from extinction is a good idea. If the process of protection hurts no one, all will accept the idea. But the elephant will want space, a specific space, not a portion of everybody's space. It will destroy a small group of trees and crops, not a portion of everyone's trees and crops. The people negatively affected by the elephant will be only those whose space overlaps with that of the elephant's range. On the other hand, if the elephant were killed, those very people whose crops and trees it would have destroyed will benefit from its tusks. It is, therefore, clear that left to events, the elephant will get killed. If society and humanity at large want the elephant protected, they must make good what those people whose territory overlaps with that of the elephant lose by its protection (NCS, Vol. I, p. 9).

The National Conservation Policy is meant to cover the following areas:

- Wildlife policy

---

[208] Chapter 2 describes the NCS process, the input from regions and the sectors involved, as well as cross-sector issues. Chapter 3 provides a broad outline of the natural resources base.

- Trees (covered by the Ethiopian Forestry Action Plan; lead agency Department of Forestry & Wildlife Conservation and Development)
- Genetic resources, particularly useful plants (covered by a genetic resources strategy; lead agency the Biological Diversity Institute)
- Fish of commercial value (lead agency Department of Fisheries)

The Ethiopian Genebank, now labelled the Biological Diversity Institute (BDI), enjoys an independent status, as some of its funding comes from external sources. Ties are now being developed with the Ethiopian Wildlife Conservation Organisation (EWCO), with a view to improving conservation in national parks. This is intended to fill some of the gaps which have previously hampered biodiversity conservation in Ethiopia. Previously, conservation was viewed mainly in terms of tourism; now the chains of ecosystems and the importance of local peoples' participation are seen as essential to conservation. This includes the realisation that revenues should be transferred back to the local communities.

The NCS is the first attempt to address all aspects of biodiversity in Ethiopia. This is the first time that wildlife and plant genetic resources are proposed to be handled under the same policy-framework. A general sentiment among my interviewees is that this element can be seen as a direct effect of the CBD.

In a discussion paper preparatory to the new NCS the relationship to local communities of people is also addressed in a novel manner:

> The protected areas had been quite alienated from the people around them and the overthrow of the repressive regime of the Derg was taken as a signal to invade the unpopular parks and slaughter the wildlife. It was not, and is not, the intention of the new government that the country's heritage should be thus devastated, but the resentment was understandable and the destruction could not be prevented. The people around the parks are, for the most part, poor and in need of resources. A lesson was learned the hard way, that wildlife conservation which ignores the needs of the people cannot work in Ethiopia.[209]

The NCS lists several guiding principles for the *conservation and management of genetic, species and ecosystem biodiversity*. Those pertaining to wildlife include:

- updating and broadening the knowledge base to strengthen the planning process for the "conservation and sustainable use of biodiversity";
- the primary conservation goal is the "maintenance of national biological resources to meet the needs and aspirations of present and future generations";

[209] Wildlife Policy Discussion Paper, the suggestions and arguments of a wildlife policy consultant, based on studies of documents and discussions with headquarters and field personnel in EWCO, Central and Regional government natural resources and environment personnel, conservation and development NGOs, the tourism sector, professional hunters, researchers, planners and others. 16 August 1995.

- "local communities inside and outside protected areas should be involved in the planning and management of such areas";
- national parks, conservation programmes etc. should "allow for a major part of any economic benefits deriving therefrom to be channelled to local communities affected by such programmes".[210]

Ethiopia has developed twelve sectoral strategies under the NCS:

Box 8.4: Twelve sectoral strategies under the NCS

1. Improved land husbandry for sustainable development
2. Improved crop and animal husbandry for improved agriculture
3. Rangeland management and pastoral development
4. Forest, woodland and tree management
5. Conservation of biodiversity
6. Water resources development and conservation
7. Energy resources
8. Minerals development
9. Sustainable human settlements and the urban environment
10. Control of industrial pollution and hazardous materials
11 Air pollution and climate change
12. Conservation of the cultural heritage.

In addition, eleven cross-sectoral areas have been identified:

Box 8.5: Eleven cross-sectoral areas under the NCS

1. Population and resources balance
2. Peoples' participation,
3. Land and resources tenure and access rights
4. Land resource use policy and strategic land use planning
5. Social, cultural and gender factors in resources management
6. Economic sustainability
7. Environmental information systems
8. Environmental research
9. Science and technology for sustainable development
10. Environmental impact assessment procedures
11. Environmental education and public awareness

---

[210] Another six principles address the manner in which *local communities* may be defined and how local people may be included in the management and planning processes.

These sector and cross-sector issues are treated at length with a view to policy objectives, principles and strategies in Volume II of the NCS.

Along with reports from the various regions these reports constitute the bases for the redrafted Vol. I and Vol. II of the NCS, finalised in December 1994. Volume III, on the institutional framework and operational arrangements for the NCS, is currently being redrafted to comply with the new political and institutional situation. Two additional volumes were also finalised in 1994: Vol. IV on a five-year action plan and Vol. V making a list of ongoing and needed projects with a bearing on the environment. These volumes have been approved by the Council of Ministers, and will eventually be transformed into legal terms.

Volume V presents a list of projects, some of which are funded and being carried out; others only proposed, with estimated costs. This is expected to become Ethiopia's environmental investment programme document. The projects have not been fully evaluated and ranked, and the document does not provide a finalised listing of projects. Only those projects that were considered of direct relevance to the sustainable development, use and management of natural resources and the environment were included. According to Volume V, this does, however, constitute the most comprehensive compilation of projects available, and it is thus the nearest to an investment programme on natural resources and the environment available. As the document does not distinguish between proposed and existing projects, it clearly has severe shortcomings with a view to portraying the economic situation within the various sectors. Nevertheless, it is the most comprehensive compilation of projects so far, and it will give an indication about the priority and weight given the various sectors.

There are two general tables in Volume V, describing the investments in the Cross Sectoral Support Programmes and the Sectoral Programmes. In addition, there are three tables on investments in Conservation of genetic resources, Species & Ecosystem Conservation and Protected Area Network, and Capacity Building and Institutional Support.[211] The tables show that most of the funding to the environmental sector comes from, or is anticipated to come from, *external sources*. This is not least true for the sub-sector of biodiversity, with only some 7% being provided by the government of Ethiopia (GoE). The related sub-sector of forestry presents much of the same picture, with only 2% GoE spending. By comparison, sustainable and improved agriculture not only account for 30% of the total budget, but 37% is expected to come from government funding. As several other sub-sectors draw relatively more on GoE resources, this could be interpreted in two ways:

- On the one hand, this may simply reflect a limited commitment to biodiversity policies on the part of the government. This would imply a lack of will and a lack of ability to pursue improved policies in the area.

---

[211] In addition to Volume V, the tables can found in Appendix 2 in the dissertation version of this book: G. Kristin Rosendal (1999) "Implementing International Environmental Agreements in Developing Countries: The Creation and Impact of the Convention on Biological Diversity" (Oslo: Unipub, Oslo University).

- On the other hand, this could mean that the GoE regards biodiversity as an area for which external funding is more likely to be forthcoming. In that case, there could still be willingness to implement the CBD, even if ability were limited.

The tables in Volume V also indicate the difficulties involved in singling out biodiversity as a specific field. While species and ecosystem conservation represents a distinct component of biodiversity policies, projects focused on forestry conservation and grain reserves are defined as components in other sub-sectors. The latter two could easily qualify as biodiversity projects, and it is symptomatic that the Ethiopian Forestry Action Plan (EFAP) is listed as a source also within the biodiversity sector. It is hence difficult to conclude that biodiversity projects represent only 5% of the total budget. This interpretation is also supported by the NCS documents themselves, where wildlife policies, forestry, fisheries and genetic resources are for the first time addressed within the same framework. This is discussed in further detail in section 8.4 on evaluation.

Finally, it must be emphasised yet again that significant uncertainties underlie this investment programme. This is best shown by comparing the figures with the estimates made by Hillman (1993) of average expenditures in the wildlife sector throughout the 1980s. In terms of US dollars, total expenditures for the wildlife sector came to about US$0.7 million. By contrast, total expenditures envisaged in the NCS compilation of projects for wildlife amount to US$139.7 million. Event though these figures relate to projects that are expected to run for several years, and also include both proposed and actually funded projects, the estimates still seem too good to be true. These differences obviously indicate the great need for caution in comparing these sources. Another significant difference between the Hillman estimates and the NCS investment plans is that the former consist of about 80% *domestic* coverage, while the latter expect 93% *external* funding. I do not have access to chronological comparable figures (similar to the Hillman study) for the plant genetic resources (pgr) sector. The NCS division between GoE money and expected external money is, however, basically similar.

- This goes to emphasise the great expectations on the part of the government of Ethiopia regarding the possibility of drawing on external donors in the biodiversity sector.
- It also raises the obvious question of whether there is a realistic basis for nurturing such high expectations.
- Another important question is how much of the increase in external funding (if any) can be put down as an effect of the CBD. I shall return to discuss these questions in greater depth, subsequent to the presentation of the remaining domestic factors.

8.3.2.1 The Seeds Sector: Content

The new legal framework for domesticated biological material is nested in a national White Paper, the so-called Mother Paper: the *National Policy and Strategy for Plant*

Genetic Resources Conservation and Development.[212] This White Paper includes an overview of gaps in conservation, as well as national plans and strategies.[213]

Box 8.6: White Paper on Plant Genetic Resources: main objectives

- to ensure that the Ethiopian plant genetic resources (pgr) are preserved, developed, managed and sustainably used;
- to integrate programmes for pgr conservation and development into national and regional development strategies and plans;
- to build scientific capacities to explore, assess, study, systematise, introduce, improve, manage and sustainably use biological resources;
- to develop capacities for the improvement, generation, development and sustainable use of biotechnology and its safe transfer;
- to encourage the participation and support of local communities in pgr conservation and development, and ensure that farmers/communities share the benefits accrued as a result of using indigenous germplasm;
- to recognise, foster and augment the traditional methods and knowledge of the local communities relevant to the conservation and sustainable use of pgr;
- to promote international and regional co-operation in pgr conservation and development; and
- to create a functional and efficient organisational set-up to facilitate co-operative action and co-ordination in pgr conservation and development.

Following the legislative framework of the CBD, the White Paper confirms that:

- Ethiopia has national sovereignty over its genetic resources, and that any exploration, survey, inventory, collection, movement, importation, exportation, exchange and use of plant genetic resources – domesticated and wild plant species, and micro-organisms – will be governed by the laws and regulations of the country (White Paper 5.06).
- Moreover, (5.09) to effectuate the national pgr policy, the adoption of the White Paper will be followed by the preparation and promulgation of necessary legislation and regulations in the areas of plant germplasm, medicinal plants, flora and natural vegetation, forest genetic resources, microbial genetic resources, biotechnology, biosafety, germplasm exchange, and plant breeders', farmers' and community rights.

[212] September 1993, Addis Ababa, The Transitional Government of Ethiopia.

[213] This is not yet an official document, but was provided with the understanding that it would probably remain very much intact, so I could make references to it. Documentation was exceptional in this area, as I was given access to the background documents for the Mother Paper; "*Highlights of current status of the agricultural sector*" and "*Gap analysis*".

There is also recognition of the need to employ strategies adapted to suit the different regions, also with a view to seeds policies. The severely degraded areas in the north, with substantial genetic erosion, are very different from those in the south. As seen from the perspective of the Genebank authorities, the challenge now is to differentiate between the various needs of conservation.[214]

In order to follow through this intended national policy, the White Paper proposes to *strengthen and upgrade some of the institutions* already in operation (5.35–5.39): the Plant Genetic Resources Centre of Ethiopia (the Genebank), the National Herbarium (a university- based agency) and the Forestry Research Centre, as well as to establish new institutions: a Microbial Genetic Resources Centre and a National Centre for Medicinal Plants. As a political and legal co-ordinating body, the White Paper envisages the creation of a National Board for Plant Genetic Resources, to be chaired by the Ministry of National Resources Development and Environmental Protection.

In the words of the Country Report for Ethiopia to FAO:

> Within the existing efforts of establishing a sound biodiversity conservation system, the already existing CPGR/E (Ethiopian Genebank) is promoted to a National Institute of Biodiversity and five plant genetic resources programmes are organised within it. These are programmes for Crop Genetic resources (already existing strong programme), Medicinal Plant Genetic Resources, National Herbarium (upgrading existing one), Forest Genetic resources (upgrading existing one) and Microbial Genetic Resources programme. This is a major step forward .. with regard to the fulfilment of global commitments such as The Convention on Biodiversity.[215]

Article 15.5 of the CBD provides the central provisions for the development of national seed legislation: "Access to genetic resources shall be subject to prior informed consent of the Contracting Party providing such resources". According to Dr. Tewolde this has had a very favourable impact on Ethiopian seeds policy, and the government is highly responsive to following it up.[216] The gist of the seeds legislation is to create an office for control – a national seeds industry agency. This has already been gazetted (i.e. registered as law) in the Negarit Gazeta (56/1993): *Proclamation to provide for the establishment of the national seed industry agency.*

---

214 Three zones have been identified: *high erosion*, most genetic variety lost due to displacement by higher yielding varieties; *initial bases lost,* but still some variability left; areas where *most variability* is contained. Different strategies are needed, and this is being surveyed at the moment in the White Paper.

215 Country Report for Ethiopia, *International Conference and Programme for Plant Genetic Resources*, FAO, Commission on Genetic Resources for Food and Agriculture, 2 August 1996.

216 This information was concurred and expanded upon by a former Director of Plant Genetic Resources Centre of Ethiopia (PGRC/E), the Genebank. He viewed the legislation process as a matter of urgency as there are now a number of groups collecting germplasm around the world. (App.1: C3).

Box 8.7 Legislation controlling access to seeds

- A main aim is to ensure benefits to which the farmers are entitled.
- No plant genetic resources (pgr) shall leave the country without legal permission, either in the form of licensing plate, or contracts. This is being developed in the law on the *National Biodiversity Board*.
- The new law will stipulate that 50% of revenues from the genetic material being used shall revert to Ethiopia, as well as 20% of royalties accruing from IPR, in cases where material has been patented.
- The Genebank has been approved by the government to control the gene flow in and out of the country. Another function of the Ethiopian Genebank is to co-ordinate African follow-up of the CBD.

8.4 Ethiopian Biodiversity Policy and the CBD: Evaluation

How then does the NCS score, in terms of consistency between domestic biodiversity policies and the objectives of the CBD? To ascertain this we need a set of measurement standards by which to structure the evaluation. As will be recalled, there are three main CBD objectives: improved conservation, sustainable use of biodiversity, and equitable sharing of the benefits arising from the utilisation of genetic resources. As argued in section 1.2.3, the main indicator of output implementation is harmonisation or integration of the objectives of the CBD in national legislation. This again is indicated by:

1. follow-up in the form of conservation plans and programmes,
2. symbolic language (i.e. using the same terms),
3. policy instruments aimed at conservation activities, and
4. related institutional changes in the government agencies.

These four indicators are relatively straightforward in terms of measurement, but there is still a need for more precise clarification. As it is impossible to go into detail here on each separate paragraph of the CBD, I need a few central aspects on which to concentrate the evaluation. Again, I take as my point of departure the first part of the analysis. In Chapter 6, I decided to concentrate on two clusters of issues, one relating to intellectual property rights (IPR) and one relating to financial aspects. The financial aspects are directly linked to implementation in terms of deliberately allocating resources to conservation activities, and these are thus incorporated in indicators (iii) and (iv) above. The IPR issue mainly concerns Ethiopia's efforts to formulate and enforce legislation controlling access and use of the country's own genetic resources. In the CBD, the equitable sharing dimension primarily aims at improved international terms of trade, but article 8 (j) concerns domestic sharing. Hence we need to look closely at access legislation, but must also bring in the element of local participation in order to study domestic sharing. Accordingly, I add to the list of indicators:

5. legislation controlling access to genetic resources, and
6. local participation.

These six evaluation criteria can be said to move on an implementation/compliance scale from weak to strong, with plans and symbolic language as the weak forms of implementation activities and e.g. the establishment of policy instruments and institutional changes as more strong forms, as they necessitate allocation of financial resources. Obviously, there is also a time component in this continuum, as plans and policy statements will usually (but not necessarily), precede the more concrete measures. Hence, lack of specific measures may be ascribed to the time factor. On the other hand, legislation for access control and local participation cannot be said to be time-dependent. If these types of legislation are found lacking, this cannot be due to the time factor; it rather indicates limited implementation score. These types of legislation will be regarded as strong and important expressions of implementation activities.

In order to pave the way for the analysis in Chapter 9, I should also point out that the six evaluation criteria can be linked up to the four institutional mechanisms identified in Chapter 7. Plans and symbolic policy obviously link up with the *moral mechanism*, while the criteria relating to policy instruments and related institutional change may necessitate financial transfers associated with the *material mechanism*. Legislation for the control of genetic resources is inherently dependent on compatible policies in other CBD Parties, thus invoking the *mutual reassurance mechanism*. The criterion of local participation is most closely connected to the *empowerment mechanism*, as this is the one mechanism directly related to non-governmental actors.

The operation in this section primarily concerns evaluating whether Ethiopian biodiversity policies represent *compliance* with the CBD. Compliance may or may not be the result of deliberate efforts to follow IEA commitments. At this point, I have not started the analysis of *what factors* may account for consistency, and whether such consistency, if found, is a result of deliberate efforts to *implement* the CBD. At the same time, these evaluation criteria are largely concerned with domestic *implementation output* ("activities moving in the right direction"), in terms of harmonisation or integration of the objectives of the CBD in national legislation. Looking back at the outline for this study in Chapter 1, we may note that the intention was to tackle the question of "whether consistency" first, and then go on to discuss "whether implementation". Finding high scores on these six evaluation criteria, however, would hardly imply compliance by default. Neither would consistent usage of similar language on criteria (i), (ii) and (v) imply unintentional compliance. High scores on (i), (ii) and (v), and low scores on the other criteria may even indicate a quite high level of implementation efforts, and a lower level of actual compliance. Hence, the findings are likely to indicate whether these are indeed deliberate efforts headed in the right direction, and aimed at following up CBD objectives.

8.4.1 The CBD and wildlife policies

Evaluation standards:

1. follow-up in the form of conservation plans and programmes,
2. symbolic language (consistent use of the same terms),
3. policy instruments aimed at conservation activities,
4. related institutional changes in the government agencies,
5. legislation controlling genetic resources,[217] and
6. local participation.

The answer to the first measurement standard is partly yes: As the NCS has a much broader scope than biodiversity, as well as constituting a process primarily initiated by external actors, it cannot in itself be linked directly to the CBD. As the components in the investment programme show, the wildlife sector is characterised by a large number of conservation plans and programmes.[218]

With regard to the second evaluation criterion there seems to be ample reason to answer in the affirmative. Many of the formulations in the guiding principles of the NCS are identical to the language used in the CBD. This is exemplified by phrases such as: updating and broadening the knowledge base to strengthen the planning process for the "conservation and sustainable use of biodiversity"; the primary conservation goal is the "maintenance of national biological resources to meet the needs and aspirations of present and future generations"; "local communities inside and outside protected areas should be involved in the planning and management of such areas"; national parks, conservation programmes etc. should "allow for a major part of any economic benefits deriving therefrom to be channelled to local communities affected by such programmes". These formulations also give clear indications regarding the fifth criterion, local participation. Another important aspect is the explicit intentions in the NCS to integrate environmental policies at all levels and across sectors. Hence on the normative (symbolic) level there are many indications of direct impact.

As far as the material aspects are concerned (evaluation criteria [iii] and [iv]), it is less easy to point out activities directly linked to the CBD. This is primarily due to the chaotic political situation that characterised the phases immediately preceding the formulation of the NCS, and the lack of substantial documentation regarding the resources allocated to this sector prior to the NCS and the CBD. In section 8.2, the highly uncertain estimates for the expenditures on and earnings from wildlife per

[217] As there is not much export of wildlife and limited scope for tourism in the parks and reserves of Ethiopia, the CBD may have little direct impact on *access to genetic resources*. In this section I will thus omit one of the evaluation standards, as the one referring to legislation controlling access to genetic resources is hardly relevant in this area.

[218] Table 8.3.2 in NCS Volume V.

average year in the 1980s were quoted (Hillman 1993): average expenditures for total wildlife support were 3,826,200 Birr (approximately US$0.54 million) from the Ethiopian government and 1,150,607 Birr (approximately US$0.16 million) from development assistance, bringing the total support for wildlife to 4,976,807 Birr (approximately US$0.7 million). Comparing this to the sectoral investment programme of the NCS, the total expenditure envisaged for the wildlife biodiversity sector is US$139.7 million. Given this large increase, it is hard to assess the credibility of the NCS figures. In order to judge whether these are realistic, I shall look further into the project portfolio in Ethiopia (section 8.6.2). Even if the figures cannot be substantiated by empirical findings, they nevertheless indicate the great expectations of the Ethiopian government concerning the economic potential of the biodiversity issue-area.

Thus it is difficult to pin down exact changes, although the NCS itself does of course represent a major step towards a more comprehensive policy in the area. It represents the first effort to address all aspects of biodiversity in a coherent manner, and it does go a long way in trying to strengthen the relevant institutions as well as initiating the process of identifying clear mandates between them. It should also be pointed out that the pre-NCS situation for EWCO could hardly have been worse, so even minor efforts would seem to improve their possibilities for executing their tasks. The way EWCO has been pulled in and out of the Ministry of Agriculture, however, illustrates the uncertainty and ambiguities still surrounding this sector.

In this sense, the institutional and legal framework proposed by the NCS may be labelled "activities going in the right direction", even though the links for the NCS as a whole may not be drawn directly to the CBD. Rather, it is an answer to general pressure by the international community, as represented by the World Bank, the IUCN and the UNDP, for DCs to formulate coherent environmental policies. The policy formulations in the biodiversity sector, however, indicate clear connections to the CBD. These are definite indications of deliberate implementation efforts, as the similarity in vocabulary can hardly be accidental.

While funding is in the process of being allocated to park infrastructure, management, and development, there are as yet no budgetary allocations for items such as community benefit-sharing, training, participation or biodiversity surveys. On the other hand, these items have been identified as part of a comprehensive package largely in line with what the CBD prescribes, and this must as such count as "implementation activities moving in the right direction".

In sum, the wildlife sector seems to have:

- A high score on consistency with the CBD in terms of plans and policy formulations (symbolic language).
- A medium score on policy instruments and related institutional change, as these are still pending both with regard to financial allocations and regarding institutional mandates. In view of the relatively short time that has elapsed since the formulation of the NCS, the latter score should not be given much weight.

- The score with regard to participation is more important; while it is clearly moving in the right direction, the pace is hardly fast enough to provide a high score on consistency.

8.4.2 The CBD and seeds sector policies

In this field the fourth criteria has particular relevance for a central question: *Will the Ethiopian administration be able to control foreign gene collectors?* The language, policy instruments and institutional set-up as well may or may not be geared towards this objective.

The first criterion (plans and programmes) has already been subject to a general discussion under the wildlife sector, but some additional comments are warranted. The sectoral investment programme directed at conservation of genetic resources indicates that the financial resources envisaged are small compared to the wildlife and ecosystem components (US$139.7 million against US$22.8 million). Only the coffee genetic resources programme (with US$14.3 million) and the farmer-based approach (with US$3.9 million) receive anything resembling substantial funding, and this is mainly expected to come from external sources.

Nevertheless, the investment programme on biodiversity conservation shows a comprehensive approach to the area. In this sector, the CBD has clearly been used as a regulatory framework within which domestic legislation is nested. This is partly reflected by the symbolic language in the objectives in the White Paper, describing the plans and programmes in this policy area, and with its emphasis on "preservation, development and sustainable use of pgr", "to integrate programs for pgr conservation and development into national and regional development strategies and plans", "to build scientific capacities to explore, assess, study, systematise, introduce, improve, manage and sustainably use biological resources", "to develop capacities for the improvement, generation, development and sustainable use of biotechnology and its safe transfer", and "to create a functional and efficient organisational set up to facilitate co-operative action and co-ordination in pgr conservation and development". The White Paper confirms that (5.06) Ethiopia has national sovereignty over its own genetic resources.

"Local participation" (criterion vi) is also comprehensively addressed in the White Paper: "to encourage the participation and support of local communities in pgr conservation and development, and ensure that farmers/communities share the benefits accrued as a result of using indigenous germplasm", and "to recognise, foster and augment the traditional methods and the knowledge of the local communities relevant to the conservation and sustainable use of pgr".[219]

---

[219] The role of the Peasant Associations is potentially important in pgr conservation and development. A major uncertainty now concerns the land tenure system; the outcome here will be of great importance for what will happen to local community rights. The Transitional Government left this point to be handled by its successor, since the political aspect of this issue so "hot".

The formulations regarding plans and local participation can all be traced back to the CBD text. Most of these objectives obviously do not go against "Ethiopian national interests", and it could hence be speculated that Ethiopia would have formulated such regulations also in the absence of the CBD. On the other hand, without the international agreement, this kind of regulatory framework would scarcely have been deemed feasible or useful, as unilateral action is clearly of little use to poor countries in the biodiversity issue-area. An exception to that rule is the formulations concerning local participation and benefit-sharing. These pertain to the domestic distribution of benefits arising from use of biological resources, and are more than what a central "national interest" would dictate in order to reap as much as possible from the new international biodiversity regime.

The symbolic terms are sought strengthened at the material level by envisaging laws and regulations to implement the national seeds policy. Within the sectoral investment programme, *access legislation* (fifth evaluation criterion) constitutes a central component in the programme, although as yet there are no resources allocated to this item. Legislation on access to genetic resources includes the following components:

- The White Paper has already gone some way in formulating the contents of such legislation with its call to "strengthen and upgrade some of the institutions already in operation" and also "to establish new institutions".
- One of the components in this process is the plan to strengthen the Ethiopian Genebank by turning it into the Biological Diversity Institute with five pgr programmes, with a view to fulfilling "global commitments such as The Convention on Biodiversity".
- A central goal in the seeds legislation is to create an office for control – a national seeds industry agency, and this has already been gazetted (registered as law) in the Negarit Gazeta (56/1993): "No plant genetic resources (pgr) shall leave the country without legal permission, either in the form of licensing plate, or contracts".
- Most significantly, the new law will stipulate that 50% of revenues from the genetic material being used is to revert to Ethiopia, as well as 20% of royalties accruing from IPR, in cases where material has been patented.

Again, these aims are completely in line with the CBD. Although they are obviously not contrary to Ethiopia's own preferences, they would hardly have been regarded as feasible or realistic without the CBD.

Feasibility is closely linked to the fourth evaluation criterion (related institutional changes), which has already been touched upon. In the course of my interviews with actors centrally placed in the legislation process, this item was subject to much interest and a mixture of hopes and doubts. There have been instances of illegal collecting of plant genetic resources, and of bribery, in the past.[220] Ethiopia is now in the process of developing directives and establishing guidelines and checkpoints to ensure that this material is not illegally transported out of the country.[221]

According to the director at the Genebank at the time, it has taken some time to convince politicians about the urgency of such matters. The former Director of the Genebank, Dr Melaku Worede, had been gaining quite a reputation internationally at that time, for keeping a firm hand on Ethiopian germplasm. Respondents in the administrative sector now agree that today there is greater awareness among politicians about the need to conserve traditional seeds and varieties, and consensus that this should take place at the local level.[222] At the University, on the other hand, one respondent pointed to the problem of controlling and monitoring deals between the government and external scientific community. In practice, my respondent noted, this would be a question of loyalty.[223]

8.4.3 Summary

The policies in the wildlife sector and the seeds sector are largely consistent with the objectives in the CBD.

Table 8.1:
Consistency between domestic output and CBD objectives: Evaluation

	i) plans & progr	ii) symbolic policy	iii) instru-ments	iv) instit. change	v) control: pgr access	vi) local particip
wildlife	high	high	medium	medium	n.a.	medium
seeds	high	high	medium	medium	high	high

i) and ii): I found evidence of a high degree of correspondence between the Ethiopian National Conservation Strategy and the Convention on Biological Diversity. At this point in time, the consistency is most outspoken at the *symbolic level*, as the NCS often employs language identical to that of the CBD. This is the case *both in the wildlife sector and in the plant genetic resources sector*. My study

220 Under the previous government there were examples of Pioneer Hi-bred (a multinational pharmaceutical company) doing corrupt deals with high-ranking politicians in order to gain access to genetic material. Pioneer Hi-bred knew, however, that they could not get access through the Genebank.

221 Interview: March 1996 (App.1, C3).

222 Interview: March 1996. (App.1, C3) This respondent believes that such corruption will not happen under the new government: "People in the Council of Ministers are directly involved in the seeds issue and they are interested in promoting it". (Ethiopian Genebank).

223 Interview: March 1996 (App. 1, N4). (University).

also affirmed that there is a *high score of consistency both in terms of symbolic language and follow-up in the form of envisaged plans and programmes.*

iii) and iv): With a view to establishing *policy instruments and related institutional frameworks* within the two sectors, there are also several indications of implementation activities. The basis for this conclusion is the comprehensiveness of Ethiopia's biodiversity policies, which largely encompass the broad spectre of activities envisaged in the CBD, including conservation, training, legislation, surveys, benefit sharing and community participation. Moreover, the NCS envisages a high level of co-ordination across relevant sectors, as indicated by the twelve sector and eleven cross-sector issues developed in the NCS documents. On the other hand, it is clear that these policy aspects are still a long way from being enforced and put into practice, as there are often *little or no financial resources* attached to the programme areas; moreover, the final distribution of *institutional mandates is still pending. With some reservations concerning future developments, it seems appropriate to grant Ethiopia a medium score on evaluation criteria three and four.*

v) and vi): The plant genetic resources sector scored high on the fifth evaluation criterion, legislative control of access to germplasm. This criterion was not regarded as relevant to the wildlife sector under the current circumstances. Local participation, the sixth evaluation criterion, had a higher score in the plant genetic resources sector compared to wildlife.

However, this is a difficult item on which to pass judgement. The *concepts of local participation and capacity-building* proliferate in international and national environmental policy formulations, projects and programmes, and at present it is hard to predict their practical implications. To some extent judgement may be based on experience from the past: The plant genetic resources sector is traditionally occupying a central place in Ethiopian politics. There are also long traditions in terms of more locally oriented practices, such as on-farm conservation. Ethiopia's wildlife conservation sector, by contrast, is inherently more inclined to create tensions between top–down conservation concerns and utilisation by local people.[224]

It is still too early to draw definite conclusions regarding compliance score on these evaluation standards, in light of the relatively short time-span. Nevertheless, as pointed out in section 1.2.2 on implementation, the time factor may go both ways: On the one hand, Ethiopian authorities should obviously be given some time to put their biodiversity policies into practice. On the other hand, it is also possible that my snapshots of the policy process have been taken at the crest of a particularly high wave in concern and attention, and that this will not be sustained in the subsequent phases.

---

[224] So far, I have been mostly concerned with implementation output (legislation) and have only briefly commented on the prospects for implementation outcome (change in the behaviour of target groups). That is as it should be; granted the limited time frame for this study, implementation outcome was not expected to have come very far. In the following sections, there is a wider scope to comment also on future prospects concerning implementation outcome.

Compliance by default or deliberate implementation?

The NCS process coincides in time with the establishment of the CBD, and this is hardly accidental. Both can be seen as expressions of greater environmental awareness at the national and international levels. The high scores on most of the evaluation criteria indicate that Ethiopian biodiversity policies are probably more a case of deliberate implementation efforts than mere coincidences. Nevertheless, we must not jump to conclusions, so I shall check for the impact of other domestic factors before passing judgement.

In the next two sections, I go deeper into the two remaining dimensions – *state–society relations,* and the role of *non-governmental actors*. Again, the assumption is that substantial changes in domestic factors may help to explain *compliance* with the CBD. This could be the case if, for instance, there were significant changes in the way non-governmental actors were brought into the legislation process. Hence, I need to control for these factors to a larger extent in order to establish whether to talk about implementation or compliance. A significant increase in NGO capacity to influence and promote biodiversity policies in Ethiopia may for instance itself be an effect of the advent of the CBD on the international arena.

8.5 State–society Relations

State–society relations are assumed to have implications for the balance between an effective central bureaucracy and local participation in policy formulation – the balance between efficiency and legitimacy. As the preceding section on the legislation process has dealt with the participatory aspects, I will here focus on *integrative capacity*, as indicated by intrapolicy co-ordination at different administrative levels and across sectors.[225] Have there been recent changes in these relationships? That might help to account for change in biodiversity policies in Ethiopia.

In my view, the situation regarding state–society relations in Ethiopia is shown by the debate on *decentralisation*. Most prominently, this concerns the disputed division of power between different levels of government in the wildlife and natural resources management sector. This debate goes a long way in revealing how the issue is approached in Ethiopian politics, and the developments so far indicate the direction of the decentralisation process. Once more, I shall look for developments on this dimension which may be said to constitute significant shifts, and which may help to account for domestic biodiversity legislation.[226]

[225] The other indicator is "participative capacity", operationalised as participation in general elections, in national environmental polices, and in long-term ecological planning. This factor was primarily dealt with in section 8.3 on the legislation process.

[226] If so, H0 is strengthened; if not, there will be added reason to examine how institutional factors have influenced domestic implementation in Ethiopia.

8.5.1 Integrative capacity: Central and local bureaucrats[227]

In 1993, a government proclamation set out proposals for the division of authority between the central and regional level in wildlife management (Negarit Gazeta, No. 41/1993). As interpreted by the authorities of Oromo Region, the proclamation confers authority to the regions over sanctuaries, reserves and controlled hunting areas, leaving only the gazetted (registered) national parks under the auspices of the central government. Initially this seemed to imply that only the two gazetted national parks would be left with the central government, and the Oromo aims were to achieve control over natural resources management in all other areas within their region. This interpretation diverged sharply from that of the Ethiopian Wildlife Conservation Organization (EWCO), as well as from that of other regions, e.g. the New Southern Region. According to EWCO, the proclamation provided the regions with control over reserves and controlled hunting areas, leaving sanctuaries and proposed as well as gazetted national parks under the authority of the central government. The latter was also the understanding with the authorities of the New Southern Region, although they were aware of the Oromo interpretation.[228] EWCO is still the natural link between government and regions, as it has a large network of contacts. In 1996, the question of division of power was resolved: all responsibility, including gazetted national parks, is now devolved to the regions.

The question of decentralisation is still under negotiation, and so far the Oromo Region has not received management control over wildlife. They have, however, received control over forestry and soil conservation – one reason for wanting to obtain an integrated responsibility for wildlife as well. The other reason, which was greatly stressed, was that it coincided with the wish of the local population that the Oromo Region should be granted management responsibility over these areas. Decentralisation would guarantee better wildlife management strategies, as the regional level was closer to the people, and had better knowledge of their culture, needs and aspirations. It was also stressed that local people must retain access to the use of natural resources; furthermore, to the extent that they are blocked from such utilisation, mechanisms must be developed to ensure that the local populace gets benefits as well as compensation for any loss. This could be in the form of access to

---

227 This section is based primarily on interview with R1 and partly R2, March 1994: The interviews were primarily concerned with the decentralisation process. I could not interview representatives from all the regions, partly because of my own financial and time-constraints and partly due to lack of organisation in several regions. Oromia represents by far the largest ethnic group, and their dealings with the central government are likely to serve as a role model for the smaller regions; hence I take their perceptions of the situation to be relevant for the relationship between the different levels of government in general. This impression is supported by my interviews with representatives from the New Southern Region (R2).

228 Interview: R2 March 1994 (Department Head of the Natural Resources Development Unit in the Wildlife Office of the New Southern Region). (App.1, R2). Here I was informed that the other regions did not have the necessary qualified staff, and hence, they did not have the same ambitions as did the Oromo region in this respect.

water, establishment of local hospitals and various forms of employment related to the management of parks and sanctuaries. The central government was criticised for neglecting local communities in their wildlife management policies in the past. The case of Senkelle, it was maintained, had failed because it was imposed from the top down without participation from the local communities.[229]

In response to whether the Oromo Region is in the process of developing wildlife management plans of their own, the respondent replied in the negative, saying that this could not be done until the local authorities knew for sure that they would receive authority over the resources. This explanation carries some ambiguity in light of the fact that the Oromo people in the Oromo Wildlife Team have previously been employed by EWCO, without succeeding in establishing better relations with the local communities during their employment there.[230] There are, however, also other factors constraining the possibilities for the Oromo office to develop plans of their own – not least the problem of very limited access to documentation from the central administration. For example, they had information about the UNDP plans for projects in the Oromo Region, but so far (as of 1994) they had had no contact with the UNDP.[231]

My respondents pointed out that the government may now be getting cold feet with regard to decentralisation, as they fear losing revenue from the resources, as well as losing employment. The employment aspect refers partly to the sheer number of positions, and partly to the government's interest of keeping employment in the urban areas. Even if they should keep their jobs, decentralisation would mean that many central government employees would have to move out to the districts. It is difficult and unpopular for the professional staff to move out into the districts, where most infrastructure is highly inadequate. The situation is similar in the wildlife and plant genetic resources sectors: Prior to the regime change in 1991, the Ministry of Agriculture concentrated people and resources in the centre, and this is cited as one reason for the lack of success of the agricultural system in Ethiopia. That is now changing; only 200 out of 3000 staff remain in the Ministry in the centre, and this has been a conscious move aimed at enhancing agricultural output in the districts.

Respondents from the central government maintained that decentralisation would be beneficial to local participation and resources management, although there will still be fear among people that they will not have access to the resources.[232]

---

[229] The "failure" of Senkelle refers to the 1991 civil war when local people turned their anger against EWCO buildings and offices within an established sanctuary.

[230] Pointed out by Per Ole Syvertsen of the University of Oslo. Syvertsen has carried out scientific work in the Ethiopian wildlife sector for six years, and has had extended contacts with EWCO during this time.

[231] Moreover, the decentralisation process is assumed to provide external donors with greater leverage: As the regions approach the donors individually, they will have a weaker bargaining position vis-à-vis donors; and at the same time there is a tendency for donors to co-ordinate their activities in Ethiopia (Hansson, 1995).

[232] Interview, March 1996, with Dr. Tewolde, the director of the NCS process.

Dr. Tewolde pointed to external actors as potential stumbling blocks in the decentralisation process. He argued that international assistance agencies generally want to put leverage on the negotiation process, and tend to favour central authority as this facilitate their dealings in Ethiopia. The European Union, embarking on a project for resources management in the south of Ethiopia, was cited as an example, as they prefer to work with the central government. This is in line with the experiences from Oromo Region, where international agencies shied away from informing and contacting the regional levels.

The decentralisation process was also discussed with interviewees in the *plant genetic resources* sector. High-ranking administrators at the Ethiopian Genebank stressed the need to gather control in the central government in order to secure an effective central bureaucracy.[233] The argument was that the government needs to collect revenue, because different regions may have different needs, and co-ordination and central planning are needed to ensure that revenue is distributed in a sensible manner to encompass these different needs. This is also the problem with the concept of *farmers' rights*: how to identify the person who did most of the work, when there are so many levels of input in this process? It was generally felt that it would be better to link the concept to *community rights*, but even then there would be a problem of distribution between the regions, and also between the local communities, due to the division of labour. Still it is *important to integrate the participation of local communities*. Thus, in the view of central officials at the Genebank, the regions go too far in aiming at separatism. On the other hand, it was conceded, some decentralisation is important for communication with the local level.

Echoing these views, a general sentiment within the scientific community was that decentralisation would have different short-term and long-term effects. In the short term, full decentralisation could prove hazardous for wildlife management, as the regions would aspire to develop at the same rate and would then have incentives to put the resources under their jurisdiction into immediate use. The government would be unable to control and set priorities within this scenario. In the long term, however, it was conceded that this might constitute the best functioning solution as it would further empower the people and give them control over their immediate environment. Local communities would benefit directly, for example from tourism, and this would give them a stake in conservation.

Are the factors associated with state–society relationship relevant for explaining consistency between Ethiopian biodiversity policies and the CBD?

The decentralisation process remains a major stumbling blocks in the development of environmental policies in Ethiopia. Here we should distinguish between the effects of *decentralisation* and the effects of the *decentralisation policy process*.

---

[233] Interview: March 1996 (Appendix 1, C3).

- The first aspect concerns uncertainty regarding the effects of *decentralisation* in itself. At present, there is great uncertainty among politicians and scientists regarding the effects of decentralisation on natural resources management. There is widespread awareness of the delicate balance between the need for effective central co-ordination and enforcement on one hand, and the need for local participation to legitimise policy formulation on the other. Giving the regions a clear mandate of resource management may in the long run enable and force them to develop the necessary *capacity*. More critical is the *uneven distribution of capacity* among the various regions, a factor which may induce unsustainable resource utilisation in less advantaged regions.
- Another aspect relates to the unresolved division of power resulting from the *decentralisation policy process*. As long as the division of power and control over natural resources remains unresolved, considerable time and energy is tied up in this issue, diverting attention from and probably postponing a final result in terms of legal *output*, including the institutional framework. The land tenure system and the national economy is still under EPRDF control. Moreover, the international donor community is primarily geared towards dealing with central authorities. As these seem to be hampering factors, they cannot be accounted for the boost in domestic biodiversity policies, and hence give no immediate support to H0.

The impact of the decentralisation process may eventually go deeper than the question of control over natural resources by central or regional levels of administration. At the core here is the political content of the decentralisation drive: Is it primarily a reaction to excessive central control of the past regimes, or is it opposed to the nation-building project of Ethiopia itself?

8.6 Non-governmental Actors

The potential effect of *societal pressure groups* was assumed to work in two directions – either *facilitating or hampering* the state's ability to comply with and enforce the regulations in question. The themes of this section are:

- whether we may identify *environmental proponents*, directly or indirectly boosting the government's policies; and
- whether it is likely that implementation of government programmes and policies may be hampered by *vertical disintegration*. This was assumed to be a possibility when government policies were found to carry explicit costs to specific sub-national target groups in the domestic implementation phase.

First, the main indicators of the factors on this dimension include the autonomy and professionalism of domestic NGOs, and the relationship between government and the international donor community. Have these factors undergone major changes coinciding with, and accounting for, changes in domestic biodiversity policies?

Second, I take non-governmental actors in Ethiopia to include the bi- and multilateral *external donor community*. These actors are not part of the Ethiopian government, but may still have an impact on Ethiopian policies, as in the development of domestic environmental policies. I will present and examine some of the major projects which are more or less externally driven.

A finding that external donors or the domestic NGO community have increased their influence on domestic policies does not, however, necessarily mean a reduced scope for effect by the new bioregime. The principles embedded in the CBD may have been adopted by non-governmental actors who advocate these in their dealing with government agencies. There is also the possibility of the fourth institutional mechanism of empowerment coming into play. This is the subject of section 9.5. Likewise, it may be difficult to distinguish between non-governmental environmental proponents being strengthened by international epistemic communities, and hence, able to put increased pressure on governmental policies, and increased influence by non-governmental environmental proponents.

The first could indicate a weakening of H0, and the second could strengthen H0. How then can we trace the rather long and complex chains and mechanisms under study here? As shown in Table 7.2, the scope for identifying impact in terms of empowerment will depend on whether we can identify a strong NGO community, and distinguish activities by an epistemic community linked to the biodiversity issue-area.

8.6.1 Domestic actors: public environmental awareness

Public environmental awareness within Ethiopia is not focused especially on biodiversity, but there is *growing concern about environmental issues*. The recurrent droughts and soil degradation have made people increasingly aware of the environmental links.

Environmental awareness and activity seems particularly outspoken in the Tigray Region in the north. This is partly because the droughts have been more serious here, but it is also because this is the home of the rebel movement that led to the formation of the new government. Some of the most important and widely acknowledged NGO work on the environment at present seems to be carried out by the Relief Society of Tigray (REST). Other parts of the NGO community generally commends REST for its work, but also distrusts (and envies) it for its affiliation with the government.[234] Apart from REST, there is no particular tradition for close working relationship between government and non-governmental organisations in Ethiopia. Since the Derg regime, people are in general insecure regarding the government, unsure as to how the new political regime will turn out.

[234] Interview with leader from NGO sector, February 1996 (App.1, N1).

Since 1991, various smaller NGOs and Community-Based Organisations have been arising throughout the country, and for the first time several are now registering as formal NGOs. The most important domestic NGOs in the wildlife sector are, the *Ethiopian Wildlife and Natural History Society* (EWNHS) and *Environment and Development Society of Ethiopia* (Lem-Ethiopia). Both EWNHS and Lem-Ethiopia have long traditions in organising green education courses among schoolchildren. These major NGOs have played very different roles in the NCS process:

- Lem-Ethiopia is registered as a formal NGO with the Ministry of Justice, but has not been directly involved in the NCS process. Registration appears to be a two-edged sword, as it gives the organisation access to duty-free import of certain products, while on the other hand opening up for government control of the organisation's work.[235]
- EWNHS has been reluctant to file for official registration due to the prospect of such government control. Nevertheless, central EWNHS members have been heavily involved in the NCS process as scientists in their personal capacity.[236]

The relationship, or lack of such, between *government and non-governmental domestic organisations* has many facets which may shed an interesting light on political decision-making in Ethiopia. *First of all,* NGOs constitute a *new phenomenon* in Ethiopia, with the exception of Lem-Ethiopia and EWNHS. *Second,* the new legislation envisaged by the NCS process opens for extended grassroots input, as the concept of *local participation* is widely recognised throughout the NCS text. In practice, however, NGO participation has so far been limited to *specialists participating* in the decision-making process in their own personal capacity, and *financial backing* of a few NGOs on the basis of ethnic and political affinity. *Third,* the government is reluctant to extend co-operation with the NGO community in general, because:

- The concept of NGO has wide connotations in Ethiopia, as it is also used to include the private business sector. During my interviews, bureaucrats frequently questioned how representative the NGOs were, as these groups, unlike local levels of the government administration, have not been democratically elected.
- There are signs of sharp competition between government agencies and NGOs in the various sectors: This uneasy relationship is partly brought about by external factors, as the environmental and developmental NGO community is increasingly receiving direct financial support from international donor agencies

---

235 As stressed by one interviewee, the formulation and enforcement of environmental policies may be difficult because ethnicity has now become more important than professionalism. Interview: N1, February 1996. During my interviews in 1996, N1 was not at all familiar with the NCS process.

236 Interview with leader from NGO sector, February 1996. (App.1, N3).

– in effect circumventing the government. On the positive side, this may be a practical way of implementing the objectives of local participation and capacity building. On the other hand, government institutions may end up having less to offer in terms of wages and opportunities compared to the NGO community, so that the final result may be a domestic *brain-drain* from government agencies to the NGO sector.

This latter effect is at present heightened by domestic *institutional factors*, as the government agencies involved in the NCS process are still uncertain about what mandates they will eventually be given. As the *legal and institutional framework* of many of the agencies involved is still unsettled, there is a wide scope for disagreement concerning authority between government and non-governmental organisations. NGOs on the other hand, in the face of pressing social and environmental needs, may wish to make use of the resources supplied to them from external donors without waiting for a legal framework.[237] Furthermore, domestic NGOs begrudge the government's inclination to favour working with external / international NGOs, rather than with them.[238]

Another aspect relating to public environmental awareness is the potential role of *mass media*. During the Mengistu regime there were two state-controlled daily newspapers. Only in 1992 did Ethiopia acquire a legally free press. Prior to this legislation there was no such thing as a free press in Ethiopia, which is reputed to have the highest score on imprisoned journalists world-wide (Murvoll, 1997:48). Even with the legislation in place, there are still obstacles to an environmental role for the media: newspaper distribution is difficult, only about half of the adult population can read and few can afford to buy a newspaper. The government still monopolises the printing offices. Since 1991, there has, however, been a sizeable increase in newspapers, with some 170 newspapers published in 1996. According to a UNESCO report (1995), the role of other types of media is equally limited, with 197 radios, 3.2 televisions and one newspaper per 1000 inhabitants in Ethiopia in 1993 (UNESCO, 1995).

8.6.2 External actors: Multi- and bilateral impact

The Marxist Derg regime was hostile towards the West, including Western donor agencies, and the number of international organisations operating in Ethiopia during this period was relatively limited. Several international donor agencies are now engaged in conservation activities in Ethiopia. Among the most active international NGOs are FarmAfrica, CARE, the Wildlife Conservation Society (WCS), the IUCN and the WWF. The most important external programmes from international donors are those run by the United National Development Programme (UNDP) and the European Union (EU). Some of the major programmes and projects will be briefly

---

[237] These sentiments were clear both with N3 and N1. Interviews February 1996.

[238] Interview: February 1996 (N1).

presented below. The main focus in this presentation is to look into the content of the various projects, in terms of their compatibility to the objectives of the CBD, as well as to get an idea of the resources being allocated to this sector.

Wildlife conservation:

CARE International is conducting a conservation and development project in Awash National Park, with a view to enhancing local communities by enabling them to manage their own natural resources and achieve food security by the year 2004.[239] The project aims to cover about 40,000 (or 55% of the) inhabitants in the area surrounding the National Park. Estimated cost of the project is US$1,963,203.

The European Union has proposed "The Southern Ethiopia Wildlife Conservation Project": a 5-year programme aimed at catalysing Ethiopia's wildlife tourism potential, while improving the security of important elements of the country's vulnerable natural resource base.[240] The ECU 16.1 million project, funded by EU Stabex and Counterpart Funds, is meant to strengthen the EWCO and assist it in the development of three protected areas in Southern Ethiopia, through infrastructural and technical inputs. The EU feasibility study report for the project has set out the following conditions for EU involvement: (i) government commitment to a central financially independent Wildlife Authority, (ii) the gazetting of all protected areas for which boundary definitions currently exist, (iii) confirmation of EWCO's status as a parastatal body empowered to retain revenue and manage its own budget and (iv) the formulation and adoption of special measures relating specifically and only to the private sector development of tourist facilities within National Parks. As the Southern Region has now been mandated by the government to administer protected areas within its jurisdiction, the EU project is currently being re-negotiated, and might well be dropped altogether.

The WWF, in collaboration with national consultants, the UNDP and the Ethiopian government is engaged in developing a Trust Fund for conservation in Ethiopia. The target for the Fund is set at US$20 million, to be provided by multilateral and bilateral development agencies, international NGOs, private foundations, and, eventually, revenues generated in-country through tourism, hunting and commercial activities. Factors used to decide where Trust Fund resources will be applied include: biodiversity importance, level of treat, importance to local or national economies, and local capacity and willingness to use the trust fund.[241]

The WWF is also engaged in establishing a Support Unit within EWCO, aimed at helping Ethiopian wildlife and conservation authorities set priorities for action, with special emphasis on forests. The Bale Mountains National Park and the Harena

---

239 Care International in Ethiopia, project proposal, February 1994.

240 "Feasibility Study for a Wildlife Conservation Project in Southern Ethiopia", Final Report, Agriconsulting SpA, Rome, 15th June 1993.

241 Development of a trust fund for conservation in Ethiopia, Consolidated report on Second Phase Mission by WWF and National Consultants, 1 August 1995.

Forest are focal areas of attention. "[The] Support Unit will work with donors and implementing agencies such as CARE, the Wildlife Conservation Society and FarmAfrica to promote, implement and disseminate effective on-the-ground strategies and methodologies".[242] The budget proposed is about 2.5 million Swiss francs, of which 1.5 is allocated to WWF staff (non-project and project location) as well as professional fees. A central aim of the programme is the involvement of local people; this, however, is not reflected in the budget.

Conservation and use of plant genetic resources:

The Ethiopian government has received financial and technical support from various external sources to develop its plant genetic resources. This started with the Federal German Republic grant of some 12 million Ethiopian Birr (approximately US$1.2 million) in 1976 – 87 to establish the Ethiopian Genebank.[243]

One of the most extensive programmes of community-based gene banks in the world has been developed in Ethiopia. The Ethiopian Genebank has formed close links with farmers over many years, and has collaborated in the maintenance and multiplication of land races (Heywood, 1995:947). A central NGO involved in this programme is the Seeds of Survival/Africa (SoS/A), involving also the Unitarian Service Committee of Canada (USC/C). This is a five-year programme involving in-situ conservation on small farms, and has been developed by USC/C, the Ethiopian Genebank and farmers themselves. Dr. Melaku Worede, former director of the Genebank, heads the programme. Community gene banks are being established in many parts of the country, with access for farmers who will get sovereignty over this material. The programme has two major parts: a seeds reserve, with locally adapted seeds that will belong to the local farmer; and a depository of germplasm, with links to the Ethiopian Genebank. The Programme will compensate the farmer for use of his/her labour by paying the difference between the amount he/she gets for the land race crop and the highest market price the farmer could have received if the land had been planted with a high-value crop (high-yield, uniform plant varieties rather than the diversity of land races).[244] At the start of this programme in 1989, 1.2 million Canadian dollars was provided.[245]

Since 1994 efforts were made to involve the UNDP/GEF in this biodiversity conservation project, focusing more specifically on plant diversity and indigenous crops maintained by farmers in situ. In their recommendation for the project, the UNDP points to both Ethiopian needs and global needs in biodiversity conservation: "Ethiopia is a poor country which is grappling with a huge problem of rehabilitating

---

242 Forest Conservation in High Priority Areas of Ethiopia, 1996-2000, Draft.

243 White Paper: *National Policy and Strategy for Plant Genetic Resources Conservation and Development*, the Transitional Government of Ethiopia, September 1993, Addis Ababa.

244 USC/SoS *Management Plan* document.

245 In this work, as well as in the organisation of workshops, the USC is aided by Rural Advancement Fund International (RAFI), an NGO with long-standing experience in the seeds issue (USC/SoS *Management Plan* document).

its basic infrastructure which has been devastated by a long civil war, and support will be required in input into human resources and infrastructure development. The international community, in the spirit of the Convention on Biological Diversity, can help insure the continuity of these valuable genetic resources for the world's food base" (UNDP, 1994:19). GEF is involved in only one similar project: In-situ conservation of wild relatives of domesticated species in natural areas in Turkey.[246] The UNDP/GEF will cover US$2,428,000 of the costs, while the Ethiopian government is expected to put in 1.3 million Birr.[247]

UNEP has in addition put up US$100,000 for an initial two-year period to support PGRC/E in their mandate of co-ordinating the African Biodiversity Network. The latter was established in 1991 by the African Ministerial Conference on the Environment (AMCEN).

In 1993/94 the World Bank proposed to start a project on screening *medicinal plants and knowledge* about this in Ethiopia. According to Dr. Tewolde, however, there are doubts about the credibility and trustworthiness of the World Bank in using this information. The project administration and scope is still being negotiated by the World Bank and the Ethiopian government. Top-level officials at the Genebank also voiced misgivings about the arrangement, fearing that the World Bank may have a hidden agenda to use the project to obtain access to valuable information.[248] Ethiopia has not rejected the proposal yet, which involves the sizeable sum of some 10 million US dollars. Indeed, the project has been incorporated in the NCS Volume V project compilation[249], but it is not yet listed in the World Bank portfolio. A major concern is for the Genebank to be appointed co-ordinating agency and secretariat in the project. As the Genebank already has the knowledge and capacity to do the job, there would be no need for foreigners to become too much involved.[250] In this respect, my informant made explicit reference to the CBD, which in his view could

---

246 GEF is expected to cover the conservation portion by financing the establishment of the local community gene banks, the USC/SoS will cover utilisation (including training of farmers and documenting their knowledge), and the Ethiopian Genebank will guide and direct in the enhancement of land races. There is also an institution-building component and an information distribution component - the African Diversity Newsletter. According to sources at the Genebank, a "hidden priority" in the new programme is to ensure conservation also in the good areas. Previous efforts have been concentrated on the drought and problem-affected areas, but it is equally important to focus on the areas with potential for future development – herein lies the future of Ethiopian agriculture.

247 The programme was expected to start in June 1994; but it is yet to be listed in the GEF project portfolio. Repeated efforts to establish the status of the project with GEF staff have not been successful so far. The project is now proceeding with UNDP/GEF funding ("A Dynamic farmer-based approach to the conservation of African Plant Genetic Resources") and the SoS has been expelled from both the project and Ethiopia. (Interview: n3, June 1998, Appendix 1)

248 Interview: March 1996 (C3).

249 Table 8.3.2, component 2.14 in the NCS Volume V.

250 Interview: C3, March 1996: "World Bank involvement would only be helping the donors, and they do not have the right to interfere."

be a very useful instrument in these negotiations, to bring leverage to bear on the discussions and to retain control in Ethiopian hands.

What do these projects tell about the financial situation? How do they score with regard to the evaluation criteria for CBD?

Since the fall of the Derg regime, the international donor community has increasingly become involved in Ethiopian projects. As illustrated in section 8.2.1, this increase is expected to make itself felt primarily in the sectors of health and education, but growth may also be anticipated in the environmental sector:

> *The general impression of the project portfolio provided by these examples seems to strengthen the probability that* significant funding *is indeed being attracted from external sources.*

The selected projects reviewed above are alone in the financial order of about US$50 million. With the exception of the World Bank project on medicinal plants, the budgets envisaged in the project proposals cannot be directly traced in the components of the NCS investment programme. Nevertheless, the scope for external funding for biodiversity-related projects seems to be headed in the direction anticipated by the government. In that case, the increase in allocations for the biodiversity sector is clearly substantial. It is difficult to judge whether this can be ascribed mainly to the effects of the new bioregime of the CBD, or whether it is basically an effect of the novel open door policy of the EPRDF government. Most likely it is a combination of the two, and in order to analyse this further we would need data from other sectors with which to compare the situation in the biodiversity area.

The projects referred to in this section give several indications concerning direct and indirect effects of the CBD: As far as the *wildlife sector* is concerned, there are few indications of direct effects. There is little evidence that the projects envisaged here would not have come about in the absence of the CBD – the increase seems more related to changes in the domestic political situation, with a more open door policy. While most of the proposed projects include the "correct" concept of local participation, this objective has hardly been followed up in practical terms:

- Large parts of the project budgets are still allocated to external experts, as shown by the example of the WWF / EWCO project.
- The underlying message in the EU project is to keep tight central control.[251]

The *plant genetic resources sector* tells a somewhat different story: Admittedly, international and domestic awareness of the value of these resources is not novel, as it can be traced back to the work of Russian botanist Vavilov and his identification

[251] It should be noted that the rationale for this policy is not necessarily a rejection of the ideal of local participation. Rather it may be built on the fear that Federated Ethiopia may split up into conflicting regions. Nevertheless, it is a decision that goes against the efforts of Ethiopian policy-makers to solve their internal affairs themselves.

of *centres of diversity* in the 1920s. Nevertheless, the CBD negotiation process and output can be said to have renewed the interest and awareness of this sector, both abroad and at home:

- The projects being developed by PGRC/E in association with GEF and SoS are largely in the spirit of the CBD, and make explicit references to its objectives (*symbolic language*).[252]
- These projects also seem to have a more credible and practical approach to the *local participation* dimension.
- An example of a direct effect of the CBD, is the reception of the World Bank initiative on medicinal plants in Ethiopia: This is not the first time Ethiopia has appeared as a tough bargaining party with regard to projects involving its national genetic heritage – as the history of the Genebank (PGRC/E) itself is ample proof. The World Bank case shows how it was possible to play this role with more credibility and legitimacy this time, since reference could be made to an international legal framework – the CBD.

With regard to *local participation*, the role of the international donors seems to be of a dual nature:

- On the one hand, there is pressure to include the new catchword of *local participation* in the overall environmental strategies of developing countries, as a precondition to fulfil funding criteria.
- On the other hand, in the actual project proposals, the external donors seem far more inclined to demand central control over natural resources.
- This points up the *important distinction between international norms emerging from the CBD regime on the one hand, and the practice by the international donor community on the other.*

There is, of course, no reason to expect compatibility between CBD objectives and the bi- and multilateral organisations. This situation indicates the potential difficulties involved in implementing IEAs in developing countries which are dominated by external actors. Despite their efforts to influence the legislation process, interference by external donor organisations have frequently been avoided and hence, the process cannot be characterised as externally-driven. This situation may or may not change as Ethiopia's need for financial resources to execute biodiversity policies becomes acute. I shall expand on this subject in Chapter 9.

[252] Ethiopia would in that case, together with Turkey, represent the only countries to receive GEF money for conservation of domesticated species of plants.

8.7 Ethiopian Biodiversity Policies and Domestic Factors: Summary

At the beginning of this chapter, I identified four central **steps** in the analysis:

1. *Inferences* about implementation based on the theoretical perspectives (Box 8.1)
2. Examining *compliance* between Ethiopian biodiversity policies and the CBD (Table 8.1).
3. Examining the *significance* of domestic factors: Did compliance came about primarily as a consequence of *change* in domestic factors (this section).
4. Examining how *links* to international institutional mechanisms may influence domestic *implementation* activities, as identified in step 1 (to be dealt with in Chapter 9).

Here let me sum up the main findings relating to step 3. In assessing the domestic factors there are three important aspects to consider:

- *Preconditions*: Do the factors indicate a certain minimum level of will and ability on the part of the Ethiopian government to formulate and enforce biodiversity policies? The assumption was that if both were found to be lacking completely, this would render the scope for implementation activities almost impossibly small (Table 8.2.).
- *Coinciding change*: Changes on the values of any of these factors might indicate that we are dealing with compliance rather than implementation activities – improvements in biodiversity policies might be due to significant changes in domestic factors, rather than being an effect of the CBD.
- *Links*: An assessment of the most significant domestic factors is important also in order to examine which factors are likely to constitute relevant links to international institutional mechanisms.

Table 8.2: Score of domestic factors on two sectors: Preconditions.

	Wildlife – preconditions	Seeds sector – preconditions
1.1 Institutional resources	low score on SOP	high score on SOP
1.2 Human resources	low commitment	high commitment and expertise
1.3 Financial resources	low capacity	low capacity
2.1 Environmental proponents	low commitment	high commitment
2.2 Environmental opponents	high score on opposition	low opposition
3.1 Participation capacity	low capacity	medium capacity
3.2 Integrative capacity	low capacity	medium capacity

Both the wildlife and the seeds policy sectors still lack actual financial resources being allocated to these sectors, but I found a high score in terms of *symbolic language and intentions regarding institutional set-up and policy instruments*. The scope of the NCS legislation furthermore indicates an explicit intention to integrate environmental concerns across a wide range of sectors. The *general lack of SOPs*, not least due to the frequently *changing institutional set-up*, may have enhanced the comprehensive approach to formulate biodiversity policies. The process of formulating and enforcing policy is likely to be easier in the seeds sector, where there are longer traditions and high awareness of the political aims.

In the following, I go through the main conclusions on each factor with regard to: whether the consistency may be accounted for by *coinciding changes* in domestic factors, and which domestic factors seem most significant in the implementation process, hence constituting potentially important *links* to the international level factors.

State capacity

The *financial* situation would seem to have experienced significant changes. However, this remains the most elusive factor, because it is very hard to judge whether the expected significant increase in external financial resources is realistic or not. If major financial increase is actually taking place, this could help to explain the *scope* of the new biodiversity policies, but it would probably still not be accountable for the specific *content*. The content itself shows too clear correspondence with the CBD text to be coincidental. Moreover, a significant increase in external *financial resources* allocated to the field of biodiversity could itself be an effect of the CBD, if the new bioregime has directed donor attention to this issue-area. This possibility will be further considered in Chapter 9.

We have seen that the NCS process has been characterised by several novel traits relating to *institutional and human resources*: First, there has been a significant change in the political system, although the democratisation process still has a long way to go. Second, there have been frequent changes in the institutional set-up dealing with the NCS legislation process. The largely opposing mandate of the Ministry of Agriculture renders a rather limited scope for state capacity within environmental policy. A third important trait of the NCS process is the great emphasis on *utilising national, domestic human resources* in the legislation process. Among the domestic factors encompassing the state capacity dimension, *human resources* seems most important in the development of Ethiopian biodiversity policies. Given the extremely low level of financial resources available and uncertainties still attending the distribution of power between the relevant institutions, Ethiopia's greatest asset would appear to relate to a *high degree of expertise and knowledge* among administrative officials.

- *Changes:* The high level of consistency between domestic biodiversity policies and the objectives of the CBD can hardly be accounted for by changes in these factors.

- *Links:* Among the state capacity factors, human resources seem to have had the most significant impact in boosting biodiversity policies – especially in the seeds sector. The institutional and financial factors are clearly also of great importance for successful implementation in the future. These three factors may hence represent important links to international institutional factors, in the analysis of whether and how the latter have influenced domestic policy (Chapter 9).

Even though economic ability is still low, the Ethiopian government seems to score high on commitment to conduct policies in this area. On the other hand, as mentioned in section 3.2.3.2, ability is a relative concept: It is up to the government to set priorities about how to spend the financial resources within its control. A low-income country may blame environmental degradation on lack of financial resources, but that justification becomes less trustworthy when the same country can afford going to war.[253]

State–society relations

As the current decentralisation process involves unresolved issues pertaining to the division of power and control over natural resources, it was expected to be more likely to represent a stumbling block than to boost the development of environmental policies in Ethiopia. *Decentralisation* represents a novelty, which will also have significant impact on the institutional framework, but, like the *democratisation process,* as yet it has hardly come far enough to account for much of the domestic legislation (*output*) score.

- *Changes:* The effect of domestic factors relating to *state–society relations* on biodiversity policy *output* was found to be small. Hence, this dimension cannot be taken to account for consistency between domestic policies and the objectives of the CBD.
- *Links:* While the factors on this dimension are clearly significant to the future enforcement of domestic biodiversity policies in Ethiopia, they seem unlikely to provide significant links to the institutional mechanisms of the CBD.

Non-governmental actors

The legislation process has been largely elite-driven, with the participation of non-governmental actor being restricted to expert individuals, acting in their personal capacity. This characterisation is, however, partly mitigated by my findings that participation at lower levels of government has clearly expanded through the NCS process. Even though these are not non-governmental, this trend towards a broader scope of participation in Ethiopian policy-making should be noted, as it could have implications also for non-governmental actors. The bi- and multilateral organisations

---

[253] As my investigation for this study ended prior to the Ethiopian-Eritrean war, I shall refrain from further comments on this.

were also examined as important factors on this dimension. A central finding was the tendency of external donors to interfere in the legislation process, which in turn have been largely resisted by Ethiopian authorities. Hence, the process cannot be characterised as externally-driven. On the other hand, as Ethiopia's need for financial resources for effectuating biodiversity policies becomes acute, this situation may change.

- *Changes:* The changes on the *non-governmental actor* dimension are too few and also too recent to account for consistency between Ethiopian policy output and the objectives of the CBD.
- *Links:* Due to the need for external funding, the role of external donors will constitute an important aspect for consideration in the next section on international and national links.

These findings are relatively straightforward, as the measurement of these factors did not present grave problems: It is hardly disputed that the institutional set-up has undergone repeated change, and there is ample evidence of the high level of awareness and knowledge among the actors involved. The one exception concerns the financial outlook, as it has been difficult to find comparable figures for the budgets over time.

Another methodological objection in this connection is that my analytical framework may over-emphasise environmental proponents. The factors and indications provided by Jänicke's model were primarily equipped with two values: presence or absence of environmentally beneficent factors. I have not identified any major public environmental opponents – short of the opposing concerns in the Ministry of Agriculture. This apparent scarcity of opponents might be due partly to my analytical framework. On the one hand, my intention was to counteract this shortcoming by the relatively flexible and intensive interviews. On the other hand, however, the interviewees could have had an interest themselves in downplaying public obstacles in the legislation process.

PART FOUR: NATIONAL AND INTERNATIONAL LINKS IN THE IMPLEMENTATION PROCESS: REGIME EFFECTS

> If enough species are extinguished, will the ecosystem collapse, and will the extinction of most other species follow soon afterward? The only answer anyone can give is: possibly. By the time we find out, however, it will be too late. One planet, one experiment.
>
> Edward O. Wilson (1992:170), *The Diversity of Life*

CHAPTER 9

THE ROLE OF INTERNATIONAL INSTITUTIONAL MECHANISMS IN ENHANCING DOMESTIC IMPLEMENTATION

9.1 Institutional Mechanisms and Theoretical Perspectives

In section 7.5, I offered four broad propositions concerning how international institutional mechanisms may have an impact on domestic implementation of IEAs in developing countries. In this chapter I shall examine how well these propositions about international and national linkages square with the empirical findings in Chapter 8. The examination of the mechanisms will be conducted mainly in separate sections, but in addition the analysis will concern the potential interplay between and among them. In this examination I shall also indicate connecting links to the conclusions in the first part of the analysis, i.e. the analysis of the regime formation process in Chapters 4 to 6.

How might causal pathways link factors at the international and national level in the process of implementing IEAs? To answer this, I will comment on the relationship between the institutional mechanisms and the three theoretical perspectives (*realist/power-based, interest-based*, and *ideational*) applied in the first part of the analysis:

The **moral mechanism** pertains to the scope for willingness to implement, assuming that an IEA may enhance domestic implementation through normative persuasion. This mechanism may come into play through two pathways: the IEA content may be perceived as legitimate in itself, or an agreed international norm may be perceived as an obligation *(pacta sunt servanta).* At the *national level,* this mechanism is assumed to affect commitment, and hence indirectly state capacity to formulate and enforce environmental policies. This mechanism is primarily linked to the *ideational perspective*, which emphasises the impact of commonly agreed norms in the international community.

The **material mechanism** concerns how the regime may affect domestic cost/benefit considerations relating to implementation through financial and technological transfers. At the *national level*, this mechanism may presumably increase commitment and capacity to enforce biodiversity policies through added financial resources. This mechanism can mostly be associated with the *interest-based perspective*. The material mechanism may also fit the assumptions linked to the *realist perspective*: Benevolent hegemonies may consent to carry a larger share of the burden so as to control developments within an issue-area.

An interplay between moral and material mechanisms could be envisaged: given that the legitimacy of the IEA content is boosted because it is *coinciding* with domestic interests. This pathway has a closer affinity to the *interest-based perspective*. Alternatively, if the correspondence between IEA content and domestic interests is due to a real *breakthrough*, the moral obligation to follow up would presumably be strengthened.[254]

Mutual reassurance focuses on the role of information, monitoring and sanctions in reducing uncertainties about free riders among negotiating parties in the implementation process. These were central factors within the *interest-based perspective*. This mechanism also has strong links to the *ideational perspective* in the sense that the negotiation forum could induce learning, representing a frame of reference for future interactions and for stabilising mutual expectations among the parties. At the *national level*, this mechanism was presumed to enhance commitment to IEA objectives, and also strengthen the development of standard operating procedures.[255]

The interface between material and mutual reassurance mechanisms gives rise to the following question: Do external donors follow up CBD objectives, not only within their own domestic biodiversity policies, but also within the arena of biodiversity policies in recipient countries?

The **empowerment mechanism** directs attention to how an international regime may foster alliances between international and non-governmental domestic actors, pushing for more far-reaching domestic implementation of IEA objectives at the *national level*. This effect was assumed to come into play primarily through international epistemic communities, which largely links this mechanism with the *ideational perspective*. The flip-side of this mechanism relates to the *realist* perspective and the possibility that central actors may be co-opted into international institutions, internalising the ideas of the prevailing hegemony.

The relationship between the four institutional mechanisms and the main assumptions associated with three theoretical perspectives may be summed up as shown in Table 9.1.

---

[254] The flip-side is authoritative persuasion – norms as part of a hegemonic agenda, and hence an explanation in line with *realist* expectations.

[255] The flip-side of this coin is implementation activities without material content, a mere "play for the grandstand".

Table 9.1:
Institutional mechanisms and theoretical perspectives – core elements

	Power-based perspective	Interest-based perspective	Ideational perspective
Moral mechanism			legitimacy: common norms obligations: *pacta sunt servanta*
Material mechanism	hegemon: sticks and carrots	cost/benefit calculations: incentive/enabling	
Mutual reassurance mechanism		information & monitoring: trust mutual compliance	learning: stabilised mutual expectations
Empowerment mechanism	norms as authoritative persuasion: co-optation		epistemic communities: coalitions[1]

1 Note the dual position of epistemic communities pointed out in section 3.2.3: First, an epistemic community may, as here, be constructive in advocating common norms and ideas. Second, it can be seen as constituting part of a specific interest coalition and, hence, fitting better into the interest-based perspective.

9.2 Moral Obligations and Enhanced Commitment?

The analysis so far has revealed a high degree of breakthrough for DC interests in the CBD, and a high degree of correspondence between the CBD and Ethiopian biodiversity policies, at least at the normative level. How may the moral mechanisms of the international institution account for this situation?

The main questions raised in this section concern how the policy formulations in the CBD are perceived from an "Ethiopian" point of view:

- Legitimacy through learning: Do Ethiopian authorities view the content of the CBD as a moral good – as *legitimate* objectives for improving conditions in the issue-area? Legitimacy through teaching: If Ethiopia views the international output as a *breakthrough* for its own interests, this was assumed to further enhance the country's moral *commitment* and inclination to implement the objectives of the CBD.
- Second, domestic implementation could come about through a sense of moral *obligations*, viewing CBD objectives in terms of *pacta sunt servanta*.[256]

256 I shall discuss potential interplay between the moral and material mechanisms in section 9.3.

9.2.1 Enhanced commitment through legitimacy

In the first (international) part of the analysis, in Chapter 5, I drew the conclusion that the objectives of the CBD largely constitute a breakthrough for DC interests. At the same time, the CBD output itself cannot be said to constitute more than a moral victory for the South. This is partly because the CBD is largely a framework convention. It must be filled with specific policy instruments for implementation if it is to fulfil its objectives in practical, material terms. Lacking this, the CBD would, it was assumed, be in danger of ending up as did UNCLOS – a hollow victory for the coastal states of the developing world.

In the second (domestic) part of the analysis, it became clear that Ethiopian biodiversity policies had a high score in terms of the symbolic (moral) level evaluation criteria – hence indicating correspondence to CBD objectives. This was apparent both in the wildlife sector and in the plant genetic resources (pgr) sector. There was a high degree of correspondence between the language used in the CBD and the NCS, and in terms of plans and programmes envisaged in the two sectors. Human resources represented an important factor in the development and improvement of biodiversity policies in Ethiopia. The potential effect of the moral mechanism was assumed to operate by way of strengthening *commitment* among officials in the state bureaucracy and non-governmental actors.

In section 3.2.3 the legitimacy of common norms was linked to two distinct processes: learning and teaching. According to the ideational perspective, DCs may themselves be *teachers* – through active participation. They may succeed in promoting and integrating their own norms and ideas on the environment and development into the international agenda and output. This was assumed to foster an increased willingness to implement IEAs. A certain level of activity was believed to be required both for there to be a breakthrough for own interests or norms, and to adapt to international norms through learning. It is thus necessary to examine the role played by Ethiopian delegations to the negotiation meetings.

During the early phases of the CBD negotiation process, Ethiopia did not play a significant role. Government concern was obviously low due to the political situation at home, with the Mengistu regime being overthrown in 1991 after a long period of civil war. Nevertheless, the issue of genetic resources had long been one of national concern, not least through the work of Dr. Melaku Worede, during his time as Director of the Ethiopian Genebank. Another central actor, Dr. Tewolde Egziabher, entered the process only during the last negotiation meeting in Nairobi, and then participated in strengthening and unifying the G–77 position on questions concerning transactions with genetic resources. In this work an alliance was established with Brazil, India, Malaysia and Mexico, among others. Alliances within the G–77 group are believed to have strengthened the DC interests in the negotiations. These interests were also strengthened by the support from the Nordic countries and from the environment and development NGO-community. This information confirms my communications with Norwegian delegates to the biodiversity negotiations, who agree that Dr. Tewolde had an entrepreneurial role in the consensus-making process.

During subsequent COP meetings, Ethiopia has played an active role in the negotiations. For instance, Ethiopia is now represented by a legal expert within the EPA. As the examination of country activity-level showed (section 5.2.1), Ethiopia scored relatively high in terms of making interventions during the negotiations. Ethiopia's activity level was equal to that of countries such as Chile, France, Germany, Italy, Japan, the Republic of Korea, Mexico, the Netherlands, Russia, Sweden, Spain, Tanzania, the UK and Zimbabwe. While the present activity-level can hardly make up for Ethiopia's limited participation during the early phases of biodiversity negotiations, it does support the impression of a high level of concern for this issue-area in Ethiopia. There are several indications that Ethiopian representatives played an entrepreneurial role in the biodiversity negotiations. This strengthens the conclusion in the first part of the analysis, that the biodiversity negotiations offered a forum in which weaker countries could get their issues and interests on the agenda. Hence, this may help to explain how Ethiopia has come to accept the CBD output and seems dedicated to following it up.

The (participating) Ethiopians are largely content with the result of the negotiations. Respondents agree that the CBD is largely regarded as *legitimate*. This is true both for those aspects corresponding to long-standing traditions in Ethiopia, and for the more novel aspects that the CBD introduces into Ethiopian biodiversity policies:

- *Legitimacy through teaching:* Some of the concepts are believed to originate, and coincide, with long-standing traditions within Ethiopia itself, such as the ideas concerning on farm, in-situ conservation of plant genetic resources, as well as the concept of local participation.
- *Adoption of new ideas (learning):* Other aspects are seen more as a direct impact from the CBD, such as the formulation of a comprehensive biodiversity policy. This constitutes a new idea on the national political agenda, which would not have happened without the CBD.[257]

The observations indicate an enhanced *commitment through legitimacy:* The comprehensive adoption in Ethiopia's NCS of the norms and principles of the new biodiversity regime implies that these norms have obtained legitimacy with national policy-makers. This is most likely to have taken place through national actors central in the international negotiation process. If this is the case, this may have fostered a higher degree of willingness – although hardly a change in ability – to follow up the policies in practical implementation activities. This indicates that the importance of the moral mechanism is mainly linked to the early phases of implementation. Finally, we must bear in mind that the CBD is a framework convention and, as such, still constitutes part of a negotiation game. As long as

[257] Interview with Dr. Tewolde Egziabher, February 1996. (App.1, C1).

negotiations continue, the negotiating parties will seldom admit that they are satisfied with the output.

9.2.2 Moral obligations – Pacta sunt servanta

The idea of *pacta sunt servanta (pss)* implies an intention or decision to implement an IEA, even though implementation is not regarded to be in the government's direct interest. In order to examine whether the *pss*-factor has come into play, we must perform several tests:

- It is necessary (but not sufficient), to point out concrete measures undertaken by the government, which would not have been undertaken in the absence of the CBD.
- Indications could also be found in oral intentions made by the implementing party.
- A stronger way of testing the *pss*-factor would in addition necessitate proof that such implementation activities were not the result of external pressure. Pressure, however, may come in many guises, and is not easily verified.

A possible indication of the first test was identified in my concluding remarks in Chapter 6: "If national legislation makes provisions for local communities of farmers and indigenous people, this could indicate an acceptance of the norms developed within the new biodiversity regime."

This supposition was based on the notion that each and every item of symbolic language in the CBD text is not necessarily due to a moral breakthrough for the South. Catch-words such as *local capacity building* and the recognition of the rights of *indigenous and local communities* may constitute part of the domestic political agendas in DC governments, but they may also be seen as part of the Northern democratisation agenda. Hence, the promotion of these principles may meet with resistance if they are interpreted as part of a conditionality scheme for achieving green external funding. These catch-words may also be distrusted by recipients as concepts which will never constitute more than a symbolic wrapping, while the contents remains the same. The supposition was, however, found to be only partly correct in the Ethiopian case, as there were differences between the two biodiversity sectors:

- The inclination to promote local participation and benefit sharing has long been an integral part of Ethiopian policies, especially in the *seeds sector*.
- The picture is different in the *wildlife sector*, where problems of vertical disintegration have an equally long history in Ethiopian conservation policies.

On the other hand, the relatively low score concerning local participation can hardly be ascribed to reluctance by the Ethiopian government alone. It is just as much a reflection of the priorities set by the *external donor community*. I shall return to this aspect in the next section.

The second test concerned oral pledges. Here, it was repeatedly pointed out (during my interviews) that Ethiopia, having ratified the CBD, intended to fulfil its international obligations by complying with the objectives of the Convention. Such oral displays of goodwill are frequent in environmental politics, where symbolic language is widely used by North and South alike. This is obviously an indication of good intentions, but it must be substantiated by specific measures if we are to conclude that the *pss*-factor has had effect. This concern for fulfilling national obligations was often cited in connection to what my respondents regarded as shortcomings of the CBD: "We must be part of the CBD process in order to improve it". Misgivings regarding the CBD include:

- *weak rules on equity*: the objective of equitable sharing is not treated adequately in the CBD, as compared to the objectives of conservation and sustainable use;
- *the gap – seeds in gene banks*: the CBD does not have retroactive effect; this means it does not include germplasm in the international gene banks, and hence, the concept of national sovereignty over germplasm is hollow;
- *distrust of the GEF*: the financial mechanisms envisaged by the CBD (the GEF) to compensate DCs for biodiversity conservation are too much in the hands of the developed countries.

When Ethiopia has chosen to embark on policy formulations in line with the CBD despite these shortcomings, this can be interpreted variously:

- It may substantiate the role of the *pss*-factor: Ethiopia chooses to honour its international obligations (even when not directly in the national interest) out of respect for commonly agreed international norms.
- The CBD is regarded as a package deal of costs and benefits: Ethiopia consents to treat it as that; – taking it on board as a whole, rather than picking out only the tasty bits.
- The policy formulations may be made as a grandstand play: implementation as a symbolic gesture to give the impression that Ethiopia is fulfilling its part of the agreement.
- The Ethiopian government may feel more or less pressured to honour its obligations: this is seen as a prerequisite for the off-hand chance that also others will fulfil their part of the agreement.

The three latter interpretations increasingly imply that Ethiopian implementation is a strategic move (rather than an effect of the pss-factor) – performed to get other parties to fulfil their obligations under the CBD, or to attract GEF funding. The forth option translates as yielding to external pressure, rather than any effect of the *pss*-factor. Whether this strategy is likely to succeed remains to be seen; this will be further discussed in the next sections. Moreover, any discussion as to whether, and if so why, Ethiopia will fulfil *all* its obligations under the CBD is also hypothetical, as its economic ability to do so is limited. This points up the distinction between implementation activities conducted on the basis of "national interest" and activities conducted with the intention of promoting implementation activities (financial transfers) in other countries.

Two preliminary conclusions may be drawn from this section:

- Respondents agree on the *legitimacy* of the norms and principles of the CBD. The norms are consistent with their own ideas. The legitimacy is of a dual nature, as it rests on both learning and teaching components: In other words, the impact of this factor can be seen both in the inclusion of novel ideas in Ethiopian biodiversity policies, as well as in aspects inherently consistent with Ethiopian norms and principles.
- There are also elements in the CBD which Ethiopian authorities regard as serious shortcomings. This may lead to diverging interpretations: Domestic implementation activities may be seen either as signifying a moral *obligation* to adhere to international conventions, or as being pursued as a grandstand play – mainly as a strategy to obtain external funding.

This state of affairs suggests that the moral mechanism have been an important factor in enhancing implementation of the objectives of the CBD – at least considering the implementation output phase. The next phase, implementation outcome, may be more dependent on the roles played by material and mutual reassurance mechanisms.

9.3 Material Mechanisms Enhancing Implementation Ability?

The material mechanism was assumed to affect domestic implementation by enhancing the state capacity to formulate and enforce environmental policy. The use of such mechanisms could be interpreted in light of the interest-based perspective, as a means of altering the cost/benefit calculations of national governments. Increased financial resources could help to mitigate the effect of vertical disintegration, as there would be greater scope for compensating target groups affected by environmental policies. Two broad scenarios may be envisaged:

- I first focus on to what extent conservation and utilisation of biological diversity in Ethiopia is supported by external factors: either by improved compensation for external use of Ethiopia's genetic resources, or by financial mechanisms targeted at biodiversity projects and policies. Financial transfers may also be seen as part of a (donor) *hegemonic scheme* to control (recipient) behaviour in a specific issue-area.
- Second, this section deals with whether and how the largely symbolic objectives in the CBD are used to increase material gains at the domestic level. Depending on the type of material mechanisms (from donors), there could be positive gains in terms of enhanced *capacity building* in general. Here, we must see whether symbolic language is utilised (by recipients) merely in order to attract external

funding: *a play for the grandstand*. If so, the effect of the moral mechanism may be drawn too far, as implementation will stop at the symbolic level. In that case, material mechanisms may still serve as an *incentive* for implementation activities, but it would also necessitate a high level of monitoring to avoid free riding. Hence, this situation would be notably different from, and more difficult than, material mechanisms being used to *enable* – to help governments execute polices to which they are highly dedicated. This section will deal with the interplay between the moral and material aspects relating to domestic implementation of international environmental agreements.

The main clues for the discussion in this section are likely to be found in the external donor community's project portfolio pertaining to the biodiversity sector in Ethiopia.

9.3.1 Financial transfers as hegemonic scheme?

At this relatively early stage, the potential of the GEF to contribute to improved implementation of the CBD objectives in developing countries is still uncertain (Chapter 6). This is due both to its relatively limited financial resources and to controversies between donors and recipients regarding its application.[258] This led to the question of how GEF money has been used in Ethiopian biodiversity projects, and with what effect.

I shall argue that Ethiopia is experiencing a "Catch-22 situation" in its dealings with the international donor community, both with regard to bi- and multilateral co-operation:

- First, projects aimed at enhancing institutional capacity and human resources in Ethiopia are likely to be turned down. Those projects that do get accepted are less likely to contribute to domestic capacity building.
- Second, even though the CBD is used and seen as an important vehicle for advocating domestic concerns for capacity building, this strategy has yet to deliver material results.
- Third, the main obstacles to change seem to lie in the SOPs of the external donor community.

The first line of argumentation rests on the finding that some of the most important projects aimed at improvements in the biodiversity sector in Ethiopia have yet to receive funding from external agencies. This was for a long time the case with the most central project in the plant genetic resources sector, the co-operation project on on-farm conservation and use. Another example is the World Bank project on medicinal plants, in which Ethiopian authorities, with reference to the CBD, strongly advocated the use of and reliance on the country's own human

---

[258] As of June 1997, about US$ 585 million, or 36% of the total GEF budget, had been spent on biodiversity projects. Only 18% of the GEF budget was allocated to projects in Africa.

resources. A third example is the EPA/DN project proposal to NORAD which has been long pending, despite NORAD's explicit emphasis on the principle of *recipient responsibility*. The special cultural and political history of Ethiopia seems to have fostered even greater distrust towards external interference here than in many other DCs. At the same time, Ethiopia ranks among the very poorest countries of the world and is in urgent need of external input. This makes Ethiopia a relevant and interesting case for studying how external relations are handled.

The second line of argumentation rests on the objective of *capacity building*. This was conceived as a central and globally acclaimed theme at UNCED, promoted in both the CBD and Agenda 21, but has had a hard time maturing. The concept itself is obviously nothing new in terms of general development aid; it has been known under labels such as institution building, institutional development and public sector management prior to its present cloak. Whatever the catch-word, development projects frequently fail to reach the goal of capacity building, as it is inherently conflicting with the short-term goal of efficiency; moreover, it can hardly be reached without long-term funding for maintenance, which is seldom part of such projects. In *Institutions for Environmental Aid*, Fairman & Ross (1996:41-44) point out two main obstacles to attaining the objective of capacity building:

- There is a tendency to rely heavily on external experts, rather than build on domestic human resources. Partly in effect –
- there is a lack of commitment to project goals in the recipient country.

Common to both the CBD objectives and the Ethiopian authorities is the acknowledgement of the need to include local people in the decision-making procedures. The involvement of local communities is regarded as a basic preconditions for improving policies on conservation and sustainable use of biological diversity.

My third line of argumentation, regarding the current lack of financial transfers to Ethiopian biodiversity projects, concerns the reluctance in the donor community to accept these kinds of project formulations. There may be various reasons for this reluctance, and the situation may certainly be dynamic, with a potential for change:

- *Northern environmental agenda*: The biodiversity project portfolio is biased towards large nature conservation projects and external efforts to establish national environmental action plans (World Bank, 1997). In this light, the North is certainly keeping a firm hand on the global environment agenda, and the long-term effect on the G–77 alliance may be divide and rule.
- *Decreasing development aid*: The case could be judged with a view to the broader setting of international capital flows: investments, loans, and development aid. The Global Development Finance (1997) report of the World Bank shows a steady decrease in the latter two, and reduced growth in the former. Within this perspective, expectations of increased financial transfers for biodiversity projects may be loosely founded.

Is the Ethiopian case simply one of bad luck, due to an inherent lack of economically interesting genetic material? Looking at the Ethiopian record with regard to plant genetic resources, this would seem hardly credible at a first glance. Ethiopia's genetic heritage is indisputably of considerable economic importance, even though parts of it may also be accessed through the international gene bank collections. On the other hand, Ethiopia differs from most other African countries with regard to the specific national capital in plant genetic resources, as well as in outspoken consciousness regarding the value of these resources. The Ethiopian case thus seems to bolster the preliminary conclusion in Chapter 6: Projects outside the traditional conservation portfolio (such as projects on sustainable use and equitable sharing in the seeds sector), tend to lose out in competition with projects focused on ecosystem and species conservation.

This line of explanation would seem to agree with the *realist perspective*, in that dominant industrialised states exert some kind of hegemonic control in the biodiversity field.[259] In this light, the industrialised countries succeed in retaining an emphasis on conservation of wild species and ecosystems, as well as keeping free access to seeds in international gene banks.

A related line of explanation is that the *standard operating procedures* of external donor agencies, both bi- and multilateral organisations, are inherently biased against effective capacity building in recipient countries. This is not a novel discovery. Several authors have pointed to the shortcomings of conditionality, which may be a major obstacle to a greater reliance on a country's own human resources. In the concluding chapter of *Institutions for Environmental Aid*, Barbara Connolly (1996) presents a similar conclusion on the basis of the examination of international environmental transfers: Donors determine priorities, and the solutions they devise, based on their SOPs, determine which problems are identified as important (1996:362). As an example the author points to the World Bank, which is more prone to grant loans to large-scale power plants than to support local projects against air pollution. Another example is the Bank's tendency to rely on expert consultants from donor countries rather than on local experts, whose employment would enhance national capacity building. This is very much in line with my own findings with regard to the EU project, among others, which has explicitly opposed the decentralisation process in Ethiopia by focusing on central government agencies. Two interpretation can be offered here:

- This can be seen as a *hegemonic scheme*, because these agencies are very much dominated by industrialised countries, and there may certainly be resistance to changing these SOPs.

[259] Whether this is benevolent or coercive hegemonic control is a matter of discussion: Some would say benevolent, as the industrialised world does pay for a common good (conservation of ecosystems and species). Some would say coercive, as the users pay only for a minimum (in which they have direct interest), while generally remaining free riders (with regard to plant genetic resources).

- On the other hand, the resistance to changing SOPs may just as well be due to *organisational inertia,* as to any deliberate hegemonic "conspiracy" on the part of donor governments and agencies. I examine donor SOPs in further detail in section 9.4.

A more general explanation for donor resistance may be found in the often delicate balance between recipients' needs and the claim for accountability in the donors' domestic constituencies. This constraint does not, however, hold up long as an excuse in donor countries that have ratified the CBD.

9.3.2 Capacity building and symbolic policy

Environmental policies were given a general boost at the international agenda in the 1980s and early 1990s, as shocking environmental reports and increased public attention made green policies a "must" for domestic policy-makers in what seems an unprecedented manner. At the same time the South argued that global environmental problems have their origin in the agenda of the North. If so, environmental awareness might be assumed to have a much smaller impact on environmental policies in developing countries. While this may be true with regard to issues such as climate change and ozone, biodiversity would seem to represent a different case. This is clearly an issue on which developing countries have been able to link essential national and local interests and concerns, like control over genetic resources and efforts to combat soil erosion. As such it may have benefited from the generally increased international attention in a similar fashion to that in the North, and there may be reason to expect increased national attention to the biodiversity issue-area also in developing countries.

My evaluation of Ethiopian biodiversity policies (Chapter 8) revealed a *medium score on the criteria pertaining to material correspondence to the CBD* in terms of establishment of policy instruments and related institutional changes. While the plans and policy formulations seemed highly courageous, it was also noted that the relatively short time span might account for the lower score on practical, material implementation activities in Ethiopia. In light of the critical economic situation in Ethiopia, however, a high score on these evaluation criteria is not likely without external allocations of resources.

There may be other than ideational (legitimacy, obligations) or power-based (hegemonic scheme) reasons for a country to implement IEAs. In this section, I focus on two interest-based rationales for Ethiopia to follow up the objectives of the CBD. The first rationale is *a play for the grandstand*; the second is *nested legislation*. I will then look at how these rationales may affect the donor-recipient relationship.

A play for the grandstand

There is a growing understanding that in order to attract international funding for conservation and development programmes, the right "catch words" must be displayed in the national policy framework. In that case, the material mechanism may bring about little or no change in willingness to implement. Nevertheless, regardless of will, an actual higher degree of external funding would help to enhance ability.

On the condition that he be kept anonymous, an interviewee from the central administration made it clear: "The language and concepts from the CBD has been integrated into the NCS. UNDP advised this, so as better to get GEF funding for projects."[260] This concern was echoed by the Ethiopian Science and Technology Commission, where one respondent pointed to the need to develop national legislation in order that Ethiopia may benefit from the CBD framework on equitable sharing and capacity building. At the same time, it was emphasised that "whether the CBD will have such beneficial impact in DCs depends on other countries implementing the CBD objectives".[261] My interviewees at the Genebank made similar comments, pointing out that "biodiversity was turning into a hot global issue and Ethiopian politicians realised that they had to prepare to accommodate the new situation: to put up a national legal framework to respond to the CBD – both because of the funding involved and because of the ideas."[262]

The question of the possible impact of the *pacta sunt servanta*-factor on Ethiopian biodiversity policies may hence be largely of academic interest. This is because the contents and concepts in the CBD generally coincide with how Ethiopia wishes to formulate its own policies in the area of biodiversity. On the other hand, the moral obligation of the *pss*-factor may have been strengthened by the "national interest", and vice versa: "The Government ratified the CBD because this was seen to serve its interests; namely the idea of equitable sharing. This (the CBD) is useful vis-à-vis GATT, the World Bank and multinational seeds corporations."[263] This points to an important aspect regarding the link between moral and material mechanisms, that of nested legislation.

Nested legislation

The idea here is that the CBD is regarded as a useful framework onto which to link national legislation, in order to provide a lever towards external donors. The CBD's utility as a framework for nesting may to a large degree explain why my centrally placed interviewees regarded the CBD as a *legitimate and materially useful* instrument for enhancing conservation and utilisation of biological diversity in Ethiopia. This was particularly so for the seeds issue, probably because politicians envisage a more immediate return from domesticated genetic materials. The CBD's call for a more equitable sharing of benefits from genetic resources is clearly in line with Ethiopian interests. The CBD is currently being used deliberately in domestic

[260] Interviews: February 1996, C7 (central administration, EPA).

[261] Interviews: February 1996, C5 (central administration).

[262] Interviews: February 1996, C3 (Ethiopian Genebank).

[263] Interview: February 1996, C5 (central administration).

policies to enhance the speed of legislation. My interviews revealed agreement that it would hardly have been possible to develop this policy without nesting it in the international framework established by the CBD. This is mainly linked to the efforts to establish agencies to control access and transactions with plant genetic resources.[264] Second, the CBD is cited as an important legal instrument for making bilateral deals on biodiversity prospecting.[265] It is hard to imagine the same incentives for this activity without the Convention. This establishes a central connection between material interests and normative legitimacy, as well as linking up with mutual reassurance.

Still, we need to distinguish between different functions of capacity building on the borderline between moral and material mechanisms. Incentive mechanisms may directly strengthen state capacity in terms of *economic resources* and enabling mechanisms may be building on domestic *human and institutional resources.*

There is a high awareness among Ethiopian policy-makers that *capacity building* represents a crucial factor for future implementation outcomes. Financial transfers as such may not, however, be very effective unless targeted at the specific capacity needed. On this basis, I will highlight two conclusions relating to the material mechanism:

- As Ethiopia is willing to pursue enhanced biodiversity policies in line with the objectives of the CBD, material mechanisms would largely be used to *enable* the government to perform implementation activities.
- The financial resources are, however, hardly sufficient to really make a difference. This aspect seems largely overrated in the literature on IEA implementation in DCs.

Both moral and material mechanisms seem to be necessary, but not sufficient, conditions for domestic implementation of CBD objectives in Ethiopia. The findings in Keohane & Levy (1996) seem to indicate that this conclusion may also apply outside Ethiopia. The window of opportunity that has been opened by the moral mechanisms to improve biodiversity policies may, however, backfire if the new bioregime fails to develop and strengthen mechanisms for mutual reassurance among the Convention Parties.[266]

---

[264] Pointed out during my interviews with staff at the Biodiversity Institute (Ethiopian Genebank), February 1996 (App.1, C3 and C4). Several interviewees made explicit reference to the CBD as an important tool for strengthening domestic control in the proposed World Bank project on medicinal plants.

[265] Interview: February 1996, C8 (central administration, EPA).

[266] Regarding the concept of "window of opportunity", see Kingdon 1984.

9.4 Mutual Reassurance: Capacity and Commitment Strengthened through Other Parties Complying?

Institutional factors were assumed to reduce uncertainty about free riders among the parties through information, monitoring and sanctions. This mechanism may be presumed to have an impact in stabilising mutual expectations among the parties, by reducing transaction costs and by providing a frame of reference for future interactions (thus enhancing implementation). There are two broad pathways through which the mutual reassurance mechanism was assumed to affect domestic implementation:

- First, if other parties are seen to honour their obligations, this would presumably enhance Ethiopian *commitment* to the CBD and increase the legitimacy for implementing CBD objectives themselves. The mutual reassurance mechanism may thus be important when compliance from other parties is regarded as a condition for own (domestic) compliance. On the same note, compatible legislation on regulating access to genetic resources is essential in both user and owner countries in order to strengthen the development of *standard operating procedures* in the area. This raises the question of how Ethiopia perceives the trends in the international community with regard to complying with the obligations on equitable sharing under the new bioregime.
- Another central aspect to be discussed further in this section concerns the link between the material and mutual reassurance mechanisms: How does the CBD affect Ethiopian biodiversity policies through external actors in the donor community? Much of Ethiopia's biodiversity policy is dependent on funding by these donors – hence, their *standard operating procedures* will also be important for how the CBD objectives will be followed up. This potential effect may work both ways, either enhancing or obstructing implementation of the CBD.

9.4.1 Mutual compliance – compatible legislation

The first question pertains to Ethiopian perceptions of other parties' behaviour vis-à-vis the new bioregime. The need for compatibility has both a moral and a material dimension:

- The *moral* aspect is rather obvious, as few parties appreciate the though of other parties choosing the free-rider strategy. Complying parties may, vice versa, represent role models and have a positive effect on overall implementation.
- The *material* aspect is somewhat harder to grasp: It is closely linked to two important rules in the CBD regarding *mutually agreed terms* and *prior informed consent*. These rules are related to the conclusion that unilateral action is hardly effective in the biodiversity issue-area. A precondition for poor countries when negotiating deals on access to biological material is that these deals can be

linked to relevant legislation in both user and owner countries. If the CBD objectives are to have any practical effects as seen from a developing country perspective, it necessitates co-operative action from the user countries as well – and these are generally the developed countries.

This may be exemplified by the international and national links: In Chapter 6, I presented some examples of how CBD objectives have affected the regulatory framework for international transactions with genetic resources. This indicated a greater acceptance of the objectives also in terms of practical policies in user countries, primarily in the private sector. Examples of this effect were the new statutes of Kew Gardens and of Biotics. The analysis in Chapter 8 showed a *high score of CBD correspondence* in Ethiopian biodiversity policies on *legislation concerning control over its own genetic material.* While it is still too early to say whether compatible regulations will be further developed in user countries, Ethiopian actors have definitely been proceeding with this objective in mind.

To judge from my interviews with central actors in the biodiversity sector in Ethiopia, anticipations for future interaction seem closely linked to perceived shortcomings of the CBD. This concerns the fact that the CBD excludes germplasm (seeds) deposited in the international gene banks from the new property rights regime for genetic resources, and also that the CBD is insufficiently specific about the rights of local communities and indigenous peoples.[267] Regarding the future effectiveness of the CBD, Dr. Tewolde believes this will depend on the effective implementation of the Convention world-wide: on the goodwill of industrialised countries, and on how the framework of the Convention will be filled in detail.

It is precisely the *lack of control over germplasm in international gene banks* that has given rise to concern among Ethiopian experts in the area. While they agree that some DCs, as indicated by the Venezuelan case (see section 6.2), may have increased their control over access to genetic material due to the CBD, this is mainly expected to pertain to *wild biological material.* Industrialised countries are expected to remain free riders with regard to *plant genetic resources*, as they believe that they have enough access to domesticated materials through the IARCs.[268] As pointed out in section 6.2.2, a retroactive clause in the CBD (concerning germplasm in international gene banks) would, however, be legally unprecedented and practically impossible.

---

267 These shortcomings were pointed out by both Dr. Tewolde Egziabher and Dr. Melaku Worede independently. (App.1, C1 and C2). They still agree that: "Conceptually the CBD is fine, and it was necessary to make it a broadly formulated framework convention."

268 Here we should note a potential methodological problem, as pointed out in Chapter 2: the danger that I may be too well versed in one interpretation of the situation in the biodiversity issue-area. Like my interviewees, I have followed the issue for several years, and our understanding of the ethical and normative aspects of the issue largely coincide.

9.4.2 Impact of external donors' SOPs

What, then, of the potential effect through the international donor community? If the objectives of the CBD are to have an impact through bi- and multilateral agencies, these agencies will have to integrate the three objectives into their *standard operating procedures*. In other words, mutuality must also include the activities of external donors in recipient countries. The record so far, judging by the external funding of the biodiversity project portfolio in Ethiopia (section 8.6.2), is not very impressive:

- Rather than supporting the CBD objectives, many of these project proposals seem to counteract successful implementation.
- The cases in which the CBD objectives are sought included are frequently left pending.

These findings are supported by the conclusions in Keohane & Levy (1996), who point to the organisational inertia which often reduces the effectiveness of financial transfers. This is because aid is transferred through existing international institutions, and these do not deal primarily with environmental protection. The lesson drawn by Keohane and Levy is the advice to direct financial transfers at concern and capacity, rather than trying to tackle contracting problems: In other words, aid should focus on moral and material aspects, and stop worrying about mutual reassurance. This may be good advice for donor countries seeking to advance their own environmental agenda, but such a strategy will hardly work with a recipient country as headstrong as Ethiopia.

On the other hand, the potential increase in funding for biodiversity projects may well be connected to the *enhanced awareness* of the importance of this issue-area in the international donor community. Such awareness may be assumed to have been fostered by the CBD. This assumption is strengthened by the dramatic increase in the budgets for this sector, as indicated by the compilation of projects for natural resources and the environment in NCS, Volume V. Even if a large part of this anticipated increase should prove not to be substantiated by actual funding, the project funding granted will still surpass the pre-CBD situation. The danger is that Ethiopia may still not be in a particularly good position to reap the benefits of this increased awareness, precisely because it has been so adamant in resisting external control.

On a more general note, the actual financial transfers from the external donor community to biodiversity projects demonstrate that the North has a vested interest in the issue-area. In this light, the need for *mutual reassurance goes both ways*. This leads to the question of who needs mutual reassurance about what.

The answer to that question would disclose a good deal about the bargaining position of the parties involved. The answer is also likely to differ from case to case, but some general elements may be discerned. So far, in line with the question of this thesis, our discussion has focused on the need to reassure the South about mutual compliance in the North. As the concept implies, mutual reassurance goes both ways. Hence, we should ask what donors expect from Ethiopia in the biodiversity field, and what Ethiopia intends to deliver.

There are various possible demand/supply scenarios:

On the donor side, project proposals reveal several different concerns.

- In the *wildlife sector*, projects are generally concerned with traditional development assistance aims. These projects are seeking to enhance living conditions of the people, or to strengthen institutional capacity building, without expecting much in return for their investments.[269] An exception is found in the World Bank project on the borderline between the wildlife and seeds sectors, in the project proposal to examine wild medicinal plants. In this case, the Bank clearly intends to get something in return, namely knowledge about medicinals which may be useful to the pharmaceutical industry in the North.
- Similar economic concerns can be deduced in most of the projects in the *plant genetic resources sector*. All the way from Germany's contributions to establish the Genebank in the 1970s to the present-day Community-gene-bank project, it is clearly possible for donors to reap benefits in terms of increased knowledge about plant genetic resources with a potentially high economic value. This hardly comes as a surprise, but it is important to note, as it is crucial to the bargaining position of Ethiopian authorities in these project proposals.

What, then, does Ethiopia intend to supply?

- One interpretation of the situation could be: *not much*. In this perspective it could seem that Ethiopia is set on remaining a free rider itself; acquiring the funds and retaining full control of the benefits derived. For this purpose, Ethiopian authorities use the CBD for what it is worth, arguing about sovereignty over their own genetic resources.
- Another interpretation is that the situation reflects the lack of mutual reassurance mechanisms in the area. The situation is one of mutual distrust, in which the parties strive to retain as much control as possible. In this light, the Ethiopian case may be seen as a demonstration of how the CBD has not come far in establishing mutual reassurance mechanisms. While countries such as Brazil and Costa Rica are clearly more successful in obtaining beneficial bargains on financial transfers and biodiversity prospecting, these cases do not necessarily shed much light on the impact of the CBD. This is because these countries would have a high probability of success also in the absence of the regime: They would probably have been able to achieve quite acceptable bilateral arrangements in the biodiversity sector, as they start out from a much

[269] This situation reflects the troubled state of the Ethiopian national parks and sanctuaries (it may also reflect the troubled economy), which makes it different from conservation projects in e.g. Brazil or Costa Rica.

stronger bargaining position compared to Ethiopia. As such, Ethiopia seems a more realistic case on which to judge the effects of the CBD in this area.

Against this backdrop, two possible conclusions may be drawn:

- On a pessimistic note, we would conclude that in order for the CBD to have a positive effect on biodiversity policies in Ethiopia, moral and material mechanisms are necessary but not sufficient conditions. In order to have a positive impact on implementation, there must also be mutual reassurance mechanisms, and these are largely lacking.
- On a more optimistic note, there are some examples of mutual reassurance mechanisms evolving in international transactions with genetic resources (bioprospecting deals). This would indicate that the situation is dynamic with a potential for improvements. Unless these mechanisms are established as multilateral regulations, however, the improvements are likely to pertain mainly to mega-diversity countries with a strong bargaining position.

9.5 Empowerment: Environmental Proponents Strengthened by Alliances with International Epistemic Communities?

The searchlight provided by the empowerment mechanism directs attention to the role of international epistemic communities in enhancing domestic implementation of international environmental agreements. Agreeing with the ideational perspective, epistemic communities were assumed to influence domestic implementation mainly through two pathways:

- The group's members may be able to create transnational and trans-governmental coalitions holding similar views, and forging consensual solutions. The main focus here is epistemic relationships with domestic non-governmental groups.
- The group's members may hold administrative positions in government, and thus encourage common policies in their respective governments and push for their enforcement. This brings the focus to individual actors in the state administration.

9.5.1 Domestic NGOs and trans-governmental coalitions

Based on the definitions in Table 7.2, the analysis in Chapter 8 did not give rise to any large scope for identifying empowerment effects. It was not possible to trace any coalition between a CBD-linked epistemic community and a strong domestic NGO community. This was basically due to a lack of a strong domestic NGO community in Ethiopia. Several distinctions were noted in this connection:

- There is the question of whether it is possible to trace a relationship between domestic NGOs and an international epistemic community in the biodiversity field, and subsequently identify an effect on domestic biodiversity policies. No such impact has been found.
- We must distinguish between non-governmental environmental proponents being strengthened by international epistemic communities (hence, able to put increased pressure on governmental policies), and increased influence by non-governmental environmental proponents in general. No impact was found on either of these.
- There is the question of the nature of the relationship between domestic NGOs and the international donor community. The principles embedded in the CBD may have been directly adopted by non-governmental actors, who advocate these in their dealing with government agencies. We will have a closer look at this below.

Local participation and capacity building constituted one of the measurement standards for evaluating whether Ethiopian biodiversity policies were in agreement with the CBD objectives (section 8.4). While it was difficult to make any definite judgement regarding this aspect in the Ethiopian wildlife sector, the plant genetic resources sector received a high score on this criteria. An interesting implication ties up to the question of whether Ethiopia will succeed in bringing its experiences and lessons from the seeds sector into the wildlife sector. This question is also related to one of the conclusions in Chapter 6 that biodiversity policies might benefit from closer co-operation between NGOs within the seeds sector and the wildlife sector respectively.

The impact from the external donor community could potentially go in two directions, either obstructing or facilitating the implementation of the CBD objectives in Ethiopia. While the expected increase in funding may to some extent be connected to the CBD, and thus interpreted as an improvement of the situation, the content of the proposed projects did not always reflect an understanding of CBD objectives. Moreover, we found the relationship between domestic NGOs and the bi- and multilateral donor community to be a potentially two-edged sword:

- The majority of externally funded projects were explicitly linked to the central authorities, thus failing to contribute to local capacity building. In the case of the EU project in the New Southern Region, this strategy was interpreted by the Ethiopian authorities as an attempt to circumvent their own objectives regarding the decentralisation process.
- On the other hand, direct co-operation between NGOs and INGOs was also regarded with suspicion by government agencies. While the federal government thus encourages co-operation between the international donor community and

lower levels of government, it seems reluctant to encourage direct links to the NGOs.[270]

The potential problems associated with external donors and domestic NGOs are addressed at length in *Institutions for Environmental Aid* (Keohane & Levy, 1996). In particular, the authors criticise the GEF for neglecting domestic needs as a building block in enhancing concern among strategic domestic groups. This relates to the conflict between global environmental concerns and local sustainable development problems. Main obstacles to successful implementation of environmental financial transfers involve the failure to build on local knowledge (society), failure to draw on domestic human resources (state), and failure to adapt capacity building to local needs (Connolly, 1996:346).

9.5.2 Individual actors linking regime and state administration

Finally, let us have a look at the relationship between individual actors and the international epistemic community.

In Chapter 7 it was assumed that DCs have less stringent SOPs, making implementation less predictable and probably less efficient. On the other hand, the lack of rigorous SOPs may enable greater flexibility in the implementation output phase. Lack of SOPs may furthermore give room for increased influence by *individual* efforts to promote certain policies. Thus, individuals in the central elite would seem to have a greater potential scope for affecting policy formulation than in developed countries. It would be easier for DC elites to formulate and pursue "national interests" in the international arena, with less disturbances from domestic opposition. These individuals may be influenced by an international institution through learning processes: via participation in transnational epistemic communities developing within the regime.

The role of individuals should not be under-estimated in the Ethiopian case. Individual Ethiopians have been central in the international debate on plant genetic resources. The success and central role of Ethiopia internationally are very much a consequence of the efforts of individuals such as Dr. Melaku and Dr. Tewolde. Ethiopia is now taking a leading role internationally, because of the groundwork done by Dr. Melaku and the international work of Dr. Tewolde. As will be recalled, Dr. Melaku Worede was also a prominent member of the Keystone Dialogues, identified as the key epistemic community debating biodiversity and plant genetic resources prior to the establishment of the CBD. Still the Director of the Ethiopian Genebank at the time, he instilled respect as a role model in the field, due to his firm handling of Ethiopian interests concerning their genetic heritage vis-à-vis external donors (GTZ). Likewise, Dr. Tewolde Egziabher is a member of the Crucible Group

[270] This two-edged sword of NGO involvement is also discussed in Keohane & Nye, who point out that NGOs play an increasingly important role in attracting external funding, while at the same time bypassing the state (1996:44).

(section 6.2), and he has a leading role in the draft convention of the Organization of African Unity on access to genetic resources.[271]

As officials with a high ranking in administrative positions, these actors seem to have had a considerable impact on the policy-making process concerning the National Conservation Strategy in Ethiopia. The divide between the bureaucratic and political sphere seems blurred, in that the actors combine a technical and academic insight in the issue-area (as shown by the various articles in international scientific journals to their names) with explicit efforts to reach political goals. Moreover, the high profile of these individuals also seems to have created role models which may have inspired the recruitment of new high-profile actors.[272]

Again, my conclusions regarding the important role of individual actors in generating and pushing for more far-reaching environmental policies find a parallel in Keohane & Levy's *Institutions for Environmental Aid*. Here also, it is asserted that the most promising findings have to do with leadership, human initiative and personality, which may come to the fore when a window of opportunity arises. These deliberations give rise to three conclusions:

- We have seen that the *moral* aspect received a high score and the *empowerment* aspect was scarcely in evidence in the Ethiopian case, except for illustrating the importance of individual actors.
- The main constraints on biodiversity policies in Ethiopia seem to be not so much a lack of willingness, as a matter of lacking capacity and financial resources. In this light the changes in the Ethiopian political system coupled with individual leadership could be used as *a window of opportunity*. The scope for material and mutual reassurance mechanisms to enable the state to provide improved biodiversity policies is probably high.
- As the material and mutual reassurance mechanisms are weak or lacking, the probability that this window of opportunity will be used, however, is accordingly small.

Hence, despite the general boost provided by the CBD towards a greater international acceptance of the need to share responsibility for improved biodiversity management, and despite the fertile recipient environment in which to execute such projects and programmes –Ethiopian biodiversity does not stand out as a candidate for reaping significant material benefits of this enhanced awareness.

---

[271] Dr. Tewolde is active in the international arena with a view to further developments of mutual legislation for access to wild and domesticated biological material. This work is carried out partly through the Crucible Group, along with GRAIN and RAFI, among others. One line of efforts concerns the need to create operative sui generis systems (alternative to the UPOV system) under the WTO TRIPs agreement.

[272] As such, both Ato Regassa Feyissa and Dr. Getachew Mengistie figure in international listings of resources persons in the issue-area (Posey & Dutfield, 1996:245). This book represents one of the most thorough treatments of this issue-area to date.

9.6 Lessons from Ethiopia and Biodiversity: Deviant Case or Role Model?

9.6.1 On generalising among DCs

Can we generalise to other countries? While I assume that the scope for generalisations between implementing the CBD in Ethiopia and in OECD countries is rather small, several lessons from the Ethiopian case may be transferred to other low income countries. Ethiopia is hardly unique among LICs. First, low income countries will have low scores on state capacity factors like institutional and financial resources, and will hence be relatively weak in terms of ability to improve their biodiversity policies. Second, many LICs have high scores on certain categories of biological diversity as well as high levels of knowledge and awareness about the value of biological resources. As these resources, as well as knowledge about their use, are now being recognised in the CBD, this would seem to provide many developing countries with a potentially high willingness to pursue biodiversity policies in line with the Convention.

Even with regard to the two aspects mentioned above, there are certainly variations among countries. For one thing, there is the *political and cultural history* to consider. Ethiopia differs from the majority of DCs in having a non-colonial historical background. As shown by the decision to turn down the World Bank offer of external experts in the NCS process and the reluctance to join the WTO, Ethiopia seem even more persistent than the majority of developing countries in guarding its sovereignty. The non-colonial past does not, however, bestow Ethiopia with a more politically tranquil past than that of many fellow DCs. Authoritarian leadership and political turmoil mixed with widespread poverty characterise the Ethiopian record. Moreover, efforts to resist conditionality have, more often than not, characterised DCs policies towards external actors. Hence, in terms of domestic factors, Ethiopia may not be a unique case among low income countries.

Second, on the *biological resources* side, Ethiopia is similar to a great number of DCs in being gene-rich, but there are important variations within this category: Ethiopia is similar to Peru and the Philippines in being particularly gene-rich in domesticated species (wheat and coffee in Ethiopia, potato in Peru, rice in the Philippines), and different from Madagascar and Brazil in being, relatively and increasingly, poor in wildlife species. In effect, the Ethiopian elite has a proud awareness concerning their plant genetic resources heritage, while predominantly lacking the means to conduct conservation and sustainable use of wildlife resources. The analysis has shown that a strong seeds sector may be conductive towards strengthening a domestic awareness in the biodiversity issue. This is due to the lessons from the plant genetic resources-debate in FAO, which conferred a symbolic momentum to this sector. On the other hand, we have also seen that strength in wildlife resources appears more useful with regard to securing material benefits in the aftermath of the CBD.

When it comes to taking advantage of the potential *material* benefits arising from the new biodiversity regime, there are obvious differences between low- and medium-income developing countries. While countries such as Brazil and Costa Rica have clearly been far more successful in obtaining materially beneficial bargains on conservation projects and biodiversity prospecting, these cases cannot shed much light on the impact of the CBD. This is because these countries would have a high probability of success also in the absence of the regime. They would probably be able to achieve quite acceptable bilateral arrangements, as they start out from a much stronger bargaining position compared to Ethiopia. As such, Ethiopia seems a more realistic case on which to judge the effects of the CBD in this area.

Hence, this seems to speak in favour of my choice of case. If a mega-diversity (in terms of wildlife) country had been chosen for the analysis, this could have led to exaggerated conclusions about the successful impact of the CBD on domestic biodiversity policies. This is because a high score on compliance here might have been the result of the country's strong bargaining position rather than an effect of the CBD. This could mean that an international environmental agreement such as the CBD will only make a difference in low income countries. However, even though the weakest seems to have a greater need for multilateral regulations of exchange and access to biological resources, all countries providing genetic resources might potentially benefit. On the other hand – and pointing to the limitations inherent in a single case-study – I have no way of knowing whether a different case would have provided different or similar findings. The Ethiopian case has revealed a number of idiosyncratic aspects, such as the pronounced emphasis on independence and the specific role played by a few individuals, both of which may demarcate the scope for generalising.

With regard to the *moral* aspects, however, we can generalise the conclusions of this study to all DCs. The CBD does represent a moral victory for the DCs. It provides a framework in which to nest domestic legislation so as to strengthen a country's bargaining position vis-à-vis external users of biological resources. So far, however, the CBD lacks the mechanisms to translate it into practical politics on a mutual basis. The general lack of *mutual reassurance* mechanisms would, however, be less problematic in medium-income countries, as these would be less dependent on a smoothly functioning multilateral system. We could for instance expect the Newly Industrialised Countries to have a higher capacity for negotiating favourable bioprospecting deals bilaterally. Against this background, the Ethiopian case may be generalised to other gene-rich, low-income countries – with two lessons learned. Ethiopia may represent both a role model and a rather dismal example for the low-income countries (LIC):

- A role model for other LIC because of Ethiopia's efforts to make the most of its genetic heritage by being at the forefront internationally with legislation controlling access to genetic resources and by seeking to rely on its own human resources.

- A dismal example as Ethiopia, like other LIC, can hardly compete with the richer, mega-diversity countries in seeking to tap international green funds.

A few remarks are in order with regard to the *empowerment* mechanism. It is not least in terms of the role of the *NGO community* that the findings in this analysis may have a limited scope for generalisations. Other studies have shown a general proliferation of green NGOs in many Asian and South American countries, which could have the potential to play a more crucial role in affecting government policies, as well as external agents in the area of biodiversity, compared to the Ethiopian situation (see Triantafillou, 1995). Similarly, Costa Rica has benefited from a disproportionate share of environmental financial transfers, because it had a large and relatively effective network of non-governmental organisations with the capacity to effectuate policy as well as put pressure on the government (Jakobeit, 1996). On the other hand, Keohane & Levy's findings on the often important role of individuals are consistent with what the present study has revealed.

The discussions of the four institutional mechanisms and the reference to *Institutions for Environmental Aid* have revealed several areas in which there seems to be *a quite wide scope for generalisations*: The most important aspect concerns the interplay between the four mechanisms. Central findings pertain to the potential obstructions posed by the SOPs of external donor agencies, and the dominance of the preferences of external donors. These two external factors may undermine domestic capacity building. The remedy proposed in *Institutions for Environmental Aid* seems to be to try to circumvent this situation by aiming at concern and capacity while downplaying contractuality (Connolly, 1996:363). To my mind, the situation rather indicates that it is not enough to focus on moral and material mechanisms; – mutual reassurance is a necessary addition in order to enhance the scope for successful implementation of IEAs in developing countries.

The difference in implications drawn by Keohane & Levy and myself may, however, partly be ascribed to differences in cases examined: The DC cases studied in Keohane and Levy's book include the GEF (general), the Montreal Protocol Multilateral Fund (general), debt-for-nature swaps (general / focus on Costa Rica) and logging reform in the Philippines and Indonesia. Issues like ozone, nature preservation and the fight against tropical deforestation are central to, and more likely to be judged from, a Northern environmental agenda perspective. The very nature of the biodiversity issue-area, with its distinct structure of resource users and resource owners, necessitates evaluation from a combination of top–down and bottom–up perspectives.

> *On the same note, the biodiversity issue puts great demands on* mutual reassurance *due to the need to develop compatible legislation in user and owner countries alike.*

The *issue of biodiversity* may to some extent be seen as deviant as regards the North-South angle: This is to say that the biodiversity issue itself may represent a deviant "case-area" compared to other global environmental issue-areas. Compared to other international environment arenas, developing countries – and Ethiopia among these – are generally held to have been more active in this particular area. Likewise, it should be recognised that the biodiversity issue puts considerable demands on the developing countries, as compared to other global environmental issues. As the main bulk of the world's terrestrial biodiversity is found in tropical countries, the South realised that they might have something to gain in the biodiversity negotiations.[273] They also realised that they might be asked to make considerable contributions, and that this could be used as leverage in the negotiations. Thus, the Ethiopian case-study may not stand out as a deviant one within this issue-area, but in relation to other areas of international environmental agreements (except the one which may be developing on forestry), the issue-area as such may be seen as deviant. This brings us to the question of generalisations across issue-areas:

9.6.2 On generalising between issue-areas

This study has been concerned with the creation and implementation of the CBD. Can we apply the conclusions in this study to other issue-areas?

There are intrinsic differences between issue-areas for which international regimes are created, and between outputs emanating from international regimes. The type of issue-area may have implications for the type of output that may be expected. We must bear in mind these regime characteristics when appraising the significance of our findings. This lesson is emphasised by the differences between the wildlife and seeds sectors within the biodiversity issue itself.

When assessing the actual and potential for successful implementation in the biodiversity sector, we must of course take into account the specific nature of the issue-area itself. For example, in *Institutions for Environmental Aid*, the authors find wide variations in the implementation success of the GEF and the Montreal Protocol Fund, with the latter achieving a significantly higher score. This variance is put down to differences in institutional traits, through the three pathways of concern, capacity and contractual environment – but this may be stretched too far: More precisely, the argument that GEF would have had a higher chance for success given an identical institutional structure as that of the Montreal Fund (Connolly, 1996:357) is not always convincing, seeing how these institutions face very different problems and issue-areas.[274]

Here, I use the division between global agreements dealing with resources- or pollution-issues, to discuss the differences and the scope for generalisation of

---

[273] It should be pointed out here that this may be a centric view of developing countries. While the central government may see it in their benefit to conserve biodiversity for future revenues from a growing tourist business, local populations of people may end up as losers in this game.

[274] The Montreal Fund aims at a much more *benign* issue-area, in the sense that problem-solving here is largely identical to, and advanced by, the interests of the target groups themselves.

lessons learned.[275] I make these comparisons for illustrative purposes and not in a systematic manner.

Among the *global environmental (resources and pollution alike) issues*, biodiversity loss and climate change are inherently difficult to tackle – not least politically, in terms of vertical disintegration. Compared to international trade in endangered species (CITES) and ozone depletion, solutions pertaining to the former issues must address larger and more sensitive sectors of society, hence adding greatly to the costs incurred. Most global conventions aim at behaviour change within sectors that are far more narrow. Unlike ozone, the biodiversity issue has no obvious focal points. Rather it pertains to a large array of interrelated sectors, and include broad human activities such as industry, agriculture and forestry. In a country like Ethiopia, with some 85% of the population engaged in the agricultural sector, coupled with critical shortcomings in agricultural production, there will be tensions between biodiversity conservation and the need to increase agricultural produce. Institutional design may improve on the situation, but hardly go all the way in mitigating the problems of complexity and huge abatement costs. In effect, the implementation problems encountered within the CBD sectors are likely to be of a more grave nature than most global environmental conventions. This underscores the added importance of controlling for third variables when seeking to generalise across issue-areas.

Another distinguishing and common trait between the biodiversity and climate change regime is the pronounced North–South cleavage line. In her study of the DCs in the climate change issue, Joyeeta Gupta (1997) makes several observations closely resembling this case. In her view, the conclusion of the UNFCCC was largely achieved by making moral concessions to the South through the inclusion of financial "sweeteners" and by accommodating DC demands "on paper". In biodiversity and climate alike, there has been a growing realisation that DC participation and behaviour change is significant for the eventual solution of the problem. In both cases, the South has sought to use this situation as a lever, while the North has been trying to persuade them to come on board by paying a symbolic entrance ticket. Both conventions state that DC implementation is dependent on industrialised countries providing their part of the deal. Judging from Gupta's conclusions, however, I would argue that there may be greater potential for domestic implementation output in the biodiversity sector. This is because climate change problems are to a much larger extent framed and perceived in line with the Northern environmental agenda. Gupta argues that the problem definition is perceived very

---

[275] There are also several important distinctions between *global* and *regional* environmental agreements. The main difference pertains to the added difficulties brought on by the large number of participants in a global issue-area, which clearly presents added constraints in terms of the moral, material and mutual reassurance mechanisms. Regarding the *moral* aspect, it is harder to engender common norms in a truly global issue-area, compared to creating consensus about the need to curb pollution in a limited, regional area. On the *material* side, the added problems are even more striking, as illustrated by the almost impossible task of providing sufficient funding for mitigating biodiversity loss.

differently in the North and the South, and that climate change is still far from being translated into the domestic agenda of developing countries. Hence, there will be little domestic support for these issues. Biodiversity seems to have a more central place on DC domestic agendas, even thought this "window of opportunity" will hardly be utilised to its full potential. On the other hand, climate change may have the potential to attract more external funding and so reach a higher score on implementation outcome and problem solving.

Next, *global resource management issues* differ from *global pollution issues* in two important ways: First, while the *polluter pays principle* has been largely accepted, there is no corresponding principle regulating, for instance, deforestation activities.[276] One obvious explanation is that resources issues are more closely connected to basic human needs for food (agriculture, forestry, fisheries), while pollution issues are one step further removed, in largely involving externalities from industrial processes.[277] Second, there are the difficulties involved in setting specific standards for regulating behaviour in global resource management issue-areas. The implementation process in pollution issues, at both the regional and global levels, may be boosted by setting specific targets for reduced emissions. At the regional level, resource management aimed at specific species has of course been aided by the formulation of scientifically based catch-quotas. Setting specific targets and time-tables is, however, hardly a feasible task in the global issue-area of biodiversity.

I would demarcate the area for which my findings may be relevant to include primarily the formation of global environmental agreements, and implementation of global environmental agreements in low income countries.

9.7 Future Prospects

In this final section I will highlight some general lessons for future implementation: What are the odds that the CBD will in fact make a difference for efforts to curb the rapid loss of biodiversity around the world? We may discern two broadly positive trends and two broadly negative ones as regards successful implementation of the CBD.

At the domestic level in owner countries, the lessons from the Ethiopian case-study are broadly optimistic:

The biodiversity issue links up with long-standing traditions, involving both the public and private sector, and fostering strong domestic commitment and awareness. The CBD embodies several principles of importance to developing countries by

[276] I am indebted to Thomas Hansteen at the Centre for Development and Environment, University of Oslo, for illuminating this point.

[277] Recent developments in biotechnology may, however, be seen to blur the boundaries between the agricultural and the industrial sector.

establishing national sovereignty over biological resources, equitable sharing of benefits from use of biological resources, financial and technological transfers tied to access to these resources, and shared responsibility for costs of conservation. There is a potential that developing countries may succeed in reaping benefits from harbouring the largest bulk of the world's species diversity. Hence, the biodiversity issue may easily be translated as a priority in domestic policy-making.

Secondly, some important users are changing their behaviour in the "prescribed" direction:

There are some indications that the message of the CBD is catching hold among central actors in international biodiversity transactions. This involves major seeds corporations, botanical gardens, and scientific institutions. A great number of these agencies are choosing to adhere to the new rules and principles provided by the CBD on equity and fair sharing in genetic resources transactions. The *private* seeds collection agencies may be more open to see the potential competitive advantage of building a fair and above-board image in germplasm transactions. This trend is far from all-inclusive, but it is nevertheless ironic that the *public* sector seems to be intimidated by what they perceive as a demand from the same private sector actors to provide biotechnology industry with strong patent systems.

On the negative side, traditional approaches to biodiversity conservation among donors may obstruct recipients' efforts to adhere to the CBD:

Even though much more than the photogenic panda is at stake, much of the international efforts to contribute to problem-solving within the biodiversity issue still focus on "survival of the cutest". Wildlife conservation still takes precedence over on-farm, in-situ plant genetic resources projects. Wildlife conservation projects and policies are more likely to be one step removed from local and indigenous community participation, and are more likely to be dominated by standard approaches within development aid. This consistent Northern conservation bias remains a central constraint on the international commitment to share responsibility for CBD implementation. While the moral principles confirmed in the CBD could be seen to open a window of opportunity in the seeds sector, combined with a great potential in terms of domestic resources, this opportunity has hardly been utilised in Ethiopia. It is important to co-operate with owner countries (principally recipients) in defining projects that fit into their own policy agenda, in order to secure commitment for domestic implementation. Hence, even though resources for overall conservation are strained, this highlights the need to make the GEF more available for projects outside the traditional conservation portfolio.

Secondly, there are overlapping and conflicting international agendas which may block efforts to ameliorate biodiversity policies:

In the face of the danger posed by biodiversity loss, there is still more prestige and influence attached to other, potentially opposing, types of international fora. The Trade-Related Aspects of Intellectual Property Rights (TRIPs) will be up for review in the new WTO ("Millennium") round. The TRIPs review is likely to reaffirm the strongly conflicting positions of the USA on the one side and DCs on the other, with European countries somewhere in-between. In effect, concern for competitiveness in the biotechnology sector may increasingly take precedence over concerns for improved conservation and equitable sharing of benefits in the biodiversity issue-area. In practical terms, this would mean that Northern governments are less likely to establish domestic legislation that recognises the rights of the providers of biological resources. This would in turn mean that the North is less likely to establish relevant, compatible legislation, such as providing the requirement in patent legislation that all genetic material must be obtained on mutually agreed terms and on prior informed consent.

Biodiversity, like climate change, provides an example of states investing large amounts of energy and resources in order to find commonly acceptable solutions to transborder environmental problems. While the optimal solution for most of these countries would still seem to be the free-rider position, the gravity of the problems at hand seems increasingly to put at least some aspects of these issues into the high salience category. At the same time, the findings in this study have indicated that, despite the investments and efforts made at international negotiation tables, the “other hands” of government – those involved with trade regimes and development aid agencies – still have a long way to go in internalising these goals and putting them into practical action.

LIST OF ANONYMOUS INTERVIEWEES

Ethiopia

Central administration:

- C1: Dr. Tewolde Egziabher, director of National Conservation Secretariat, Ministry of Natural Resources Development and Environmental Planning (MNRDEP), also Head of the Ethiopian delegation during the latter phase of the Biodiversity negotiations. Since 1994 Head of the National Herbarium. Member of Crucible Group.
- C2: Dr. Melaku Worede, former director of the Ethiopian Genebank (PGRC/E), now scientific advisor to the USC Canada (Unitarian Service Committee of Canada), the Seeds Of Survival programme. Like Dr. Tewolde, a central figure in the international debate on property rights to genetic resources (through i.a. the Keystone Dialogues).
- C3: Director of Plant Genetic Resources Centre of Ethiopia (PGRC/E) / the Biodiversity Institute (Ethiopian Genebank).
- C4: Head of Division: Community Biodiversity Conservation and Development, the Biodiversity Institute (Ethiopian Genebank).
- C5: Leading official, "Patent, Technology Transfer and Development Department", Ethiopian Science and Technology Commission.
- C6: Leading official, Ethiopian Wildlife Conservation Organisation (EWCO).
- C7: Leading official, EPA, IUCN expert advisor to the NCS, (experience through three regimes in the administration) (former in MNRDEP).
- C8: Team Leader, Policy and Legislation, Environmental Protection Authority.

Regional:

- R1: Leading official, Oromia Natural Resources Development and Environmental Protection Bureau, Oromia Region.
- R2: Leading official, Natural Resources Development Unit in the Wildlife Office of the New Southern Region.

Non-governmental:

- N1: Leading official, Consortium of Ethiopian Voluntary Organisations (CEVO).
- N2: Resident Representative, Centre for International Environment and Development Studies, NORAGRIC, Norway.
- N3: Leading official, Ethiopian Wildlife & Natural History Society (EWNHS), member of Wildlife Steering Committee, and former university professor at Addis Ababa University (Biology Dept).
- N4: Dr., Dept. of biology, Addis Ababa University

- N5: Dr., LEM – The Environment and Development Society of Ethiopia.
- N6: Leading official, CARE Ethiopia.

About LEM-Ethiopia:

Established in 1992. Seven employees, and 15 voluntary members assisting the organisation. Has environmental clubs in schools all over the country. Funded by the Henrich Bull Foundation.

About EWNHS:

Established in the late sixties, primarily an expat. organisation with involvement from the UN. In 1991 it was "domesticated" and Dr. Shibru Tedle took over as leader. This change involved a new role in enhancing environmental concern, environmental movement. Rather "competing" than co-operating with LEM. Also environmental school clubs, as well as their own environmental magazine in English and Amharic. Compared to LEM, EWNHS has a more clear profile, more projects and more competent experts with closer ties to professional groups in Ethiopia. Financed by SIDA and EU.

Norwegian interviewees (confidential):

- n1: Leading official, Norwegian Directorate for Development Aid (NORAD), Environment Project.
- n2: Leading official, Norwegian Directorate for Nature Management (DN).
- n3: Dr., senior research fellow, NORAGRIC, NLH-Norway.
- n4: Leading official, Norwegian Directorate for Nature Management (DN).

GLOSSARY OF TERMS

Biological diversity (biodiversity): the variability among all living organisms and the ecological complexes of which they are part; this includes diversity within species, between species and between ecosystems.

Biological resources: includes genetic resources, organisms or parts thereof, populations, or any other biotic component of ecosystems with actual or potential value for humanity.

Biotechnology: means any technological application that uses biological systems, living organisms, or derivates thereof, to make or modify products or processes for specific use. Production may be carried out by using intact original or modified organisms, such as yeast and bacteria, or by using active cell components, such as enzymes from organisms.

Country of origin of genetic resources: means the country which possess those genetic resources in in-situ conditions.

Country providing genetic resources: means the country supplying genetic resources collected from in-situ sources, including populations of both wild and domesticated species, or taken from ex-situ sources, which may or may not have originated in that country.

Domesticated or cultivated species: species in which the evolutionary process has been influenced by humans to meet their needs.

Ecosystem: a dynamic complex of plant, animal and micro organism communities and their non-living environment interacting as a functional unit.

Endemic species: a species whose natural geographic distribution is restricted to a specific area or state.

Ex-situ conservation: used about conservation away from the natural environment, primarily for domesticated plants and animals; e.g. in gene banks, botanical gardens and zoos.

Gene: a segment of DNA carrying, due to its base sequence, a very specific information. Some genes carry the information for the synthesis of proteins (structural genes), others carry information for regulatory functions (regulatory genes).

Genetic engineering: a technology used to alter the genetic material of living cells through direct interference with the genome in order to make them capable of producing substances or performing functions alien to the unmanipulated cells.

Genetic material: hereditary material found in living organisms or parts thereof, responsible for the characteristics of an organism.

Genetic resources: means genetic material of actual or potential value.

Germplasm: the total genetic variability available to a particular population of organisms.

Habitat: the place or type of site where an organism naturally occurs.

Herbicide: more or less specific plant poison.

Indigenous species: a species occurring in the wild in a particular area or state, within its known natural range.

In-situ conservation: means the conservation of ecosystems and natural habitats and the maintenance and recovery of viable populations of species in their natural surroundings and, in the case of domesticated or cultivated species, in the surroundings where they have developed their distinctive properties.

Land race (or cultivar): plant varieties developed over thousands of years, adapting to local environments. Landraces are generally less specialised and systematically bred than the new, pure line, high-yielding varieties, but they are improved by humans and they contain a larger genetic diversity.

Micro-organism: any organism that can be seen only with the aid of a microscope. Also called microbe.

Wild species: a species which has not been deliberately modified by humans.

REFERENCES:

Allison, Graham (1971) Essence of Decision: Explaining the Cuban Missile Crisis (Boston, MA: Little and Brown).

Amin, Samir (1976) Unequal development (Hassocks: Harvester Press).

Andresen, Steinar (1989) "Science and Politics in the International Management of Whales", Marine Policy 13:99-117.

Andresen, Steinar & Jørgen Wettestad (1995) "International Problem-Solving Effectiveness: The Oslo Project So Far", International Environmental Affairs, 7:127-149.

Andresen, Steinar, J.B. Skjærseth, & J. Wettestad (1995) Regime, the State and Society: Analyzing the Implementation of International Environmental Commitments, Working Paper: WP-95-43 (Laxemburg, Austria: IIASA).

Agrawala, S. & S. Andresen (1999) "Indispensability and Indefensibility? The United States in the Climate Treaty Negotiations", Global Governance 5, pp. 457-482.

Ashton, P.S. (1988) "Conservation of Biological Diversity in Botanical Gardens", in Wilson (ed) Biodiversity (Washington D.C.: National Academy Press).

Axelrod, Robert (1984) The Evolution of Cooperation (New York: Basic Books).

Barton, J.H. & W.E. Siebeck (1992) "Intellectual Property Issues for the International Agricultural Research Centres. What are the Options?", Issues in Agriculture, no. 4 (Washington D.C.: Consultative Group on International Agricultural Research).

Bent, S.A., R.L. Schwaab, D.G. Conlin & D.D. Jeffery (1987) Intellectual Property Rights in Biotechnology Worldwide (New York: Stockton Press).

Berg, Trygve (1995) "Kulturplanter og husdyr", in Nils Christian Stenseth, Kjetil Paulsen, & Rolf Karlsen (eds) Afrika - natur, samfunn og bistand (Oslo: Gyldendal, ad Notam).

Bergesen, H.O., G. Parman & Ø. B. Thommessen (eds) (1998) Yearbook of International Co-operation on Environment and Development (London: EarthScan Press Ltd.).

Bergesen, H.O., G. Parman & Ø. B. Thommessen (eds) (1999/2000) Yearbook of International Co-operation on Environment and Development (London: EarthScan Press Ltd.).

Bernauer, Thomas (1994) Explaining the consequences of international environmental institutions, Theory and research strategy (Center for International Affairs, Harvard University).

Bernauer, Thomas (1995) "The effect of international environmental institutions: how we might learn more", International Organization, 49, 2, Spring 1995, pp. 351-77.

Bilder, R.B. (1980) "International law and natural resources policies", Natural Resources Journal, vol. 20, July, pp. 451–86.

Brown Weiss, Edith, & Harold K. Jacobson (1994) "Strengthening Compliance with International Environmental Accords: Some Preliminary Observations from a Collaborative Project", paper presented at the 35th Convention of the International Studies Association, Washington D.C., 29 March - 3 April.

Bull, Hedley, Benedict Kingsbury & Adam Roberts (eds) (1990) Hugo Grotius and International Relations (Oxford: Oxford University Press).

Carr, E.H. (1962) The Twenty Years' Crisis 1919-1939 (1939) (London: Macmillan).

CGIAR (1992) Working Document on Genetic Resources and Intellectual Property. Draft, Istanbul, May 22 (Washington D.C.: Consultative Group on International Agricultural Research).

Chayes, Abram, & Antonia Handler Chayes (1991) "Compliance Without Enforcement: State Behaviour Under Regulatory Treaties", Negotiation Journal, July, pp. 311-30.

Chayes, Abram, & Antonia Handler Chayes (1993) "On Compliance", International Organization, vol. 47, no. 2, Spring, pp. 175-205.

Cheru, Fantu (1992) Constraints for a conservation-based agricultural development policy in Ethiopia (Bergen: Chr. Michelsen Institute).

Connolly, B. (1996) "Increments for the Earth: The Politics of Environmental Aid", in Keohane & Levy (eds), Institutions for Environmental Aid. Pitfalls and Promise (Cambridge, MA: MIT Press).

Cox, Robert W. (1983) "Gramsci, Hegemony, and International Relations: An Essay in Method", Millennium: Journal of International Studies vol. 12, no. 2, pp. 162-75.

Crespi, R.S. (1988) Patents: a Basic Guide to Patenting in Biotechnology (Cambridge: Cambridge University Press).

Crucible Group (1994) People, plants and patents. The Impact of Intellectual Property on Biodiversity, Conservation, Trade and Rural Society (Ottawa: International Development Research Centre).

Dorm-Adzobu, Clement, & Allen Hober (1995) Institutional mechanisms for environmental management in Africa. Ethiopian National Conservation Strategy, Network for Environment and Sustainable Development in Africa.

Edwards, G.C. (1980) Implementing public policy (Washington D.C.: Congressional Quarterly Press).

El-Gewely, M.R. (ed) (1990) Genetic Engineering and Biotechnology Monitor, Unido; Issue no. 29, May.

Elmore, R.F. (1978) "Organizational models of social program implementation", Public Policy, 26, 185-228.

Elmore, R.F. (1979) "Backward Mapping: Implementation research and policy decision", Political Science Quarterly, 94, 606-16.

Elster, Jon (1989) Nuts and Bolts for the Social Scientist (Cambridge: Cambridge University Press).

Etzioni, Amitai (1961) A Comparative Analysis of Complex Organizations (New York: Free Press).

Evans, Peter (1989) "Predatory, developmental and other apparatuses: A comparative economic perspective on Third World state", Sociological Forum, vol. 4, no. 4.

Evans, T. & P. Wilson (1993) "Regime Theory and the English School of International Relations: A Comparison", Journal of International Studies, vol. 21, no. 3, pp. 329-51.

Fairman, David & Michael Ross (1996) "Old Fads, New Lessons: Learning from Economic Development Assistance" in Keohane & Levy (eds) (1996).

Fowler, Cary (1993) "Biological Diversity in a North–South context", in Helge Ole Bergesen & George Parman (eds) Green Globe Yearbook (Oxford: Oxford University Press) pp. 35–44.

Franck, Thomas M. (1990) The Power of Legitimacy among Nations (New York: Oxford University Press).

Frank, André Gunder (1967) Capitalism and Underdevelopment in Latin America: Historical Studies of Chile and Brazil (New York: Monthly Review Press).

Fraz, Wendy E. (1996) "The Scope of Global Financing: the Cases in Context", in Robert O. Keohane & Marc A. Levy (eds) Institutions for Environmental Aid. Pitfalls and Promise (Cambridge,MA: MIT Press).

Fried, Robert (1990) "Comparative Public Administration", in Lynn, N.B. & Aa. Wildawsky (eds) Public Administration. The State of the Art (Clapham House).

Galtung, J. (1971) "A Structural Theory of Imperialism", Journal of Peace Research, vol. 8, no. 2, pp. 81-117.

GATT (1989) Communication from India: Standards and principles concerning the availability, scope and use of Trade-Related Intellectual Property Rights, GATT Secretariat, MTN GNG/NG11/W/37, July.

GEF (1992) Summary Report on the Review of the GEF Biodiversity Portfolio, November (Washington).

GRAIN (1997) Signposts to Sui Generis Rights. Resource material from the international seminar on sui generis rights; Bangkok, 1-6 December 1997 (Bangkok: BIOTHAI & GRAIN).

Grajal, Alejandro (1999) "Biodiversity and the Nation State: Regulating Access to Genetic Resources Limits Biodiversity Research in Developing Countries", Conservation Biology, Vol. 13, No. 1, pp. 6-10.

Grieco, Joseph M. (1988) "Anarchy and the limits of cooperation: a realist critique of the newest liberal institutionalism", International Organization, 42, 3, Summer 1988, pp. 485-507.

Grindle, Merilee (1980) Politics and Policy Implementation in the Third World (Princeton, NJ: Princeton University Press).

Grotius, Hugo (1609) Mare Liberum sive de iure quod Batavis competit ad Indicana commercia dissertatio (Leiden: Elsevier).

Gupta, Joyeeta (1997) The Climate Change Convention and Developing Countries. From Conflict to Consensus? (Dordrecht: Kluwer Academic Publishers).

Haas, Ernst B. (1982) "Words can hurt you; or, who said what to whom about regimes", International Organization, vol. 36, no. 2, pp. 207-244.

Haas, Peter M. (1990) Saving the Mediterranean. The Politics of International Environmental Cooperation (New York: Columbia University Press).

Haas, Peter M. (ed.) (1992) "Knowledge, Power, and International Policy Coordination", Special Issue of International Organization, 46(1), pp. 1-390.

Haas, Peter M. (1993) "Epistemic Communities and the Dynamics of International Environmental Cooperation", in V. Rittberger (ed) Regime Theory and International Relations (Oxford: Clarendon Press).

Haas, Peter M., Robert O. Keohane & Marc A. Levy (eds) (1993) Institutions for the Earth. Sources of Effective International Environmental Protection (Cambridge, MA: MIT Press).

Hanf, K. & A. Underdal (1998) "Domesticating International Commitments: Linking National and International Decision-Making", in A. Underdal (ed) (1998b) The Politics of International Environmental Management.

Hansenclever, Andreas, Peter Mayer, & Volker Rittberger (1996) "Interests, power, Knowledge: The Study of International Regimes", Mershon International Studies Review, 40, pp. 177-228.

Hansson, Göte (1995) Ethiopia 1995, Consolidation of Reforms - Implications for Foreign Aid. Macroeconomic Studies: 64 (Stockholm: SIDA).

Hardin, Gareth (1968) "The Tragedy of the Commons", Science, 162, pp. 1243–48.

Harrison, Paul (1987) The greening of Africa, Breaking through in the battle for land and food., An IIED-earthscan study, Paladin.

Hendrickx, Frederic, Veit Koester & Christian Prip (1993) "Convention on Biological Diversity. Access to genetic resources: A Legal Analysis", Environmental Policy and Law, 23/6.

Hernes, Gudmund (1975) Makt og Avmakt (Oslo: Oslo University Press).

Heywood, V.H. (ed.) (1995) Global Biodiversity Assessment, UNEP (Cambridge: Cambridge University Press).

Hillman, Jesse C. (1993) Ethiopia: Compendium of Wildlife Conservation Information (Addis Ababa: EWCO).

Hindar, K., G.K. Rosendal & H.N. Trønnes (1990) Bioteknologi og norsk tilpassning til EFs indre marked (Trondheim: Norsk Institutt for Naturforskning, NINA).

Hjern, B. & D. Porter (1981) "Implementation structures: A new unit of administrative analysis", Organization Studies, 2, pp. 211-27.

Hollis, Martin, & Steve Smith (1990) Explaining and understanding international relations (Oxford: Oxford University press).

Human Rights Report, No. 4 (1994) The 1994 Election and Democracy in Ethiopia, Siegfried Pausewang (Oslo: Norwegian Institute of Human Rights).

Human Rights Report, No. 5 (1995) The Process of Democratisation in Ethiopia. - An Expression of Popular Participation or Political Resistance?, Kjell Tronvoll & Øyvind Aadland (Oslo: Norwegian Institute of Human Rights).

Huntley, B.J. (1988) "Conserving and Monitoring Biotic Diversity. Some African Examples", in Wilson (ed), Biodiversity.

Hurrell, Andrew, & Benedict Kingsbury (eds) (1992) "The international politics of the environment - introduction", in The International Politics of the Environment (Oxford: Oxford University Press).

Hveem, Helge (1996) Makt og velferd i det globale samfunn. Teorier i internasjonal politisk økonomi (Oslo: Oslo University Press).

IMF (1994) Ethiopia: Recent Economic Developments, August 29 (Washington D.C.: IMF, African Department).

Jakobeit, Cord (1996) "Nonstate Actors Leading the way: Debt-for-nature swaps", in Robert O. Keohane & Marc A. Levy (eds) Institutions for Environmental Aid. Pitfalls and Promise (Cambridge, MA: MIT Press).

Jänicke, Martin (1995) "The political system's capacity for environmental policy", enlarged text to a key note address to "Nordic Research Conference on Environment and Society", Gothenburg, June 13-15, 1995; later published in Martin Jänicke & Helmut Weidner (eds) (1997) National Environmental Policies. A Comparative Study of Capacity Building, (Berlin: Freie Universität Berlin).

Juma, C. (1989) The Gene Hunters (Princeton, NJ: Princeton University Press).

Kapuscinski, R. (1984) the Emperor (London: Pan Books Lmt).

Keohane, Robert O. & Joseph S. Nye (1977) Power and Interdependence (Boston, MA: Little, Brown & Co).

Keohane, Robert O. (1984) After Hegemony. Cooperation and Discord in the World Political Econom, (Princeton, NJ: Princeton University Press).

Keohane, Robert O. (1989) "Neoliberal Institutionalism: A Perspective on World Politics", in Robert O. Keohane (ed), International Institutions and State Power: Essays in International Relations Theory, pp. 1-20 (Boulder, CO: Westview Press).

Keohane, Robert O. (1993) "The Analysis of International Regimes: Towards a European-American Research Programme", in Rittberger (ed) Regime Theory and International Relations (Oxford: Clarendon Press).

Keohane, Robert O. & Marc A. Levy (eds) (1996) Institutions for Environmental Aid. Pitfalls and Promise (Cambridge, MA: MIT Press).

Keystone International Dialogue Series on Plant Genetic Resources (1990) Final Consensus Report, Madras Plenary Session, 14 February (Colorado: the Keystone Center).

Keystone International Dialogue Series on Plant Genetic Resources (1991) Final Consensus Report: Global Initiative for the Security and Sustainable Use of Plant Genetic Resources. Oslo Plenary Session. Third Plenary Session, 31 May – 4 June (Colorado: the Keystone Center).

King, Gary, Robert O. Keohane & Sidney Verba (1994) Designing Social Inquiry. Scientific Inference in Qualitative Research (Princeton, NJ: Princeton University Press).

Kingdon, J.W. (1984) Agendas, Alternatives, and Policy Alternatives (Boston: Little and Brown)

Kiss, Agnes (ed) (1990) Living with Wildlife. Wildlife Resource Management with Local Participation in Africa, Technical Paper No. 130 (Washington D.C.: World Bank).

Kloppenburg, J.R. & D.L. Kleinman (1987) "The Plant Germplasm Controversy", BioScience. vol. 37, no. 3.

Kloppenburg, J.R. (1988) First the seed (Cambridge: Cambridge University Press).

Koester, Veit (1994) "Implementation Issues in the Biodiversity Convention" in Implementing Environmental Conventions, Scandinavian Seminar Publications, Papers from the Second High-level Nordic Policy Seminar (Copenhagen: Erling Olsens Forlag).

Koester, Veit (1997) "The Biodiversity Convention Negotiation Process and some Comments on the Outcome", Environmental Policy and Law, 27/3, pp. 175-192.

Krasner, Stephen (1981) "Transforming international regimes: What the Third World wants and why", in Hollist, W.L. & Rosenau, J.N (eds) World System Structure (London: SAGE).

Krasner, Stephen (1982) "Structural causes and regime consequences: regimes as intervening variables", International Organization, vol. 36, no. 2, Spring 1982, pp. 185-205.

Krasner, Stephen (ed) (1983) International Regimes (Ithaca, NY: Cornell University Press).

Kratochwil, Friedrich, & John G. Ruggie (1986) "International organization: a state of the art on an art of the state", International Organization 40, 4, Autumn, pp. 753-75.

Laird, Sarah A. (1993) "Contracts for biodiversity prospecting", pp. 99-130, in WRI, Biodiversity Prospecting (Baltimore: World Resources Institute Publications).

Laugen, Torunn (1995) Compliance with International Environmental Agreements. Norway and the Acid Rain Convention, Report:003-1995 (Lysaker: Fridtjof Nansen Institute).

Levy, Marc A. & Oran R. Young (1994) "The Effectiveness of International Regimes", paper presented at Annual Convention of the International Studies Association, Washington D.C., 29 March - 1 April.

Levy, Marc A., Oran R. Young & Michael Zürn (1995) "The study of international regimes", European Journal of International Relations, vol. 1, no. 3, pp. 267-330.

Leys, Colin (1996) "The Crisis in 'Development Theory'", New Political Economy, vol. 1, no. 1, pp. 41-58.

Lijphart, Arend (1969) "Typologies of Democratic Systems", in A. Lijphart (ed) Politics in Europe (Englewood Cliffs, New Jersey: Prentice Hall).

Lijphart, Arend (1971) "Consociational Democracy" in R.J. Jackson & M.B. Stein (eds) Issues in Comparative Politics: A Text with Readings (London: Macmillan).

Lijphart, Arend (1975) "The Comparable-Case Strategy in Comparative Research", Comparative Political Studies, vol. 8, no. 2.

Lipschutz, Ronnie D., with Judith Mayer (1996) Global Civil Society & Global Environmental Governance. The Politics of Nature from Place to Planet (New York: State University of New York Press).

Litfin, Karen T. (1994) Ozone discourses. Science and politics in global environmental cooperation (New York: Columbia University Press).

Luard, E. (1984) "Who owns the Antarctic?", Foreign Affairs, pp. 1175–94.

Lunde, Leiv (1991) Science or Politics in the Global Greenhouse? The Development Towards Scientific Consensus on Climate Change, A Fridtjof Nansen Institute, Center for Economic Analysis (ECON) and Energi Data (ED) Report: 1991/8 (Lysaker: Fridtjof Nansen Institute).

Lyster, Simon (1985) International Wildlife Law. An analysis of international treaties concerned with the conservation of wildlife (Cambridge: Grotius Publications Limited).

Malnes, Raino (1990) The Environment and Duties to Future Generations, An Elaboration of 'Sustainable Development', Report 1990/2 (Lysaker: Fridtjof Nansen Institute).

Malnes, Raino (1995) "'Leader' and 'Entrepreneur' in International Negotiations: A Conceptual Analysis", in European Journal of International Relations, vol. 1(1), pp.87-112.

March, James G. & Johan P. Olsen (1989) Rediscovering Institutions. The Organizational Basis of Politics (New York: Free Press).

Martinussen, John Degnbol (1994) Samfund, stat og marked (København: Mellomfolkeligt Samvirke).

Masood, Ehsan (1998) "A formula for indigenous involvement", Nature 392, 539.

Mazmanian, D.A. & P.A. Sabatier (1981) Effective Policy Implementation (Lexington, MA: Lexington Books).

Mazmanian, D.A. & P.A. Sabatier (1983) Implementation and Public Policy (Glenview, IL: Scott, Foresman).

Midgaard, Knut, Halvor Stenstadvold, & Arild Underdal (1973) "An Approach to Political Interlocutions", Scandinavian Political Studies, vol. 8, pp. 9- 36 (Oslo: Oslo University Press).

Migdal, Joel (1988) Strong societies and weak states (Princeton, NJ: Princeton University Press).

Miles E., Andresen S., Lee K., Underdal A., & Wettestad J. (1994) Science, Technology and International Collaboration: Conditions for Effective Global and Regional Action

Concerning the Problem of Global Climate Change, research prospectus (Lysaker: Fridtjof Nansen Institute).

Miles, E. et al. (eds) (1998) Explaining Regime Effectiveness: Confronting Theory with Evidence (Cambridge, MA: MIT Press) (forthcoming).

Mittermeier , R.A. (1988) "Primate Diversity and the Tropical Forests", in Wilson (ed) Biodiversity.

Mooney, Pat R., Cary Fowler, Eva Lachkovics & Hope Shand (1988) "The Laws of Life. Another Development and the New Biotechnologies", Development Dialogue, Dag Hammarskjöld Foundation, Uppsala, 1–2.

Morgenthau, Hans (1960) Politics Among Nations, 6th edition 1985 (New York: Knopf).

Murvoll, Mariann (1997) Miljøpolitisk kapasitet: En case studie av Etiopias evne til å løse jorderosjonsproblemet for perioden 1972 til 1996, Hovedoppgave i statsvitenskap (Oslo: Oslo University Press).

Najam, Adil (1994) Are developing countries "different"?, Annotated outline of Think Piece (Laxenburg: IIASA, International Institute of Applied Systems Aanalysis).

Najam, Adil (1995) Learning from the literature on implementation: A synthesis perspective, Working paper WP-95-61 (Laxenburg, Austria: International Institute of Applied Systems Aanalysis).

Onwuka, Ralph I., & Timothy Shaw (eds) (1989) Africa in World Politics (London: Macmillan Press).

Ostrom, Elinor (1990) Governing the Commons: the evolution of institutions for collective action (Cambridge: Cambridge University Press).

Pierce, F. (1992) "Brazil, where the ice cream comes from", New Scientist, 17 July, p. 47.

Popper, Karl R. (1979) Objective Knowledge. An Evolutionary Approach (Oxford: Oxford University Press).

Porter, G. (1993) "The United States and the Biodiversity Convention", International Conference on the Convention on Biological Diversity, African Centre for Technology Studies, Nairobi, Kenya, January 26-29.

Posey, Darrel A. & Graham Dutfield (1996) Beyond Intellectual Property. Toward Traditional Resource Rights for Indigenous Peoples and Local Communities (Ottawa: International Development Research Centre).

Pressman, J.L. & A. Wildavsky (1973) Implementation (Berkeley: University of California Press).

Prescott-Allen, C. & R. Prescott-Allen (1988) Genes from the Wild, Using Wild Genetic Resources for Food and Raw Materials (London: Earthscan Publications).

Principe, P.P. (1989) "Valuing the biodiversity of medicinal plants", in Akerele, O., Heywood, V., & Synge, H., (eds) The Conservation of Medicinal Plants (Cambridge: Cambridge University Press).

Purdue, Derrick (1995) "Hegemonic Trips: World Trade, Intellectual Property Rights and Biodiversity", Environmental Politics, vol. 4, Spring 1995, no. 1, pp 88-107.

Putnam, R.D. & C.R. Henning (1989) "The Bonn Summit of 1978: A Case Study in Coordination", in Cooper et al. (1989) Can Nations Agree? Issues in International Economic Cooperation (Washington, D.C.: Brookings Institution).

Reid, Walter (1992) "Bush Biodiversity Policy Risks Dangerous Side Effects", Wall Street Journal, October 8.

Riggs, Fred W. (1964) Administration in Developing Countries. The Theory of Prismatic Societies (Boston: Houghton Mifflin Co).

Rittberger, Volker (ed) (1993) Regime Theory and International Relations (Oxford: Clarendon Press).

Rosendal, G. K. (1989) A Sustainable Development for Plant Genetic Resources: The Output of the Debate in FAO; a Sisyphean Victory for an Environmental Organisation? R:010-1989 (Lysaker: Fridtjof Nansen Institute).

Rosendal, G. K. (1991) International Conservation of Biological Diversity: the quest for effective solutions, R:012-1991 (Lysaker: Fridtjof Nansen Institute).

Rosendal, G. K. (1995) "The politics of patent legislation in biotechnology: an international view", in El-Gewely, M.R. (ed) Biotechnology Annual Review (Amsterdam: Elsevier), pp. 453–76.

Rosendal, G. K. (1995b) "Genbanker: Bevaring av biologisk mangfold" in Nils Christian Stenseth, Kjetil Paulsen, & Rolf Karlsen (eds) Afrika - natur, samfunn og bistand (Oslo: Gyldendal, ad Notam), pp. 375-92.

Rosendal, G. K. (1995c) "The Biodiversity Convention: An Analysis of the Background for and Implications of the US "no" in Rio", in Calestous Juma & Vincente Sànchez (eds) Biodiplomacy (Nairobi: ACTS Press).

Rosendal, G.K. (1995d) "The Convention on Biological Diversity: a viable instrument for conservation and use?" in Helge Ole Bergesen & George Parman (eds) Green Global Yearbook 1995 (Oxford: Oxford University Press), pp. 69-81.

Rosendal, G.K. (1995e) "The forest issue in post-UNCED international negotiations: conflicting interests and fora for reconciliation", Biodiversity and Conservation, vol. 4, pp. 91-107.

Ruggie, John G. (1975) "International Responses to Technology: Concepts and Trends", in J.G. Ruggie and Ernst B. Haas (eds) "International Responses to Technology", International Organization 29(3): 557-83.

Ruggie, John G. (1982) "International Regimes, Transactions, and Change: Embedded Liberalism in the Postwar Economic Order, in Steven Krasner (ed.) International Organization, vol. 36, no. 2.

Raustiala, Kal (1997) "Domestic Institutions and International Regulatory Cooperation", World Politics 49, pp. 482-509.

Sabatier P.A. & D.A. Mazmanian (1980) "Policy implementation: A framework for analysis", Policy Studies Journal, 8, 538-60.

Sabatier, Paul (1986) "What can we learn from implementation research?", in F-X Kaufman, V. Ostrom & G. Majone (eds), Guidance, Control and Performance Evaluation in the Public Sector (Berlin: De Gruyter), pp. 313-25.

Sabatier, Paul A., & Hank C. Jenkins-Smith (1993) Policy Change and Learning. An Advocacy Coalition Approach (Boulder, CO: Westview Press).

Scheiber, Harry, N. (1998) "Ocean Law, the Biodiversity Convention, and Access to Marine Genetic Materials" in Davor Vidas & Willy Østreng (eds) Order for the Ocean at the Turn of the Century (the Hague: Kluwer Law International).

Shiva, Vandana (1994) "The need for sui generis rights", Seedling, March, pp. 11–15.

Simon, Herbert A. (1947) Administrative Behaviour (New York: Macmillian).

Skjærseth, J.B. & G. K. Rosendal (1995) "Norges miljø-utenrikspolitikk" in Torbjørn L. Knutsen, Gunnar M. Sørbø, Svein Gjerdåker (eds) Norges Utenrikspolitikk (Oslo: Cappelen Akademisk Forlag), pp. 161-80.

Slaughter Burley, Anne-Marie (1993) "International law and international relations theory", American journal of international law, vol. 87, no. 2, p. 205-39.

Smartt, J. & N.W. Simmonds (eds) (1995) Evolution of crop plants, 2. ed. (London: Longman Scientific & Technical).

Snidal, Duncan (1985) "The Limits of Hegemonic Stability Theory", International Organization (Autumn), vol. 39, no. 4.

Sprinz, Detlef & Tapani Vaahtoranta (1994) "The interest-based explanation of environmental policy", International Organization 48, 1, Winter 1994, pp. 77-105.

Strange, Susan (1983) "Cave! hic dragones: A Critique of Regime Analysis", in Stephen D. Krasner (ed.), International Regimes (Ithaca, NY: Cornell University Press), pp. 337-54.

Strange, Susan (1988) States and Markets (London: Pinter Publishers).

Stokke, Olav Schram & Davor Vidas (eds) (1996) Governing the Antarctic. The effectiveness and legitimacy of the Antarctic Treaty System (New York: Cambridge University Press).

Svarstad, Hanne (forthcoming), Exchanges in Bioprospecting: Shaman Pharmaceuticals in Tanzania (PhD dissertation in progress).

Svensson, Ulf (1993) "The Convention on Biodiversity - A New Approach", in Sjøstedt, Svedin and Hägerhäll Aniansson (eds) International Environmental Negotiations; Process, Issues and context (Stockholm: FRN Utrikespolitiska Institutet).

Triantafillou, Peter (1995) Environmental Politics in Asia. Framing the environmental policy process, CDR Working Paper (Denmark: Centre for Development Research).

UNEP (1996) Conference on biological diversity and exotic species, Trondheim, June 1996. (Trondheim: NINA) Forthcoming Report.

Underdal, Arild (1980) The Politics of International Fisheries Management: The Case of the Northeast Atlantic (Oslo: Universitetsforlaget).

Underdal, Arild (1984) "Can we in the study of international politics, do without the model of the state as a rational unitary actor? A discussion of the limitations and possible fruitfulness of the model, and its alternatives", Internasjonal politikk, temahefte I (Oslo: NUPI).

Underdal, Arild (1993) The study of international regimes, review essay: a review of three contributions from Tübingen.

Underdal, Arild (1997a) Modelling the International Climate Change Negotiations: A Non-Technical Outline of Model Architecture, CICERO Working Paper 1997:8, ISSN: 0804-452X.

Underdal, Arild (1997b) "Patterns of Effectiveness: Examining Evidence from 13 International Regimes", paper presented at the 38th Annual Convention of the International Studies Association, Toronto, 19–22 March 1997.

Underdal, Arild (1998a) "Explaining Compliance and Defection: Three Models", European Journal of International Relations, vol. 4(1): 5-30.

Underdal, Arild (ed) (1998b) The Politics of International Environmental Management (Dordrecht: Kluwer Academic Publishers).

UNDP (1994) ETH/93/G31 - A Dynamic Farmer-Based Approach to the Conservation of Ethiopian Plant Genetic Resources.

UNESCO (1995) Statistical Yearbook. USA: UNESCO Publishing & Bernan Press, United Nations Educational, Scientific and Cultural Organization.

Van Meter, Donald S., & Carl E. van Horn (1975) "The policy implementation process - A Conceptual Framework", Administration and Society, vol. 6, no. 4, February, pp. 445-88.

Van Wijk, J. (1991) "Diminishing National Sovereignty in Intellectual Property Protection: Enforced Global Recognition of Patents in Biotechnology", paper presented at the International Symposium and Experts Workshop: Property Rights, Biotechnology and Genetic Resources, African Centre for Technology Studies and World Resources Institute, Nairobi, 10-14 June.

Vogel, David, & Veronica Kuhn (1987) "The comparative study of environmental policy: a review of the literature", in Meinholf Dierkes, Hans N. Weiler, Ariane Berthoin Antal (eds) Comparative Politics Research: Learning from Experience (Aldershot: Gower).

Waltz, K.N. (1959) Man, the State and War (New York: Columbia University Press).

Wapner, Paul (1995) "The state and environmental challenges: A critical exploration of alternatives to the state-system", Environmental Politics, Vol. 4, Spring 1995, No 1, pp. 44-69.

WCED, World Commission for Environment and Development (1987) Our Common Future (Oxford: Oxford University Press).

Weale, Albert, & Andrea Williams (1998) "National Science and International Policy", in Arild Underdal (ed) The Politics of International Environmental Management (the Netherlands: Kluwer Academic Press).

Wettestad, Jørgen, & Steinar Andresen (1991) The Effectiveness of International Resource Cooperation: Some Preliminary Findings (Lysaker: Fridtjof Nansen Institute).

Wettestad, Jørgen (1995) Nuts and bolts for the Environmental Negotiators? Designing Effective International Regimes (Lysaker: Fridtjof Nansen Institute).

Wettestad, Jørgen (1999) Designing Effective Environmental Regimes (London: Edward Elgar Publishing).

Wildavsky, Donald P. (1982) Bitter Pills: Population Policies and their Implementation in Eight Developing Countries (Cambridge: Cambridge University Press).

Williams, Marc (1993) "Re-articulating the Third World Coalition: the role of the environmental agenda", Third World Quarterly, vol. 14, no. 1, pp. 7-29.

Wilson, E.O. (ed) (1988) Biodiversity (Washington D.C.: National Academy Press).

Wilson, E.O. (1992) The Diversity of Life (Cambridge, MA.: Harvard University Press).

Wilson, James Q. (1973) Political Organization (New York: Basic Books).

Wood, D. (1993) "An Enhanced Role for Agrobiodiversity in Global Conservation Policy", paper presented at the International Conference on the Convention on Biological Diversity: National Interests and Global Imperatives, organised by UNEP and ACTS, Kenya, January 26-29.

Woodford, J.D. (1990) "The Status of Wildlife in Ethiopia, its Conservation and Utilisation", in Hillman (1993).

World Bank (1990) The long-term perspective study of Sub-Saharan Africa. Volume 1. Country Perspective, Background Papers.

World Bank (1997) Environmental Matters, Annual review, fall 1997.

World Conservation Monitoring Centre (1991) Biodiversity Guide to Ethiopia, Cambridge.

WRI, World Resources 1990-1991, A Report by the World Resources Institute, in collaboration with UNEP and UNDP (Oxford: Oxford University Press).

WRI, World Resources Institute (1993) Biodiversity Prospecting (Baltimore: WRI Report).

Yamin, Farhana (1995) "Biodiversity, ethics and international law", International Affairs, Vol. 71, No. 3, 1995, 529-546.

Yin, Robert K. (1991) Case study research (London: Sage Publications).

Young, Oran R. (1982a) “Regime Dynamics: the rise and fall of international regimes”, International Organization, 36, 2, Spring, p. 277-97.

Young, Oran R. (1982b) Resource Regimes: Natural Resources and Social Institutions (Berkeley: University of California Press).

Young, Oran R. (1989a) “The Politics of International Regime Formation: Managing Natural Resources and the Environment”, International Organization 43(3):349-75.

Young, Oran R. (1989b) International Cooperation: Building Regimes for Natural Resources and the Environment (Ithaca, NY: Cornell University Press).

Young, Oran R. (1991) “Political leadership and regime formation: on the development of institutions in international society”, International Organization, 45, 3, Summer, pp. 281-308.

Young, Oran R. (1994) International Governance – Protecting the Environment in a Stateless Society (Ithaca, NY: Cornell University Press).

Young, Oran R. (1996) “Institutional Linkages in International Society: Polar Perspectives”, Global Governance, vol. 2, no. 1, pp. 1-24.

Young, Oran R. (ed) (1997) Global Governance: Drawing Insights from the Environmental Experience (Cambridge: MIT Press).

Young, Oran R. (ed) (forthcoming) International Society: Enhancing the Effectiveness of Environmental Governance Systems.

Young, Oran R. & Arild Underdal (forthcoming) “Institutional Dimensions of Global Change: A Full Scoping Report”, IHDP: International Human Dimension Programme.

Zacher, Mark W. & Brent A. Sutton (1991) “Mutual Interests, Normative Continuities and Regime Theory: Cooperation in International Transportation and Communication Industries”, European Journal of International Relations, 1991, vol. 2 (1): 5-46.

Zürn, Michael (1993) “Bringing the second image (back) in. About the domestic sources of regime formation”, in Rittberger (ed) Regime Theory and International Relations, pp. 282-311.

Østerrud, Øyvind (1978) Utviklingsteori og historisk endring (Oslo: Gyldendal Norsk forlag).

Aakvaag, Ruth Kleppe (1998) “Biopiratene slår til igjen”, Genialt, no. 2, p. 5.

ENVIRONMENT & POLICY

1. Dutch Committee for Long-Term Environmental Policy: *The Environment: Towards a Sustainable Future*. 1994 ISBN 0-7923-2655-5; Pb 0-7923-2656-3
2. O. Kuik, P. Peters and N. Schrijver (eds.): *Joint Implementation to Curb Climate Change*. Legal and Economic Aspects. 1994 ISBN 0-7923-2825-6
3. C.J. Jepma (ed.): *The Feasibility of Joint Implementation*. 1995 ISBN 0-7923-3426-4
4. F.J. Dietz, H.R.J. Vollebergh and J.L. de Vries (eds.): *Environment, Incentives and the Common Market*. 1995 ISBN 0-7923-3602-X
5. J.F.Th. Schoute, P.A. Finke, F.R. Veeneklaas and H.P. Wolfert (eds.): *Scenario Studies for the Rural Environment*. 1995 ISBN 0-7923-3748-4
6. R.E. Munn, J.W.M. la Rivière and N. van Lookeren Campagne: *Policy Making in an Era of Global Environmental Change*. 1996 ISBN 0-7923-3872-3
7. F. Oosterhuis, F. Rubik and G. Scholl: *Product Policy in Europe: New Environmental Perspectives*. 1996 ISBN 0-7923-4078-7
8. J. Gupta: *The Climate Change Convention and Developing Countries: From Conflict to Consensus?* 1997 ISBN 0-7923-4577-0
9. M. Rolén, H. Sjöberg and U. Svedin (eds.): *International Governance on Environmental Issues*. 1997 ISBN 0-7923-4701-3
10. M.A. Ridley: *Lowering the Cost of Emission Reduction: Joint Implementation in the Framework Convention on Climate Change*. 1998 ISBN 0-7923-4914-8
11. G.J.I. Schrama (ed.): *Drinking Water Supply and Agricultural Pollution*. Preventive Action by the Water Supply Sector in the European Union and the United States. 1998 ISBN 0-7923-5104-5
12. P. Glasbergen: *Co-operative Environmental Governance: Public-Private Agreements as a Policy Strategy*. 1998 ISBN 0-7923-5148-7; Pb 0-7923-5149-5
13. P. Vellinga, F. Berkhout and J. Gupta (eds.): *Managing a Material World*. Perspectives in Industrial Ecology. 1998 ISBN 0-7923-5153-3; Pb 0-7923-5206-8
14. F.H.J.M. Coenen, D. Huitema and L.J. O'Toole, Jr. (eds.): *Participation and the Quality of Environmental Decision Making*. 1998 ISBN 0-7923-5264-5
15. D.M. Pugh and J.V. Tarazona (eds.): *Regulation for Chemical Safety in Europe: Analysis, Comment and Criticism*. 1998 ISBN 0-7923-5269-6
16. W. Østreng (ed.): *National Security and International Environmental Cooperation in the Arctic – the Case of the Northern Sea Route*. 1999 ISBN 0-7923-5528-8
17. S.V. Meijerink: *Conflict and Cooperation on the Scheldt River Basin*. A Case Study of Decision Making on International Scheldt Issues between 1967 and 1997. 1999 ISBN 0-7923-5650-0
18. M.A. Mohamed Salih: *Environmental Politics and Liberation in Contemporary Africa*. 1999 ISBN 0-7923-5650-0
19. C.J. Jepma and W. van der Gaast (eds.): *On the Compatibility of Flexible Instruments*. 1999 ISBN 0-7923-5728-0
20. M. Andersson: *Change and Continuity in Poland's Environmental Policy*. 1999 ISBN 0-7923-6051-6

ENVIRONMENT & POLICY

21. W. Kägi: *Economics of Climate Change: The Contribution of Forestry Projects*. 2000
ISBN 0-7923-6103-2
22. E. van der Voet, J.B. Guinée and H.A.U. de Haes (eds.): *Heavy Metals: A Problem Solved?* Methods and Models to Evaluate Policy Strategies for Heavy Metals. 2000
ISBN 0-7923-6192-X
23. G. Hønneland: *Coercive and Discursive Compliance Mechanisms in the Management of Natural Resourses*. A Case Study from the Barents Sea Fisheries. 2000
ISBN 0-7923-6243-8
24. J. van Tatenhove, B. Arts and P. Leroy (eds.): *Political Modernisation and the Environments*. The Renewal of Environmental Policy Arrangements. 2000
ISBN 0-7923-6312-4
25. G.K. Rosendal: *The Convention on Biological Diversity and Developing Countries*. 2000 ISBN 0-7923-6375-2

KLUWER ACADEMIC PUBLISHERS – DORDRECHT / BOSTON / LONDON